GEOLOGICAL HAZARDS

Recent Titles in
Sourcebooks on Hazards and Disasters

Biological Hazards: An Oryx Sourcebook
Joan R. Callahan

GEOLOGICAL HAZARDS

A SOURCEBOOK

Timothy M. Kusky

Sourcebooks on Hazards and Disasters

An Oryx Book

GP

Greenwood Press
Westport, Conn. • London

Library of Congress Cataloging-in-Publication Data

Kusky, Timothy M.
Geological hazards : a sourcebook / Timothy M. Kusky.
p. cm.—(Sourcebooks on hazards and disasters)
Includes bibliographical references and index.
ISBN 1–57356–469–9 (alk. paper)
1. Geodynamics. 2. Natural disasters. I. Title. II. Series.
QE501.3.K795 2002
363.34—dc21 2002192773

British Library Cataloguing in Publication Data is available.

Library of Congress Catalog Card Number: 2002192773
ISBN: 1–57356–469–9

First published in 2003

Greenwood Press, 88 Post Road West, Westport, CT 06881
An imprint of Greenwood Publishing Group, Inc.
www.greenwood.com

Printed in the United States of America

The paper used in this book complies with the Permanent Paper Standard issued by the National Information Standards Organization (Z39.48–1984).

10 9 8 7 6 5 4 3 2 1

Contents

Preface		vii
Acknowledgments		ix
Chapter 1	Introduction to Geological Hazards	1
Chapter 2	Earthquakes	21
Chapter 3	Volcanic Eruptions	49
Chapter 4	Tsunami	75
Chapter 5	Mass Wasting	97
Chapter 6	Streams and Floods	121
Chapter 7	Coastal Hazards	145
Chapter 8	Deserts, Drought, and Wind	169
Chapter 9	Glaciers and Glaciation	199
Chapter 10	Hazards Associated with Geologic Materials	221
Chapter 11	Natural Geologic Subsidence Hazards	245
Chapter 12	Hazards of Sudden Catastrophic Geologic Events	267
Index		293

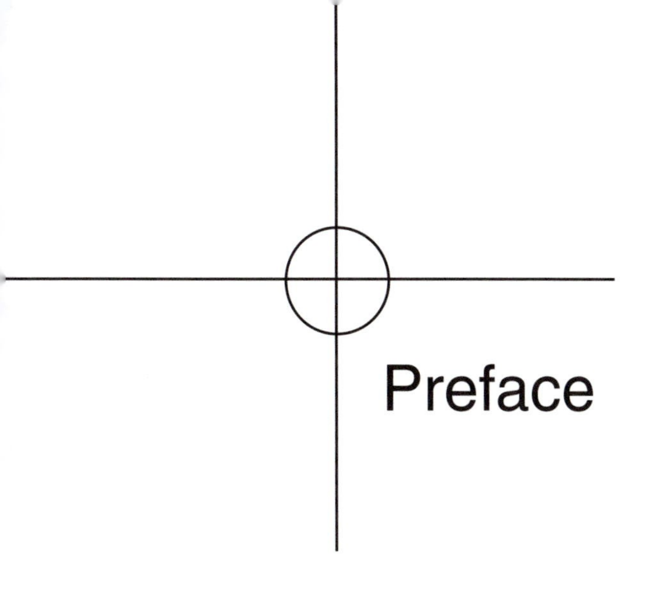

Preface

Geological hazards include processes that are harmful to humans or to our way of life. There are three major types of hazards. Some, such as volcanic eruptions or huge earthquakes that may involve vast regions of the planet, are catastrophic and kill or affect up to hundreds of thousands of people. Some single earthquake events have killed nearly a million people, whereas volcanic eruptions have caused entire regions to be abandoned and, in some cases, may have wiped out entire cultures. Both volcanic eruptions and earthquakes are associated with tsunami, giant killer waves that travel rapidly across ocean basins and rise unexpectedly on distant shores, washing away all that is not secured.

Other geologic hazards are slow acting, such as the gradual but steady downhill creep of soil that causes power poles, fences, and foundations to gradually tilt downhill. The imperceptibly slow dissolution of bedrock by water in the ground may form large subterranean caverns that eventually collapse, swallowing any structures that were unsuspectingly built above them. Some natural geologic materials are hazardous, including types of clays that expand and contract dramatically with changes in ground moisture. Other minerals constitute health hazards, such as asbestos, dissolved metals in ground water, and odorless radon gas that seeps into homes.

A final class of geologic hazard is related to the interaction of the atmospheric system with other systems. It includes destruction from coastal storms, including hurricanes; long periods of drought that may lead to famine and desertification; and regional floods.

Geological Hazards: A Sourcebook is designed to give its readers an understanding of how Earth's natural processes are sometimes hazardous to humans. The book is intended as a reference resource for high school and college students, teachers and professors, scientists, librarians, journalists, general readers, and specialists looking for information outside their specialty. It will also give readers an understanding of the forces and processes behind natural disasters. This book achieves this goal by first analyzing geological processes, then discussing how aspects of these processes may be dangerous. It examines the scientific principles behind geological hazards and discusses specific examples of when, where, and why certain hazardous and disastrous events have taken place. Each chapter includes an extensive source list where the reader may find additional information, including relevant organizations, print resources, web resources, and other types of media.

Chapter 1 discusses basic geological processes, and Chapters 2–5 then present information on geologic hazards that are associated with the movement of tectonic plates. The next section (Chapters 6–9) focuses on hazards associated with interactions of the atmosphere, ocean, and land, and a third section (Chapters 10–11) includes aspects of geologic materials that are hazardous when humans come in contact with them. A final chapter (Chapter 12) discusses extraterrestrial hazards and how Earth's interactions with extraterrestrial materials, such as asteroids, have influenced life here. Discussions throughout the book focus on what people may have been able to do to foresee or reduce the devastation from past disasters, and what we have learned that may help reduce the cost of similar disasters in the future.

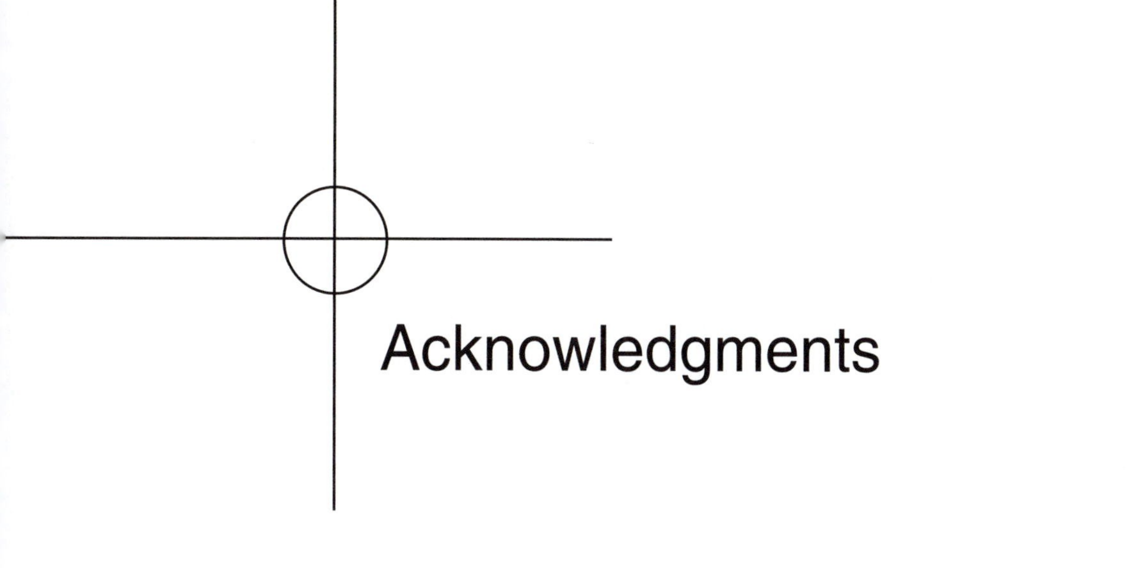

Acknowledgments

I would like to thank my wife, Carolyn, and my children Shoshana and Daniel for their patience during the long hours spent at my desk preparing this book. I would also like to thank Henry Rasof, John Wagner, and Nicole Cournoyer for their help with editorial and review aspects of the book. Finally, I thank Angie Bond and Soko Made for help with the indexing and figures.

CHAPTER 1

Introduction to Geological Hazards

INTRODUCTION

The earth is a naturally dynamic world, with volcanic eruptions spewing lava and ash and earthquakes pushing up mountains, shaking the surface, and forming tsunami that sweep across ocean basins at hundreds of miles per hour and rise in huge waves on distant shores. Mountains may suddenly collapse and bury entire villages, and slopes are gradually creeping downhill, moving everything built on them. Storms sweep coastlines and remove millions of tons of sand from one place and deposit it on another in a single day. Large parts of the globe are turning into desert, and glaciers that once advanced are rapidly retreating. Sea level is beginning to rise faster than previously imagined. Not long ago, society viewed these seemingly unrelated events as acts of God and did little to try to understand them or reduce their lethality. All these natural phenomena are, in fact, expected consequences of the way the planet works, and as scientists understand these geological processes better, they are able to better predict when and where natural geologic hazards could become disasters and then take preventative measures.

Advances in science and engineering in recent decades have dramatically changed the way we view natural hazards. In the past, we viewed destructive natural phenomena, including earthquakes, volcanic eruptions, floods, landslides, and tsunamis, as unavoidable and unpredictable. Our society's attention to basic scientific research has changed that view dramatically, and we are now able to make general predictions of when, where, and how severe such destructive natural events may be, reducing their consequences significantly. We are therefore able to plan evacuations, strengthen buildings, and make detailed plans of what needs to be done before natural geological disasters occur to such a degree that the costs of these disasters have been greatly reduced. This greater understanding has come with increased governmental responsibility. In the past, society placed little blame on its governments for the consequences of natural disasters. For instance, in September 1900 nearly 10,000 people perished in a hurricane that hit Galveston, Texas, yet, since there were no warning systems in place, no one was blamed. In 2001, two feet of rain with consequent severe flooding hit the same area, and nobody perished. However, residents filed billions of dollars worth of insurance claims. Now, therefore, things are different, and few disasters go without blame being placed on public officials, engineers, or planners. Our extensive warning systems and building codes and our increased understanding have certainly prevented the loss of thousands of lives, yet they also give us a false sense of security. When an earthquake or other disaster strikes, we expect our homes to be safe, yet they are only built to be safe to a certain level. When a natural geological hazard exceeds that expected level of damage, it may result in great destruction, and we blame the government for not anticipating the event. What can be done? Our planning and construction efforts are only designed to meet certain levels of force for earthquakes and other hazards, and planning for the rarely occurring stronger events would be exorbitantly expensive.

We still have far to go in mitigating natural geologic hazards. We need a better understanding of the natural processes that cause geologic hazards, an understanding that is achieved through basic scientific research. This book provides an introduction to the physical processes behind natural geologic hazards as well as an extended discussion of specific catastrophes and what may be done to mitigate their effects. Understanding the forces and processes behind hazardous geologic phenomena is the first step for anyone trying to reduce their destructive consequences.

WHAT ARE GEOLOGICAL HAZARDS?

Geological hazards take many shapes and forms, from earthquakes and volcanoes to the slow downhill creep of material on a hillside and the expansion of clay minerals in wet seasons. Natural geologic processes are in constant operation on the planet. These processes are considered hazardous when they go to extremes and interfere with the normal activities of society. For instance, the surface of the earth is constantly moving through plate tectonics, yet we do not notice this process until sections of the surface move, suddenly causing an earthquake.

In this book, natural geologic hazards are examined by first discussing the physical processes behind each kind of hazard, then discussing when and how these processes may be hazardous to humans. Geologic hazards can be extremely costly in terms of price and human casualties. With growing population and wealth, the cost of natural disasters has grown as well. The amount of property damage (measured in constant U.S. dollars) has doubled or tripled every decade, with individual disasters sometimes costing tens of billions of dollars. A recent report to the U.S. Congressional Natural Hazards Caucus estimated the costs of some recent disasters—Hurricane Andrew in 1992 cost $19 billion, the 1993 Midwest floods cost $21 billion, and the 1994 Northridge earthquake cost $14 billion. In contrast, the entire Persian Gulf War cost the United States and its allies $65 billion. The costs of natural geologic hazards are now similar to the costs of warfare, demonstrating the importance of understanding their causes and potential effects.

There are several causes of geologic hazards, and each of these topics is discussed in this book. First, the slow but steady movement of tectonic plates on the surface of the earth causes many geologic hazards either directly or indirectly. Plate tectonics controls the distribution of earthquakes, the location of volcanoes, and the uplifting of mountain ranges. A second group of hazards is related to earth surface processes, including river flooding, coastal erosion, and changing climate zones. We will see that many earth surface processes are parts of natural earth cycles but are considered hazardous to humans because we have not adequately understood the cycles before building on exposed coastlines and in areas prone to shifting climate zones. A third group of hazards is related to materials, such as clay that dramatically expands when wetted and sinkholes that develop within limestone. Still other hazards are extraterrestrial in origin, such as the occasional impact of meteorites and asteroids with Earth. Earth's exponentially growing human population worsens most of these hazards: Human dominance on the planet has plunged the earth into a mass extinction event, the severity of which has not been seen since 66 million years ago, when natural disasters killed the dinosaurs and many of the planet's other species.

NATURAL HAZARDS, MYTH, AND RELIGION

Natural geological hazards and catastrophes may have been the sources of many traditional myths and biblical odysseys. Since early peoples did not fully understand many of the processes associated with the causes of natural disasters, they often invoked supernatural explanations. Many peoples on the planet still attribute the causes of natural disasters to animistic gods, and many of the world's major religions ask believers to accept divine intervention as the triggering mechanism for many natural disasters. Moreoever, early peoples and the religiously devout today often regard natural phenomena as the direct wrath of gods in response to some human behavior.

Among the most famous examples are the biblical stories of the great flood, drought, days of darkness, seven plagues, parting of the Red Sea for Moses and the Israelites, and destruction of many temples and cities by earthquakes. Of these, the biblical great flood is perhaps the best known natural disaster of all time, being found in the traditions of several of the world's major religions including Judaism, Christianity, early Babylonian and Egyptian religions, and stories from the Euphrates Valley. Similar legends pervade the early history of other peoples, including the Zapotec and other Indians of Mexico and Central America, early Hawaiians, Aboriginal Australians, Chinese, and Malaysians. Other stories of floods, principally from Pacific Rim cultures, seem to relate to tsunami (see Chapter 4) because they describe floods as great swellings from the ocean.

The Tigris and Euphrates Rivers that flow out of present-day Turkey into Iraq, which is in the same region as the biblical-era Mesopotamia was, are prone to regular severe floods and are the likely location of many of the early biblical legends. Anecdotes that the entire world was flooded are scientifically unrealistic, but may refer to the entire world known to these early peoples. These floods have left thick deposits of silt and mud in the Tigris-Euphrates Valley and are the source of the region's fertility, development of early agriculture, and civilization.

Most of these early legends have similarities, such as some premonitory event warning of the upcoming deluge; the flooding of the entire world; and the survival of a chosen few, often on a boat, raft, or other craft stranded on some mountaintop. These legends relate the survival of the few to some inherent righteousness, forming the basis for many of our world's cultural values. Despite this, there is no evidence that all of these catastrophic floods were contemporaneous with each other or that they represented anything divinely different from natural geologic processes that had been operating on Earth for billions of years before people even inhabited these regions. This is not to deny any divine role in the formation of Earth and the universe and in the establishment of fundamental physical relationships, nor does it address the question of why specific events happened, but simply identifies natural physical processes associated with specific events.

Large historical volcanic events may have also prompted biblical and other legends. For instance, the eruption of Santorini, Greece, in 1628 B.C. (see Chapter 3) destroyed much of the previously densely populated island when the volcano collapsed and formed a huge caldera complex that was filled with water from the surrounding Aegean Sea. The huge tsunami generated by this eruption was catastrophic for many Mediterranean cultures; it was hundreds of feet tall near the volcano and swept across much of the eastern Mediterranean, affecting the modern-day coasts of Lebanon, Israel, Egypt, and Libya with a twenty-foot-high crest. The tsunami even moved 200 miles inland along the Nile River, much to the awe and confusion of local people who thought the river was running backwards. It is widely believed that this eruption destroyed the previously flourishing Minoan culture on the island of Crete, and it may also be the cause of many Greek legends, including Plato's lost continent of Atlantis.

The eruption of Santorini produced such a large ash cloud that it covered much of the eastern Mediterranean, including Sinai, northern Egypt, and the Levant, with dense ash fallout. The timing of this eruption, the largest-known historical eruption of all time, may account for several other biblical legends. For example, the ash clouds may account for the biblical three days of such "darkness throughout the land of Egypt" that the darkness could be felt. These passages accurately describe a volcanic ash cloud. Also, the parting of the Red Sea for the Israelites and subsequent drowning of the Egyptian army by returning waters might have been a real effect of the eruption, known as a seiche wave (Chapter 4). These are similar to tsunami, but in this case are caused by a huge atmospheric blast or pressure wave from the eruption. Seiche waves can have initial drawdowns of water in one place, followed soon (typically hours later) by huge crashing inflows.

THE GEOLOGICAL FRAMEWORK OF THE EARTH

Geology is the study of the earth, the physical processes that operate on the planet, and the materials that make up the earth. Many natural geologic processes are hazardous to humans, and these processes are the focus of this book.

The branch of geology that deals with the relationships between people and their natural physical environment is called *environmental geology*. Some common ways that environmental geology impacts the lives of most people are through the development and exploitation of mineral resources and through other, less fortunate circumstances such as earthquakes, volcanic eruptions, landslides, coastal erosion, and floods. To understand environmental geology and geological hazards, it is essential to fully grasp the basic principles of physical geology. Therefore, much of this introductory chapter will be devoted to developing this understanding and discussing the implications that these geological phenomena have on our lives.

To understand geological hazards, we must consider biological and social factors as well as the natural physical environment. The physical environment includes things such as water, air, soil, rock, temperature, humidity, light, and how these factors vary with time. The earth can be thought of as a very complex system that includes many smaller systems interacting with each other. Some of these systems are the hydrosphere, the atmosphere, the biosphere, and the lithosphere. It is no simple task to understand this Earth, for it is an ever-changing and dynamic system.

The External Layers of the Earth and Earth Systems

The external layers of the earth include the *hydrosphere*, which consists of the ocean, lakes, streams, and the atmosphere. The air/water interface is very active, for here erosion breaks rocks down into the lithosphere's upper unconsolidated part, loose debris at the earth's surface called the *regolith.*

The hydrosphere is a dynamic mass of liquid that is continuously on the move. It includes all the water in oceans, lakes, streams, glaciers, and groundwater, although most water is in the oceans. The hydrologic cycle (Figure 1.1) describes both long- and short-term changes in the earth's hydrosphere. The hydrosphere is powered by heat from the sun, which causes evaporation and transpiration. Evaporated water moves into the atmosphere and precipitates as rain or snow, which then drains off in streams, evaporates, or becomes groundwater, to begin the cycle over and over again.

The *atmosphere* is the sphere around the earth that consists of the mixture of gases we call air. It is hundreds of kilometers thick, and it is always moving because more of the sun's heat is received per unit area at the equator than at the poles. The

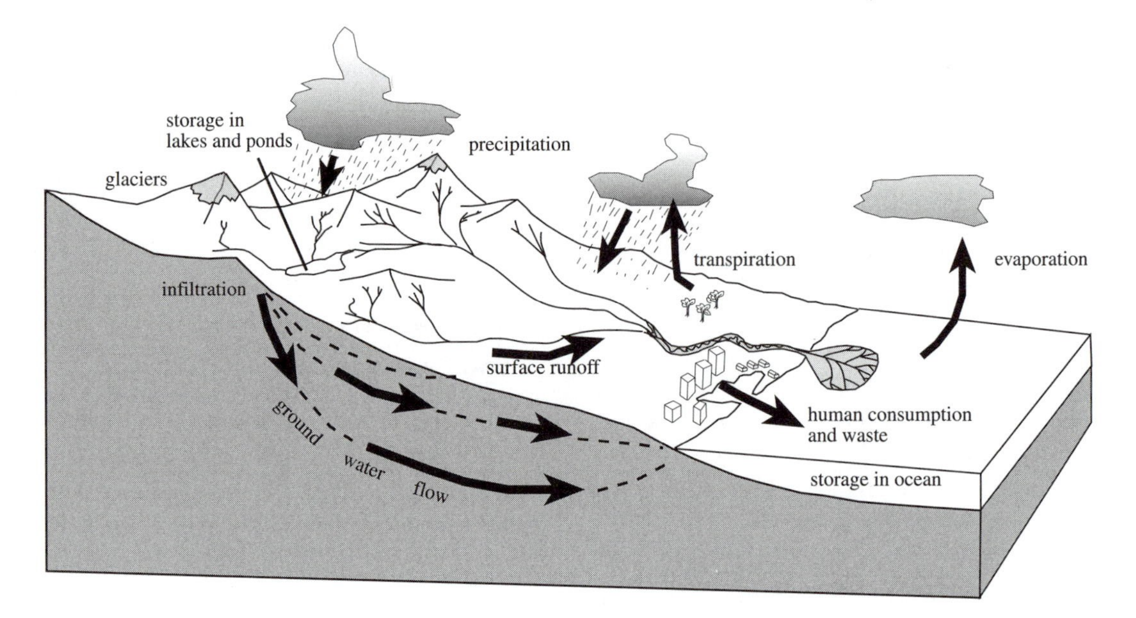

Figure 1.1. The Hydrologic Cycle. Water evaporates from the oceans, forming clouds that release precipitation that falls on the land. This precipitation fills streams and rivers, forms glaciers, and works its way into the ground water system. The water eventually returns to the ocean, completing the cycle.

heated air expands and rises to where it spreads out, then it cools, sinks, and gradually returns to the equator. The effects of the earth's rotation, called the Coriolis force, modify this simple picture of the atmosphere's circulation by causing any freely moving body in the northern hemisphere to veer to the right and any in the southern hemisphere to veer to the left (Figure 1.2).

Climate is the average weather of a place or area and its variability over a period of years. The average temperature, precipitation, cloudiness, and windiness of an area determine a region's climate. Factors that influence climate include latitude; proximity to oceans or other large bodies of water that could moderate the climate; topography, which influences prevailing winds and may block precipitation; and altitude. All of these factors are linked together in the climate system of any region on the earth. The global climate is influenced by many other factors: Chemical interactions between seawater and magma significantly change the amount of carbon dioxide in the oceans and atmosphere, and may change global temperatures. Pollution from humans also changes the amount of greenhouse gases in the atmosphere, which may be contributing to global warming.

The biosphere is the totality of Earth's living matter and its partially decomposed dead plants and animals. The biosphere is

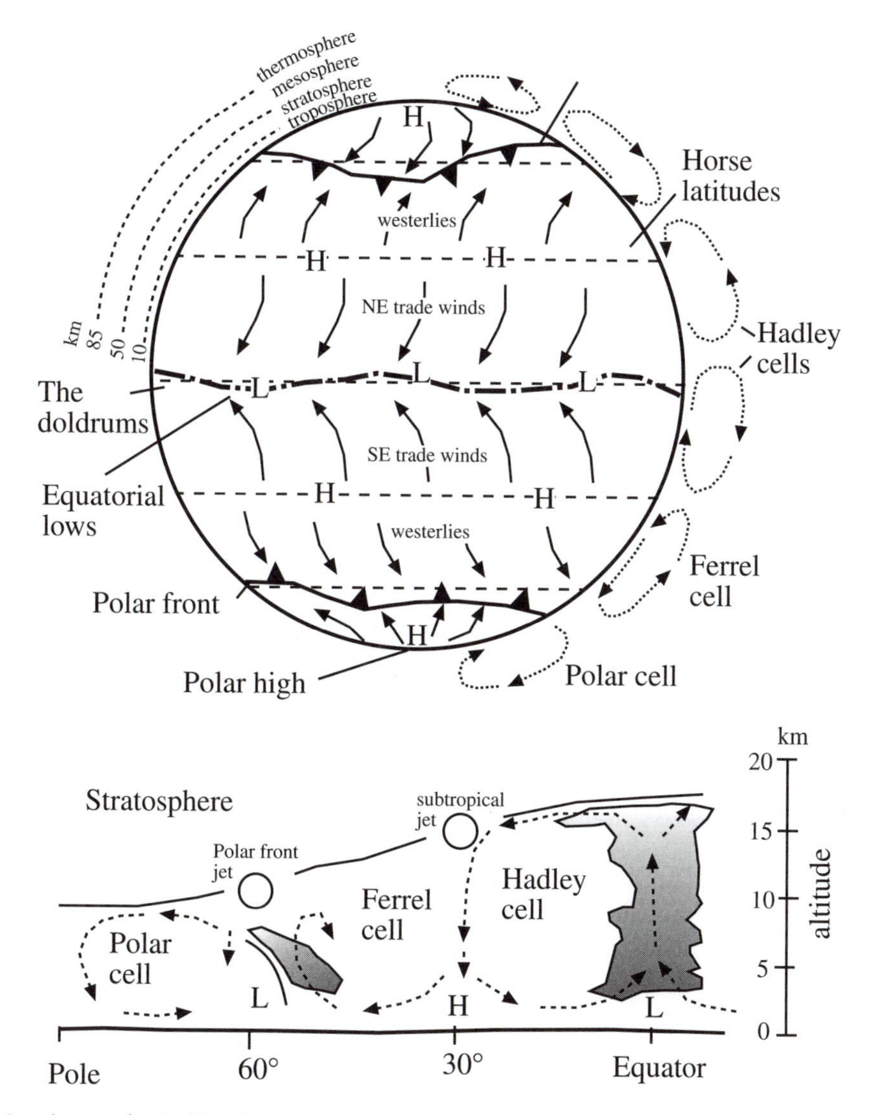

Figure 1.2. Atmospheric Circulation patterns showing predominant wind directions on the globe. Cross-section shows different layers of atmosphere (scale exaggerated). Detail in lower part of figure shows how upwelling areas near the equator and between 45–60° form belts of high-precipitation, whereas cold, dry, downwelling air between 20–40° contributes to the formation of most of the world's deserts.

made up largely of the elements carbon, hydrogen, and oxygen. When these organic elements decay, they may become part of the regolith and be returned through geological processes back to the lithosphere, atmosphere, or hydrosphere.

Rocks and the Rock Cycle

There are three major families of rocks: igneous, metamorphic, and sedimentary. The cooling, crystallization, and consolidation of magma form *igneous rocks. Sedimentary rocks* are products of weathering and deposition, and/or chemical precipitation. *Metamorphic rocks* are rocks whose original form has been changed from physical and chemical reactions at high temperatures and pressures. Most crustal rocks initially formed from magmas. Ninety-five percent of all rock in the earth's crust is either igneous or was derived from igneous rocks. However, sedimentary rocks cover about 75 percent of the surface of the earth, forming a thin cover resting on top of the igneous rocks. *The rock cycle* describes how rocks change with time. Igneous rocks may form at the earth's surface, then interact with external processes through erosion until eventually they become part of the regolith. With deep burial, these sediments become rocks again, and if buried deeply enough they may melt and form magmas, which again rise to the surface, restarting the cycle. This rock cycle has occurred over and over again throughout the history of the earth, in a process called crustal recycling. Continents are recycled on time scales of about 1 billion years or more, whereas oceans are recycled in periods of about 60–80 million years.

THE NATURE OF SCIENTIFIC INQUIRY (THE SCIENTIFIC METHOD)

All science, including the study of natural hazards in environmental geology, proceeds by an orderly set of rules. This is based on the assumption that the natural world behaves in a consistent and predictable manner and that the same set of laws that operates on small atomic particles also operates on large scales, spanning galaxies. Uniformitarianism is a principle, widely held in geology, which holds that the same processes that operate today also operated in the past. This principle is widely used to interpret the history of the planet preserved in the rock record.

In science, we strive to discover underlying patterns or laws to be able to predict the outcome of future events, given a specific set of circumstances. The first step in the scientific method includes the observation and measurement of a phenomenon and the collection of facts. In this step, scientists search for patterns that are repeatable. In the second step, scientists form a working hypothesis, or tentative explanation, that attempts to explain the facts collected in the first step. Many of the best scientists operate with multiple working hypotheses at one time, each of which may be a viable explanation for a phenomenon. The third step is to test the hypotheses with continued observation or experiments. If a hypothesis survives many tests, it may be elevated to the status of theory. If it does not, it is rejected or modified. Theories may eventually be discarded, but most are generally accepted as nearly fact. After continued rigorous testing and observation, a theory may become a scientific law, from which there has been no known deviation even after numerous observations and experiments.

GEOLOGIC TIME SCALES

The rates at which most geologic processes happen are very slow compared to what we experience in our day-to-day lives. For instance, plate tectonics is moving North America away from Europe at a few millimeters per year, which seems slow, yet only a short instant of geologic time has passed since North America and Europe were connected as part of the same land mass. Other geologic processes are more rapid, such as earthquakes, which may last only a minute or two. These short but dramatic events are only part of a much longer and slow geologic process that lasts millions of years.

The discovery of radioactivity and radioactive decay in the late 1800s has led to very precise ways to measure the ages of rocks and events by measuring the ratios of parent and daughter products of radioactive decay. We now know that the universe is about 13 billion years old, the earth is approximately 4.6 billion years old, and the oldest known rocks on the earth are about 4 billion years old. Before geologists could date rocks this way, however, they were able to only assign relative ages by looking at which rocks were laid down on top of one another, or which ones intruded into another one. To do this, they used the *law of superposition*, which states that in an undeformed sequence of sedimentary rocks or lava flows, each layer is older than the one above it and younger than the one below it. First established in the mid-1600s, this law represented a milestone in scientific reasoning and led to the first rational way to establish the relative ages of strata. For instance, if sequence ABCD outcrops in one area and CDEFG in another, the law of superposition made it possible to say that B is older than F, even though the two rock units do not occur in the same sequence. Similarly, by recognizing that any igneous rock that cuts across or intrudes into another rock must be younger than the rock into which it intrudes, early geologists were able to establish a relative chronology of events for specific areas. Another law, the *principle of faunal succession*, states that fossil organisms succeed one another in a definite and determinable order, and different time periods can be recognized by their fossil content. This law allowed geologists to recognize the relative ages of strata in widely separated locations. Together, these laws and radioactive dating have led to the construction of the geologic time scale (Figure 1.3), which serves as the fundamental way in which geologists discuss the different distinctive periods in the history of the earth, especially in historical geology.

BASIC EARTH STRUCTURE AND TECTONICS

Earth is one of a group of nine planets that condensed from a solar nebula about 5 billion years ago. The process was like a great big swirling cloud of hot dust, gas, and protoplanets that collided with each other, eventually forming the main planets. The accretion of the earth was a high-temperature process, which allowed the early earth to melt, with heavier metallic

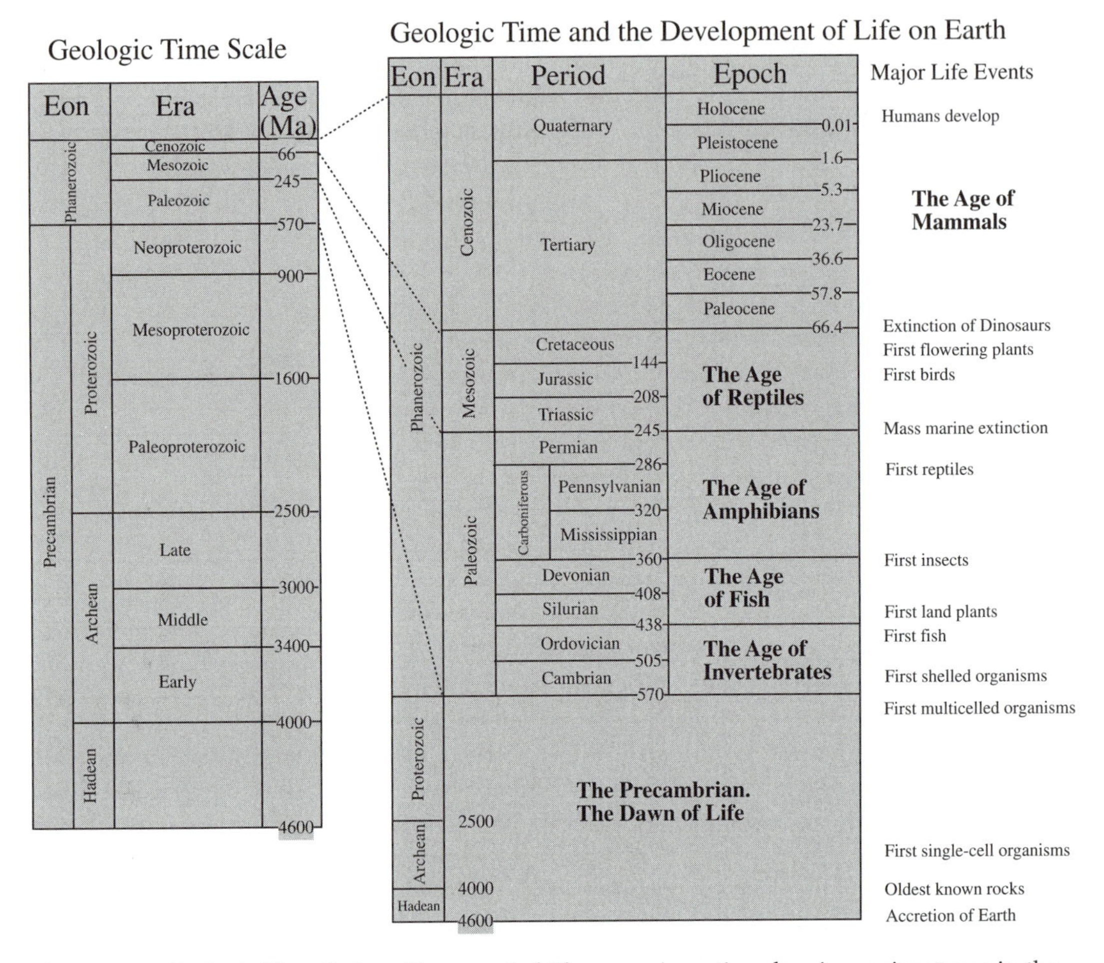

Figure 1.3. Geologic Time Scale, with expanded Phanerozoic section showing major stages in the development of life.

elements such as iron (Fe) and nickel (Ni) segregating and sinking to the earth's core and lighter rocky elements floating upward. This process led to the differentiation of the earth into several different concentric shells of contrasting density and composition (Figure 1.4), and was the main factor in constructing the earth's large-scale structure.

These main chemically defined shells of the earth include the crust, a light outer shell 5–70 km thick. This is followed inward by the mantle, a solid rocky layer extending to 2900 km beneath the earth's surface. The outer core is a molten metallic layer extending to 5100 km depth, and the inner core is a solid metallic layer extending to 6370 km (Figure 1.4).

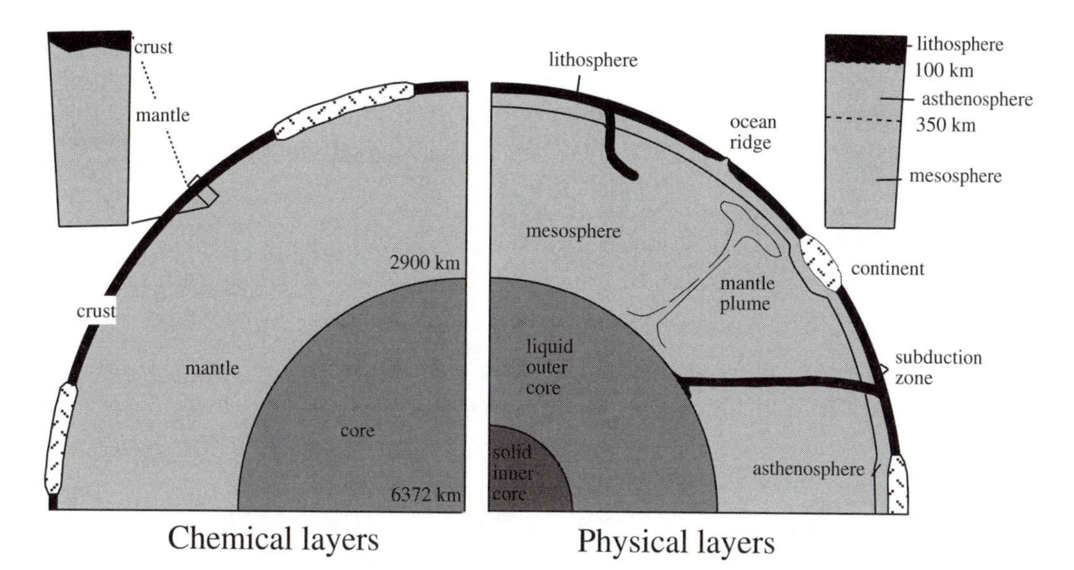

Figure 1.4. Basic Earth Structure, showing different ways of considering divisions in the Earth's interior. The chemical layers (left) show a basic layering into crust, mantle and core, whereas the physical layers (right) show layers that are solid, liquid, and have different mechanical properties. The strong lithosphere consists of the mobile tectonic plates, which move over the softer and easily deformable asthenosphere.

With the recognition of plate tectonics in the 1960s, geologists recognized that the outer parts of the earth were also divided into several zones that had very different mechanical properties. It was recognized that the outer shell of the earth was divided into many different rigid plates, all moving with respect to each other and some carrying continents in continental drift. This outer rigid layer became known as the *lithosphere*, which is Greek for "rigid rock sphere" and which ranges from 75 to 150 km thick, and includes the crust and upper part of the mantle. The lithosphere is essentially floating on a denser but partially molten layer of rock in the upper mantle known as the *asthenosphere*, or weak sphere. It is the weakness of the asthenosphere that allows the plates to move about on the surface of the earth.

Physiography of the Planet

The most basic division of the earth's surface shows that it is divided into continents and ocean basins, with oceans occupying about 60 percent of the surface and continents 40 percent. A transect or cross-section across any given continent to the ocean shows some major physiographic divisions (Figure 1.5). Mountains are elevated portions of the continents. Shorelines are where the land meets the sea. Continental shelves are areas underlain by continental crust and covered by shallow coastal water; they can be broad or narrow.

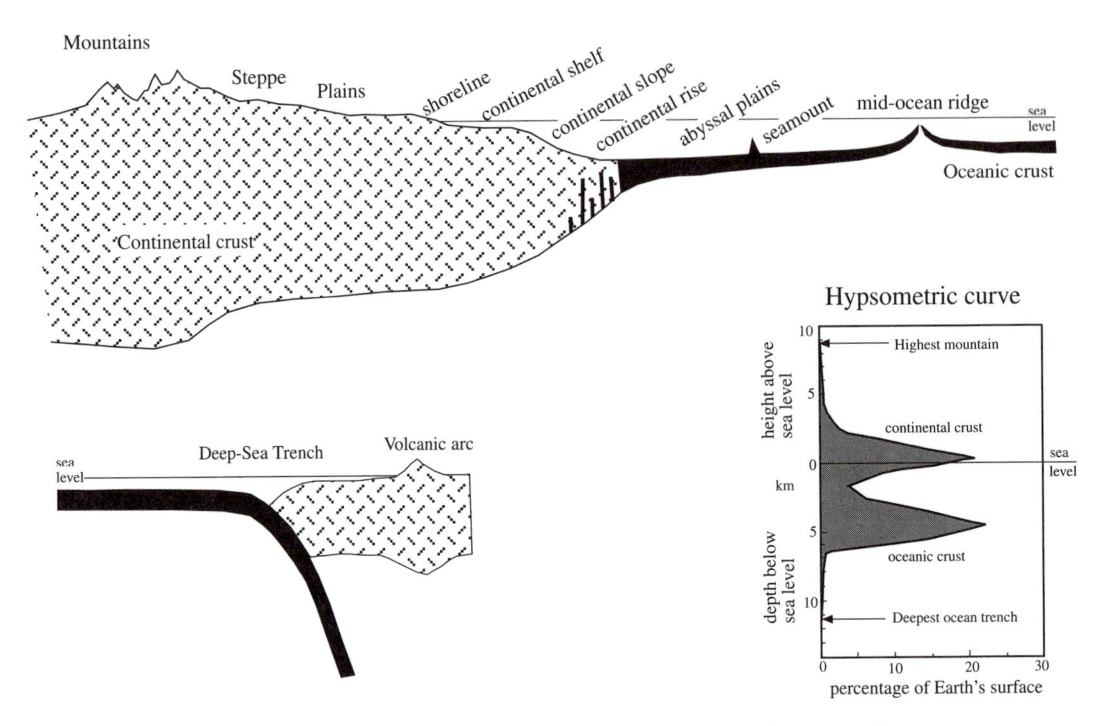

Figure 1.5. Physiography of the Earth. A cross section of a continental margin shows physiographic changes from mountains, steppes, plains and shoreline environment, to offshore and underwater environments of continental shelf, slope and rise, abyssal plains, and mid-ocean ridges. The continental crust is on average 35 kilometers (21 miles) thick beneath continents, but may be up to 70 kilometers (42 miles) thick beneath mountain ranges. Oceanic crust is typically 7 kilometers (4.2 miles) thick. Cross section of a convergent margin shows deep-sea trench. The hypsometric curve shows the percentages of crust at different elevations and clearly shows the planet's division into relatively deep, dense oceanic crust, and relatively high, light continental crust.

Continental slopes are steep drop-offs from the edge of a continental shelf to the deep ocean basin, and continental rises are where slopes flatten to merge with the deep ocean's abyssal plains. Ocean ridge systems are subaquatic mountain ranges where new ocean crust is being created by sea floor spreading.

Mountain belts on the earth are of two basic types. Orogenic belts are linear chains of mountains, largely on the continents, that contain highly deformed, contorted rocks that represent places where lithospheric plates have collided or slid past one another. The mid-ocean ridge system is a 65,000-kilometer mountain ridge that represents vast outpourings of young lava on the ocean floor; it represents places where new oceanic crust is being generated by plate tectonics. After it is formed, it moves away from the ridge crests and fills the space created by the plates drifting apart.

The oceanic basins also contain long, linear, deep ocean trenches that are up to several kilometers deeper than the surrounding ocean floor and can reach depths of fourteen kilometers below the sea surface. These represent places where the oceanic crust is sinking back into the mantle of the earth, completing the plate tectonic cycle for oceanic crust.

Plate Tectonics

In the 1960s a revolution shook up the earth sciences and resulted in the acceptance of the plate tectonic paradigm, which states that the earth's outer shell, or lithosphere, is broken into several rigid pieces, called plates, that are all moving. These plates are rigid and do not deform internally when they move, but only deform along their edges. The edges of plates are therefore where most mountain ranges are located, where most of the world's earthquakes occur, and where most active volcanoes are located. The plates are moving as a response to radioactive decay heating the mantle underneath, and they are much analogous to lumps floating on a pot of boiling stew.

We can define three fundamental types of plate boundaries (Figure 1.6). Divergent boundaries are where two plates move apart, creating a void that is typically filled by new oceanic crust that wells up to fill the progressively opening hole. Convergent boundaries are where two plates move toward each other; they result in one plate sliding beneath the other when a dense oceanic plate is involved or in collision and deformation when continental plates are involved. Transform boundaries form where two plates slide past each other, such as along the San Andreas Fault in California.

Since all plates are moving with respect to each other, the surface of the earth is made up of a mosaic of various plate boundaries, and the geologist has an amazing diversity of different geological environments to study. Every time one plate moves, the others must move to accommodate this motion, creating a never-ending saga of different plate configurations.

Where plates diverge, new oceanic crust is produced by *sea floor spreading,* in which volcanic basalt pours out of the depths of the earth and fills the gaps generated by the moving plates. Examples of this can be seen on the earth's surface, including Iceland along the Rekjanes Ridge.

In a process called *subduction,* oceanic lithosphere is being destroyed by sinking back into the mantle at the deep ocean trenches. As the oceanic slabs go down, they experience higher temperatures that cause melts to be generated, which then move upward to intrude the overlying plate. Since subduction zones are long narrow zones where large plates are being subducted into the mantle, the melting produces a long line of volcanoes above the down-going plate and forms a volcanic arc; this arc can form on either a continent or an oceanic plate, depending on what type of lithosphere the overlying plate is composed.

Plate Tectonics and Geologic Hazards

Plate tectonics and tectonic boundaries are important for understanding geologic

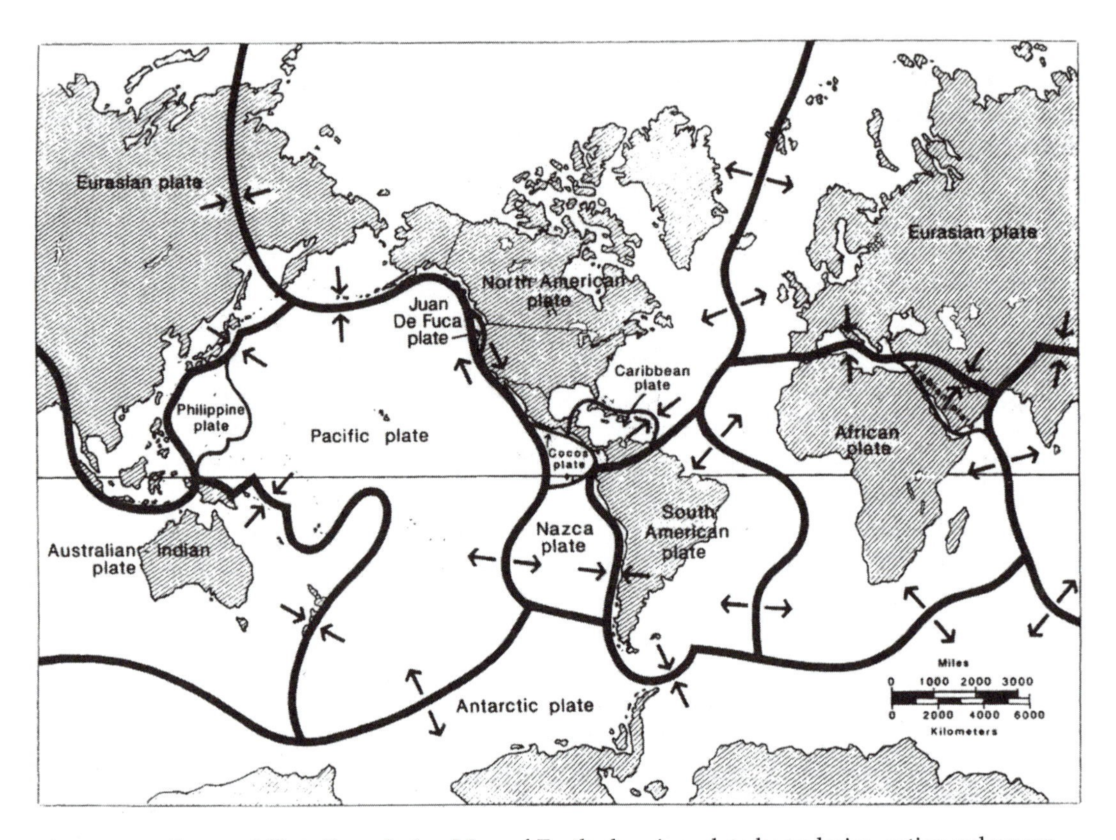

Figure 1.6. Types of Plate Boundaries. Map of Earth showing plate boundaries, active volcanoes, and earthquakes. Divergent boundaries (where two plates are moving apart) include the mid Atlantic ridge and the east Pacific rise. Convergent boundaries (where two plates are moving toward each other) include the Aleutian subduction zone, and the Alpine-Himalayan mountain chain. Transform or strike-slip boundaries (where two plates are sliding past each other) include the San Andreas fault in California, and many smaller faults on the ocean basins (too small to show on map). Figure Credit: U.S. Geological Survey

hazards because most of the planet's earthquakes, volcanic eruptions, and other hazards are located along and directly created by the interaction of plates; also, the plate tectonic setting controls the concentration of economic minerals (including petroleum). Thus, an understanding of plate tectonics is essential for planning for geologic hazards.

Plate tectonics can be thought of as the surface expression of great levels of energy loss from deep within the earth. Single earthquakes (Chapter 2) have killed as many as hundreds of thousands of people, such as the 1976 Tangshan earthquake in China that killed a quarter million people. Earthquakes also cause enormous financial and insurance losses. For instance, as mentioned earlier, the 1994 Northridge earthquake in California caused more than $14 billion in losses. Most of the world's volcanoes (Chapter 3) are also associated with plate boundaries. Thousands of volcanic vents are located along the mid-ocean ridge system, and most of the volume of magma produced on the earth is erupted through these volcanoes. However, volcanism associated with the mid-ocean ridge system is rarely explosive, hazardous, or even noticed by humans. In contrast, volcanoes situated above subduction zones at convergent boundaries are capable of producing tremendous explosive eruptions with great devastation of local regions. Volcanic eruptions and associated phenomena have killed tens of thousands of people this century, including the massive mudslides at Nevada del Ruiz in Colombia that killed 23,000 in 1985. Some of the larger volcanic eruptions cover huge parts of the globe with volcanic ash and are capable of changing the global climate.

Earth Surface Processes and Geologic Hazards

Plate tectonics is also responsible for uplifting the world's mountain belts, which are associated with their own sets of hazards, particularly landslides and other mass wasting phenomena (Chapter 5). These geologic hazards are associated with steep slopes and the effects of gravity moving material down these slopes to places where people live. Landslides and the slow downhill movement of earth material occasionally kill thousands of people in disasters, such as when parts of a mountain collapsed in 1970 in the Peruvian Andes and buried a town village dozens of miles away, killing 60,000 people. More typically, downhill movements are more localized and destroy individual homes, neighborhoods, roads, or bridges. Some downslope processes are very slow and involve the gradual, inch-by-inch creeping of soil and other earth material downhill, taking everything with it. This process of creep is one of the most costly of natural hazards, costing U.S. taxpayers billions of dollars per year.

Many geologic hazards are driven by energy from the sun and reflect the interaction of the hydrosphere, lithosphere, atmosphere, and biosphere. Heavy or prolonged rains can cause river systems to overflow, flooding low-lying areas, destroying towns and farmlands, and even changing the courses of major rivers (Chapter 6). Floods are of several types, including flash floods in mountainous areas and regional floods in large river valleys such as the great floods of the Mississippi and Missouri Rivers in 1993. Coastal regions may also experience floods, sometimes the result of typhoons, hurricanes, or coastal storms that bring high tides, storm surges, heavy rains, and deadly winds. Coastal storms may cause large amounts of coastal erosion, including cliff retreat, beach and dune migration, the opening of new tidal inlets, and the closing of old inlets (Chapter 7). Hurricane Andrew caused more than $19 billion of damage to the southern United States in 1992. These are all normal beach processes, but they have become hazardous since so many people have migrated into beachfront homes.

Deserts and dry regions are associated with their own set of natural geologic hazards (Chapter 8). Blowing winds and shifting sands make agriculture difficult, and deserts have a very limited capacity to support large populations. Some of the greatest disasters in human history have been caused by droughts, some associated with the expansion of desert regions into areas that previously received significant rainfall and supported large populations dependent on agriculture. In this century, the sub–Saharan Sahel region of Africa has been hit with drought disaster several times, affecting millions of people and animals. A particularly severe drought in 2003 affected millions of people in Ethiopia with rampant famine and starvation.

Desertification is but one possible manifestation of global climate change. Planet Earth has fluctuated in climate extremes from hot and dry to cold and dry or cold and wet, and it has experienced several periods when much of the land's surface was covered by glaciers (Chapter 9). Glaciers have their own set of local-scale hazards that affect those living or travelling on or near them. Crevasses can be deadly if fallen into, glacial meltwater streams can change their course so quickly that encampments on their banks can be washed away without a trace, and icebergs present hazards to shipping routes. Glaciers may, more importantly, reflect subtle changes in global climate: When glaciers are retreating, climate may be warming and becoming drier. When glaciers advance, the global climate may be getting colder and wetter. Glaciers have advanced and retreated over northern North America several times in the past 100,000 years; we are still in an interglacial episode and will see the start of the return of the continental glaciers over the next few hundred or thousand years.

Geologic materials can be hazardous (Chapter 10). Asbestos, a common mineral, is being removed from thousands of buildings in the nation because of the perceived threat that airborne asbestos fibers present to human health. In some cases for certain types of asbestos fibers, this perceived threat is real because the particles are already airborne. In other cases, the asbestos would be safer if it were left where it is rather than disturbing it and making its particles airborne. Natural radioactive decay is releasing harmful gases including radon that creep into our homes, schools, and offices and cause cancer in numerous cases every year. This hazard is easily mitigated, and simple monitoring and ventilation can prevent many health problems. Other materials can be hazardous even though they seem inert. For instance, some clay minerals expand by hundreds of percent when wetted. These expansive clays rest under many foundations, bridges, and highways, and cause billions of dollars of damage every year in the United States.

Sinkholes have swallowed homes and businesses in Florida and other locations in recent years. Sinkhole development and other subsidence hazards (Chapter 11) are more important than many people realize. Some large parts of southern California near Los Angeles have sunk tens of feet due to pumping groundwater and oil out of underground reservoirs. Other buildings that rest above former mining areas have begun sinking into collapsed mine tunnels. Coastline areas that are experiencing subsidence have the added risk of having the ocean rise into former living space. Coastal subsidence, coupled with gradual sea level rise, is rapidly becoming one of the major global hazards that the human race is going to have to deal with in the next century, since most of the world's population lives near the coast in the reach of the rising waters. Cities may become submerged and farmlands covered by shallow salty seas. An enormous amount of planning is needed, as soon as possible, to begin to deal with this growing threat.

Hazards from Out of This World

Occasionally in Earth's history, the planet has been hit with asteroids and meteorites from outer space, and these have completely devastated the biosphere and climate system (Chapter 12). Many of the mass extinctions in the geologic record are now thought to have been triggered, at least in part, by large impacts from outer space. For instance, the death of the dinosaurs and a huge percent of other species on Earth 66 million years ago is thought to have been caused by a combination of massive volcanism from a flood basalt province preserved in India, coupled with an impact with a six-mile-wide meteorite that hit the Yucatán Peninsula of Mexico. When the impact occurred, a 1,000-mile-wide fireball erupted into the upper atmosphere and a tsunami hundreds or thousands of feet high washed across the Caribbean, southern North America, and across much of the Atlantic. Huge earthquakes accompanied the explosion. The dust blown into the atmosphere immediately initiated a dark global winter, and as the dust settled months or years later, the extra carbon dioxide in the atmosphere warmed the Earth, for many years forming a greenhouse condition. Many forms of life could not tolerate these rapid changes and perished. Similar impacts have occurred at several times in Earth's history and have had a profound influence on the extinction and development of life on Earth.

Hazards of Accelerating Population Growth

The human population is growing at an alarming rate, with the population of the planet currently doubling every fifty years. At this rate, there will only be a three-by-one-foot space for every person on Earth in 800 years. Our unprecedented population growth has put such a stress on other species that we are driving a new mass extinction on the planet. We do not know the details of the relationships between different species and scientists fear that destroying so many other life forms may contribute to our own demise. In response to the population explosion, people are moving into hazardous locations including shorelines, riverbanks, along steep-sloped mountains, and along the flanks of volcanoes. Populations that grow too large to be supported by the environment usually suffer some catastrophe, disease, famine, or other mechanism that limits growth, and we as a species need to find ways to limit our growth to sustainable rates to ensure our very survival on the planet.

RESOURCES AND ORGANIZATIONS

Print Resources

Abbott, P. L. *Natural Disasters.* 3rd ed. Boston: McGraw Hill, 2002, 422 pp.
Bryant, E. A. *Natural Hazards.* Cambridge: Cambridge University Press, 1993, 294 pp.

Eldredge, N. *Life in the Balance.* Princeton, N.J.: Princeton University Press, 1998, 224 pp.

Erikson, J. *Quakes, Eruptions, and Other Geologic Cataclysms: Revealing the Earth's Hazards.* The Living Earth Series. New York: Facts on File Science Library, 2001, 310 pp.

Griggs, G. B., and Gilchrist, J. A. *Geologic Hazards, Resources, and Environmental Planning.* Belmont, Calif.: Wadsworth Publishing Co., 1983, 502 pp.

Keller, E. A. *Environmental Geology.* 8th ed. Englewood Cliffs, N.J.: Prentice Hall, 2000, 562 pp.

Mackenzie, F. T., and Mackenzie, J. A. *Our Changing Planet, An Introduction to Earth System Science and Global Environmental Change.* Englewood Cliffs, N.J.: Prentice Hall, 1995, 387 pp.

Murck, B. W., Skinner, B. J., and Porter, S. C. *Dangerous Earth: An Introduction to Geologic Hazards.* New York: John Wiley and Sons, 1997, 300 pp.

Skinner, B. J., and Porter, B. J. *The Dynamic Earth: An Introduction to Physical Geology.* New York: John Wiley and Sons, 1989, 541 pp.

Non-Print Sources

Web Sites

EROS Data Center Web site:
http://edcwww.cr.usgs.gov/
Lists satellite images, land cover maps, elevation models, maps, and aerial photography useful for natural hazards studies.

NASA's Web site on natural hazards:
http://earthobservatory.nasa.gov/NaturalHazards/natural_hazards_v2

The Natural Hazards Observer Web site:
http://www.colorado.edu/UCB/Research/IBS/hazards/o/o.html
This site, the online version of *The Natural Hazards Observer,* contains features about various hazards and disasters.

U.S. Geological Survey (USGS) Web site for natural hazards:
http://www.usgs.gov/themes/hazard.html

WeatherMatrix Web site:
http://www.weatherwatchers.org/
WeatherMatrix is a worldwide organization of over 3,000 amateur and professional weather enthusiasts—meteorologists, storm chasers and spotters, and weather observers from all parts of the globe. WeatherMatrix was formerly the Central Atlantic Storm Investigators (CASI). The site has frequently updated news about weather-related disasters.

Organizations

Congressional Natural Hazards Work Group
This group is a cooperative endeavor between a group of private and public organizations, whose goal is to develop a wider understanding within Congress of the value of reducing the risks and costs of natural disasters. The work group supports the effort of the Congressional Natural Hazards Caucus, led by co-Chairs Senator Ted Stevens (R-AK) and Senator John Edwards (D-NC). Information on the Natural Hazards Caucus Work Group can be found at http://www.agiweb.org/workgroup/. Some of the lead organizations include the American Meteorological Society and University Corporation for Atmospheric Research (http://www.ucar.edu), and the National Science Foundation (http//www.nsf.gov).

Federal Emergency Management Agency (FEMA)
500 C Street, SW
Washington, D.C. 20472
202–646–4600
http://www.fema.gov
FEMA is the nation's premier agency that deals with emergency management and preparation and that issues warnings and evacuation orders when disasters appear imminent. FEMA maintains a Web site that is updated at least daily and includes information of hurricanes, floods, fires, national flood insurance, and information on disaster prevention, preparation, and emergency management. Divided into national and regional sites. Also contains information on

costs of disasters, maps, and directions on how to do business with FEMA.

The National Drought Mitigation Center (NDMC)
University of Nebraska-Lincoln
http://enso.unl.edu/ndmc/
The NDMC helps people and institutions develop and implement measures to reduce societal vulnerability to drought. The NDMC stresses preparation and risk management rather than crisis management.

National Oceanographic and Atmospheric Association (NOAA)
http://www.noaa.gov/
NOAA conducts research and gathers data about the global oceans, atmosphere, space, and sun, and applies this knowledge to science and service that touch the lives of all Americans. NOAA's mission is to describe and predict changes in the earth's environment and to conserve and wisely manage the nation's coastal and marine resources. NOAA's strategy consists of seven interrelated strategic goals for environmental assessment, prediction, and stewardship. These include 1) providing advance short-term warnings and forecast services, 2) implementing seasonal-to-interannual climate forecasts, 3) assessing and predicting decadal-to-centennial change, 4) promoting safe navigation, 5) building sustainable fisheries, 6) recovering protected species, and 7) sustaining healthy coastal ecosystems. NOAA's Web site includes links to current satellite images of weather hazards, issues warnings of current coastal hazards and disasters, and has an extensive historical and educational service. The National Hurricane Center, located at http://www.nhc.noaa.gov/, is a branch of NOAA, and it posts regular updates of hurricane paths and hazards.

U.S. Geological Survey (USGS)
U.S. Department of the Interior
345 Middlefield Road
Menlo Park, CA 94025
650–329–5042
http://www.usgs.gov/
The USGS, which also has offices in Reston, Virginia, and Denver, Colorado, is responsible for making maps of many of the different types of hazards discussed in this book, including earthquake and volcano hazards, tsunami, floods, landslides, and radon. The National Landslide Information Center Web site is http://landslides.usgs.gov/html_files/nlicsun.html.

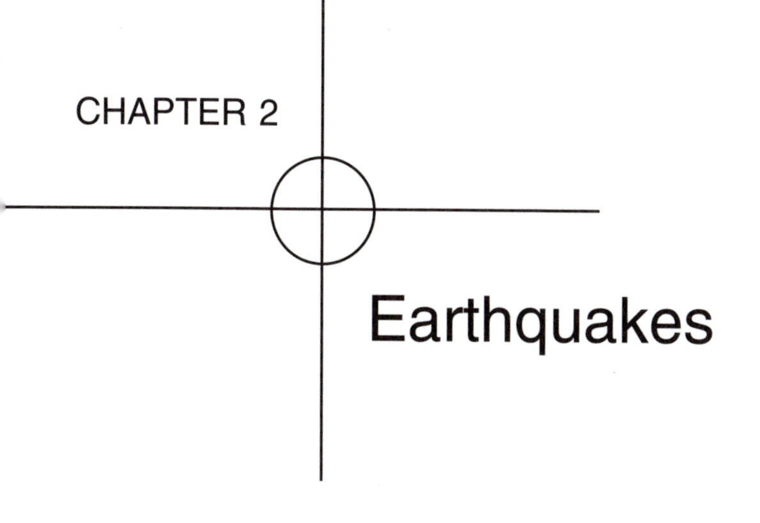

CHAPTER 2

Earthquakes

Earthquakes can be extremely devastating and costly events that sometimes kill hundreds of thousands of people and level entire cities in a matter of seconds. Recent earthquakes have been covered in detail by the news media, and the destruction and trauma of those affected is immediately apparent. A single earthquake may release the energy equivalent to thousands of nuclear blasts and may cost billions of dollars in damage, not to mention the toll in human suffering. Earthquakes are also associated with secondary hazards, such as tsunami, landslides, fire, famine, and disease, which also exert their toll on humans. In this chapter, the causes and consequences of earthquakes are examined in detail. This provides a context for a discussion of where earthquakes are most likely to occur, how they are studied, the hazards associated with earthquakes, their cost, and what might be done to reduce their devastating effects.

The lithosphere (or outer rigid shell) of the earth is broken into seven large tectonic plates, each moving relative to the other, and many smaller plates. Most of the world's earthquakes happen where two of these plates meet and are moving past each other, such as in southern California. Recent earthquakes in Turkey, Taiwan, Greece, and Mexico have also been located along plate boundaries. Figure 2.1 shows the plate boundaries of the earth, with dots showing where significant earthquakes have occurred in the past fifty years. Most really big earthquakes occur at boundaries where the plates are moving toward each other (as in Alaska) or sliding past one another (as in southern California). Smaller earthquakes occur where the plates are moving apart, such as along mid-oceanic ridges where new magma rises and forms ocean-spreading centers.

In the conterminous United States, the area that gets the most earthquakes is southern California along the San Andreas Fault, which is where the Pacific plate is sliding north relative to the North American plate. In this fault, the motion is characterized as a "stick-slip" type of sliding, where the two plates stick to each other along the plate boundary as they slowly move past each other and where stresses

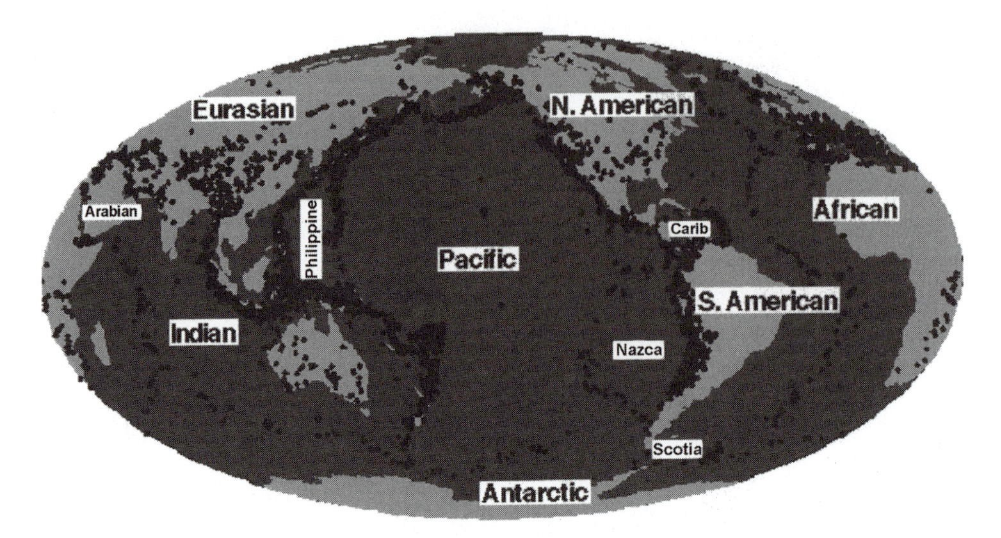

Figure 2.1. Plate boundaries of the Earth, with dots showing where significant earthquakes have occurred in the past 50 years. Note how the earthquakes correspond to the plate boundaries shown in Figure 1.6. Figure Credit: U.S. Geological Survey

rise over tens or hundreds of years. Eventually, the stresses along the boundary rise so high that the strength of the rocks is exceeded and the rocks suddenly break, causing the two plates to dramatically slip, moving up to a few meters in a few seconds. This sudden motion of previously stuck segments along a fault plane is an *earthquake*.

The severity of an earthquake is determined by how large of an area breaks in the earthquake, how far it moves, how deep within the earth the break occurs, and the length of time that the broken or slipped area along the fault takes to move. The *elastic rebound theory* states that recoverable (also known as elastic) stresses build up in a material until a specific level or breaking point is reached. When the breaking point or level is attained, the material suddenly breaks and releases the stresses in an earthquake. Initially straight lines formed by rows of fruit trees, fences, roads, and railroad lines that cross an active fault line and became gradually bent as the stresses built up are typically noticeably offset after an earthquake. When the earthquake occurs, the rocks snap along the fault and the bent rows of trees, fences, roads, or rail lines become straight again, but displaced across the fault. Figure 2.2 shows an orchard from southern California in which the trees were planted in straight rows but are now offset across a strand of the San Andreas Fault.

Some areas that are distant from active plate boundaries are also occasionally prone to earthquakes. Even though earthquakes in these areas are uncommon, they can be very destructive. Places including Boston, Massachusetts, Charleston, South Carolina, and New Madrid, Missouri (near

Figure 2.2. A fault trace crosses a cultivated field near El Centro California. The surface rupture on the Imperial Fault extended from about 2.5 miles (4 km) north of the International Border to about 2.5 miles south of Brawley. Maximum lateral displacement was about 22 inches (55 cm) in Heer Dunes and the maximum vertical displacement was 7.5 inches (19 cm) southeast of Brawley. Photo Credit: University of Colorado and NOAA/NGDC

St. Louis) have been sites of particularly bad earthquakes. For instance, in 1811 and 1812 three large earthquakes with magnitudes of 7.3, 7.5, and 7.8 were centered in New Madrid and shook nearly the entire United States, causing widespread destruction. Most buildings were toppled near the origin of the earthquake and several deaths were reported (the region only had a population of 1,000 at the time, although it is now densely populated). Damage to buildings was reported from as far away as Boston and Canada, where chimneys toppled, plaster cracked, and church bells were set to ringing by the shaking of the ground.

Many earthquakes in the past have been incredibly destructive, killing hundreds of thousands of people, like the ones in Armenia, Iran, and Mexico City in recent years (see Table 2.1). Some earthquakes have killed nearly a million people, such as one in 1556 in China that killed 800,000–900,000 people and one in Calcutta, India, in 1737 that killed about 300,000 people.

ORIGINS OF EARTHQUAKES

Earthquakes can originate from sudden motion along a fault, from a volcanic eruption, or from bomb blasts. Not every fault

Table 2.1
The Ten Worst Earthquakes in Terms of Loss of Life

Place	Year	Deaths	Estimated magnitude
Shaanxi, China	1556	830,000	
Calcutta, India	1737	300,000	
T'ang Shan, China	1976	242,000	m. 7.8
Gansu, China	1920	180,000	m. 8.6
Messina, Italy	1908	160,000	m. 7.5
Tokyo, Japan	1923	143,000	m. 8.3
Beijing, China	1731	100,000	
Chihli, China	1290	100,000	
Naples, Italy	1693	93,000	
Gansu, China	1932	70,000	m. 7.6

is associated with active earthquakes. In fact, most faults are no longer active but were at some time in the geologic past. Of the faults that are active, only some are characterized as being particularly prone to earthquakes. Some faults are slippery, and the two blocks on either side just slide by each other passively without producing major earthquakes. In other cases, however, the blocks stick to each other and deform like a rubber band until they reach a certain point where they suddenly snap, releasing energy in an earthquake event.

Rocks and materials are said to behave in a *brittle* way when they respond to built-up tectonic pressures by cracking, breaking, or fracturing. Earthquakes represent a sudden brittle response to built-up stress and are almost universally activated in the upper few kilometers of the earth. Deeper than this, the pressure and temperature are so high that the rocks simply deform like silly putty and don't snap, but are said to behave in a *ductile* manner.

An earthquake originates in one place and then spreads out. The *focus* is the point in the earth where the earthquake energy is first released. The *epicenter* is the point on the earth's surface that lies vertically above the focus.

When big earthquakes occur, the surface of the earth actually forms into waves that move across the surface, just as in the ocean. These waves can be pretty spectacular and also extremely destructive. When an earthquake occurs, these seismic waves move out in all directions, just like sound waves or ripples that move across water after a stone is thrown in a still pond. After the seismic waves have passed through the ground, the ground returns to its original shape, although buildings and other human constructions are commonly destroyed. Imagine experiencing a really large earthquake and actually seeing waves of rock, several feet high, moving toward you at very high speeds.

During an earthquake, these waves can either radiate underground from the focus—called *body waves*—or aboveground from the epicenter—called *surface waves*. Body waves travel through the whole body of the earth and move faster than surface waves, whereas surface waves cause most of the destruction associated with earthquakes because they briefly change the shape of the surface of the earth when they pass. There are two types of body waves: P (primary) or compressional waves and S (secondary) or shear waves. *P waves,* or *compressional waves,* deform material through a change in volume and density, and these waves can pass through solids, liquids, and gases. A back-and-forth type of motion is associated with passage of a P wave. P waves move with high velocity, about 6 km/second, and are thus the first to be recorded by seismographs; this is why they are called primary waves. P waves cause a lot of damage because they temporarily change the area and volume of ground that humans built things on or modified in ways that require the ground to keep its original shape, area, and volume. When the ground suddenly changes its volume by expanding and contracting, many of these constructions break. For instance, if a gas pipeline is buried in the ground, it may rupture or explode when a P wave passes because of its inability to change its shape along with the earth. It is common for fires and explosions originating from broken pipelines to accompany earthquakes.

The second kind of body waves is known as shear waves or S (secondary) waves because they change the shape of a material but not its volume. Shear waves can only pass through solid materials. Shear waves move material at right angles to the direction of wave travel and thus consist of an alternating series of sideways motions. Holding a jump rope so that one end is on the ground and moving it rapidly back and forth can simulate this kind of motion. Waves form at the end being held and move the rope sideways as they move toward the loose end of the rope. A typical shear wave velocity is 3.5 km/second. These kinds of waves may be responsible for knocking buildings off foundations when they pass, since their rapid sideways or back-and-forth motion is difficult for buildings to withstand. The effect is much like pulling a tablecloth out from under a set table—if done rapidly, the building (like the table setting) may be left relatively intact, but detached from its foundation.

Surface waves can also be extremely destructive during an earthquake. These waves have complicated types of twisting and circular motions, much like the circular motions you might feel while swimming in waves out past the surf zone at the beach. Surface waves travel slower than either type of body waves, but because of their complicated types of motion they often cause the most damage. This is a good thing to remember during an earthquake, because if you realize that the body waves have just passed your location, you may have a brief period of no shaking to get outside before the very destructive surface waves hit and cause even more destruction.

MEASURING EARTHQUAKES

How is the shaking of an earthquake measured? Geologists use *seismographs,* which display earth movements by means of an ink-filled stylus on a continuously turning roll of graph paper. When the ground shakes, the needle wiggles and leaves a characteristic zigzag line on the paper. Many seismograph records clearly show the arrival of P and S body waves, followed by the surface waves (Figure 2.3).

Seismographs are built using a few simple principles. To measure the shaking of the earth during a quake, the point of reference must be free from shaking, ideally on a hovering platform. However, since building perpetually hovering platforms is impractical, engineers have designed an instrument known as an inertial seismograph. These make use of the principle of *inertia,* which is the resistance of a large mass to sudden movement. When a heavy weight is hung from a string or thin spring, the string can be shaken and the big heavy weight will remain stationary. Using an inertial seismograph, the ink-filled stylus is attached to the heavy weight and remains stationary during an earthquake. The continuously turning graph paper is attached to the ground, and moves back and forth during the quake, resulting in the zigzag trace of the record of the earthquake motion on the graph paper.

Seismographs are used in series, some set up as pendulums and others as springs, to measure ground motion in many directions. Engineers have made seismographs that can record motions as small as one-hundred-millionth of an inch, about equivalent to being able to detect the ground motion caused by a car driving several blocks away. The ground motions recorded by seismographs are very distinctive, and geologists who study them have methods of distinguishing between earthquakes produced along faults, earthquake swarms associated with magma moving into volcanoes, and even explosions from different types of construction and nuclear blasts. Interpreting seismograph traces has therefore become an important aspect of nuclear test ban treaty verification.

Earthquake Magnitude

Earthquakes vary greatly in intensity, from undetectable ones to ones that kill millions of people and wreak total destruction. For instance, a bad earthquake in 1999 killed several thousands of people in Turkey, yet several thousand earthquakes that do no damage occur every day throughout the world. The energy released in large earthquakes is enormous, up to hundreds of times more powerful than large atomic blasts. Strong earthquakes may produce ground accelerations greater than the force of gravity, enough to uproot trees or send projectiles right through buildings, trees, or anything else in their path.

Earthquake magnitudes are most commonly measured using the *Richter scale.* The Richter scale gives an idea of the amount of energy released during an earthquake and is based on the *amplitudes* (half the height from wave-base to wave-crest) of seismic waves at a distance of 100 km (sixty-one miles) from the epicenter. The Richter scale magnitude of an earthquake is calculated using the zigzag trace produced on a seismograph, once the epicenter has been located by comparing signals from several different, widely separated seismographs (see Figure 2.4). The Richter scale is logarithmic, which means that each increasing point on the scale corresponds to a tenfold increase in amplitude. This is necessary because the energy of earthquakes changes by factors of more than 100 million.

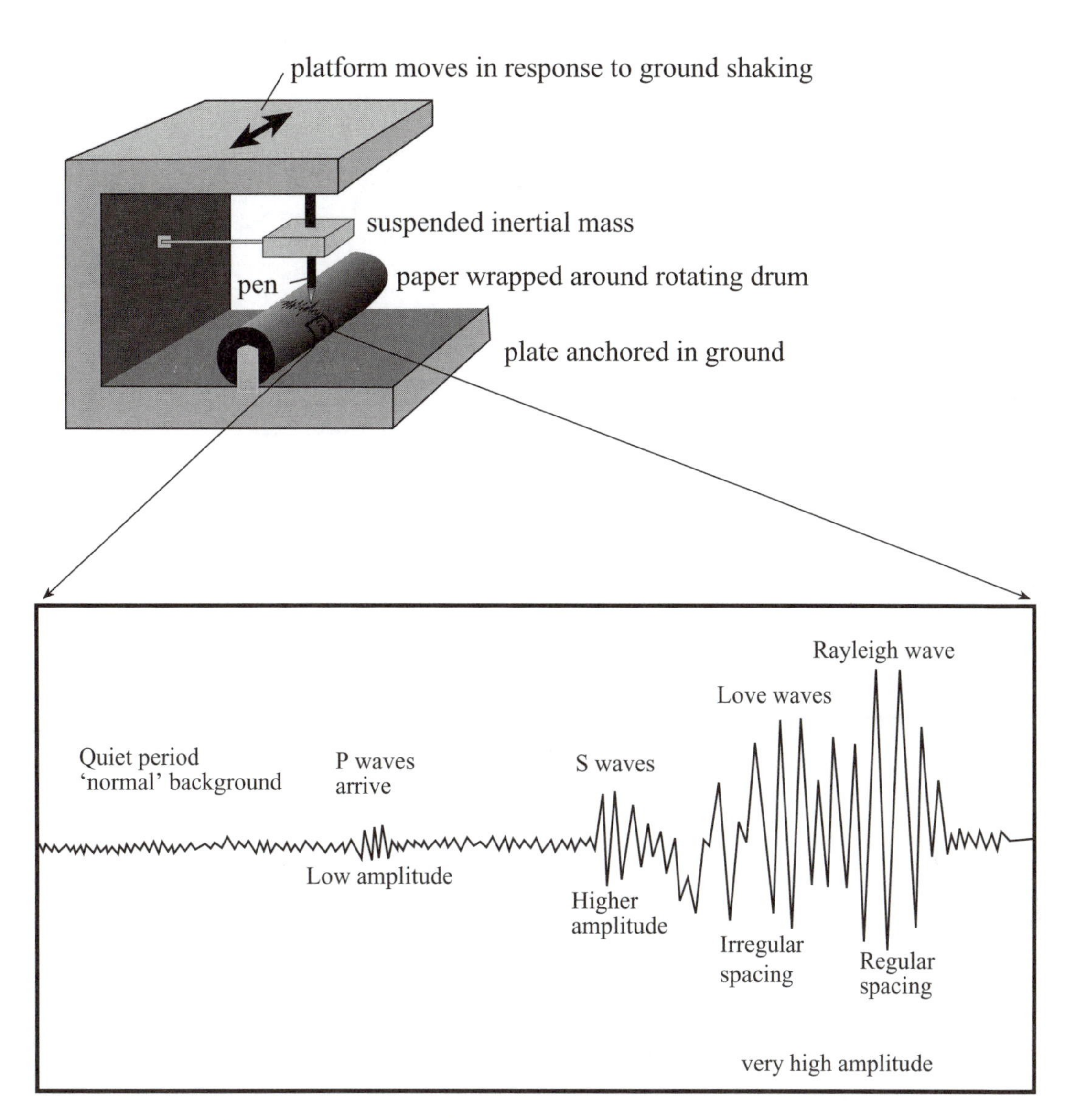

Figure 2.3. Inertial seismograph principle. A heavy inertial mass is suspended from a spring or hinge attached to a fixed platform. When an earthquake shakes the platform, the inertial mass stays in one place, swinging from the hinge. A paper wrapped around a rotating drum vibrates with the shaking platform, and a pen attached to the suspended inertial mass records the back and forth motions of the platform, recording the amplitude of the seismic waves. Since the drum is rotating at a set rate, the time of arrival of different seismic waves can be measured. Lower part of diagram shows a seismograph record indicating the arrival of P and S body waves, followed by the surface waves.

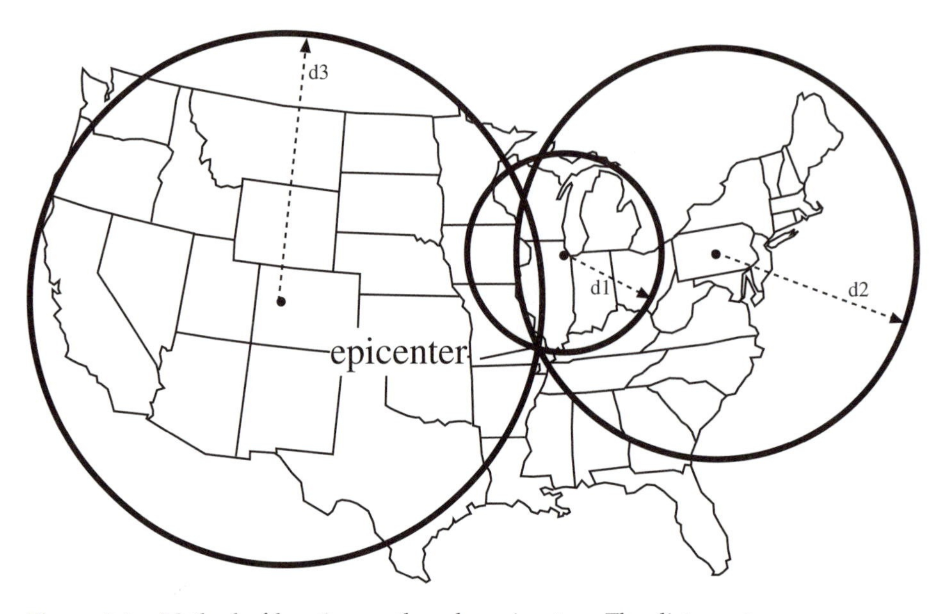

Figure 2.4. Method of locating earthquake epicenters. The distance to an earthquake from any point can be calculated using the time difference of arrival of the P and S waves, which increases as the distance increases since P waves travel much faster than S waves (6 kilometers/sec vs. 3.5 kilometers/sec). This distance can be drawn as a circle around the point at which the time difference is measured, and the epicenter must lie on this circle. If this procedure is done for three different seismic stations, the three circles must intersect at one point, which is the earthquake epicenter (in this case, New Madrid Missouri).

The energy released in earthquakes changes even more rapidly with each increase in the Richter scale because the number of high amplitude waves increases with bigger earthquakes and also because the energy released is according to the square of the amplitude. Thus, it turns out in the end that an increase of 1 on the Richter scale corresponds to a thirtyfold increase in energy released. The largest earthquakes so far recorded are the 9.2 Alaskan earthquake of 1964 and the 9.5 Chilean earthquake of 1960, each of which released the energy equivalent to approximately 10,000 nuclear bombs the size of the one dropped on Hiroshima.

Before the development of modern inertial seismographs, earthquake intensity was commonly measured using the *modified Mercalli intensity scale.* This scale, named after Father Giuseppi Mercalli, was developed in the late 1800s; it measures the amount of vibration people remember feeling for low-magnitude earthquakes and measures the amount of damage to buildings in high-magnitude events. Table 2.2 compares the Richter and modified Mercalli scales. One of the disadvantages of the

Table 2.2
Modified Mercalli Intensity Scale Compared to Richter Magnitude

Mercalli Intensity	Richter Magnitude	Description
I–II	< 2	Not felt by most people.
III	3	Felt by some people indoors, especially on high floors.
IV–V	4	Noticed by most people. Hanging objects swing. Dishes rattle.
VI–VII	5	All people feel. Some building damage, especially to masonry. Waves on ponds.
VII–VIII	6	Difficult to stand, people scared or panicked. Difficult to steer cars. Moderate damage to buildings.
IX–X	7	Major damage, general panic of public. Most masonry and frame structures destroyed. Underground pipes broken. Large landslides.
XI–XII	8 and higher	Near total destruction.

Mercalli scale is that it is not corrected for distance from the epicenter. Therefore, people near the source of the earthquake may measure the earthquake as a IX or X, whereas people further from the epicenter might only record a I or II event. However, the modified Mercalli scale has proven very useful for estimating the magnitudes of historical earthquakes that occurred before the development of modern seismographs, since the Mercalli magnitude can be estimated from historical records.

EARTHQUAKE HAZARDS

Earthquakes are associated with a wide variety of specific hazards, including primary effects such as ground motion, ground breaks (or faulting), mass wasting, and liquefaction. Secondary and tertiary hazards are indirect effects that are caused by events initiated by the earthquake. These may include seiche waves and tsunami; fires and explosions caused by disruption of utilities and pipelines; and changes in ground level that may cause disruption of habitats, changes in groundwater level, displacement of coastlines, loss of jobs, and displacement of populations. Financial losses to individuals, insurance companies, and loss of revenue to businesses can easily soar into the tens of billions of dollars for even moderate-sized earthquakes.

Ground Motion

One of the primary hazards associated with earthquakes is ground motion caused by the passage of seismic waves through populated areas. The most destructive waves are surface waves, which in severe earthquakes may visibly deform the surface of the earth into moving waves. Ground motion is most typically felt as shaking and causes the familiar rattling of objects off shelves reported from many minor earthquakes. The amount of destruction associated with given amounts of ground motion depends largely on the design and construction of buildings and infrastructure according to specific codes.

The amount of ground motion associated with an earthquake generally increases with the magnitude of the earthquake, but depends also on the nature of the substratum—loose, unconsolidated fill tends to shake more than solid bedrock. This was dramatically illustrated by the 1989 Loma Prieta earthquake in California, where areas built on solid rock vibrated the least (and saw the least destruction) and areas built on loose clays vibrated the most. Much of the San Francisco Bay area is built on loose clays and mud, including the Nimitz Freeway, which collapsed during the event. The area that saw the worst destruction associated with ground shaking was San Francisco's Marina district. Even though this area is located far from the earthquake epicenter, it is built on loose unconsolidated landfill that shook severely during the earthquake, causing many buildings to collapse and gas lines to rupture and initiate fires. More than twice as much damage from ground shaking during the Loma Prieta earthquake was reported from areas over loose fill or mud than from areas built over solid bedrock. Similar effects were reported from the 1985 earthquake in Mexico City, which is built largely on old lakebed deposits.

Additional variations in the severity of ground motion are noted in the way that different types of bedrock transmit seismic waves. Earthquakes that occur in the western United States generally affect a smaller area than those that occur in the central and eastern parts of the country. This is because the bedrock in the West (California, in particular) is generally much softer than the hard igneous and metamorphic bedrock found in the East. Harder, denser rock generally transmits seismic waves better than softer, less dense rock, so earthquakes of a given magnitude may be more severe over larger areas in the East than in the West. Fortunately, at least from the perspective of ground motion intensity, more large earthquakes occur in the West than in the East.

Ground motions are recorded as *accelerations,* which measure the rate of change of motion. This type of force is the same as accelerating in a car, where you feel yourself being pushed gently back against the seat. This is a small force compared to another common acceleration force, gravity. Gravity is equal to 9.8 meters per second squared, or 1 g (this is what you would feel if you jumped out of an airplane). People have trouble standing up and buildings begin falling down at 1/10 the acceleration of gravity (0.1 g). Large earthquakes can produce accelerations that greatly exceed and even double or triple the force of gravity. These accelerations are able to uproot large trees and toss them into the air, shoot objects through walls and buildings, and cause almost any structure to collapse.

Some of the damage typically associated with ground motion and the passage of seismic waves includes swaying and pancaking of buildings. During an earthquake, buildings may sway with a characteristic frequency that depends on the building's height, size, construction, and underlying material, and on the intensity of the earthquake. This causes heavy objects to rapidly dash from side to side inside the buildings, which can be quite destructive. The shaking generally increases with height, and in many cases the shaking causes concrete floors at high levels to separate from the walls and corner fastenings, causing the floors to progressively fall or pancake upon another, crushing all in between. With higher amounts of shaking, the entire structure may collapse.

Ground Breaks

If you've ever watched an action movie about an earthquake, you've probably seen pictures of great ruptures opening in the ground, swallowing up all in their path, and then closing again. Although these scenes are far from reality, ground breaks or ruptures are a serious hazard associated with earthquakes.

Ground breaks form where a fault cuts the surface, and they may also be associated with mass wasting, or the downhill movements of large blocks of land. These ground breaks may have horizontal, vertical, or combined displacements across them and may cause considerable damage. Fissures that open in the ground during some earthquakes are mostly associated with the mass down slope movement of material and not with the fault trace itself breaking the surface. For instance, in the 1964 Alaskan earthquake, ground breaks displaced railroad lines by several yards; broke through streets, houses, storefronts, and other structures; and caused parts of them to drop by several yards relative to other parts of the structure. Ground breaks are also one of the causes of the rupture of pipelines and communication cables during earthquakes.

Mass Wasting

Mass wasting (Chapter 5) is the downhill movement of material. In most instances, mass wasting occurs by a slow gradual creeping of soils and rocks downhill, but during earthquakes large volumes of rock, soil, and all that is built on them may suddenly collapse in a landslide. Earthquake-induced landslides occur in areas with steep slopes or cliffs, such as parts of California, Alaska, South America, Turkey, and China. One of the worst recorded earthquake-induced landslides occurred in Peru during the 1970 m 7.8 earthquake, in which at least 18,000 people were killed.

In the 1964 m 9.2 earthquake in Alaska, landslides destroyed power plants, homes, roads, and railroad lines. Some landslides even occurred undersea and along the seashore. Large parts of the towns of Seward and Valdez in Alaska were sitting on the top of large submarine (underwater) escarpments; during the earthquake, large parts of these towns slid out to sea in giant submarine landslides and were submerged. Another residential area near Anchorage, Turnagain Heights, was built on top of cliffs with fantastic views of the Alaska range and Aleutian volcanoes. When the earthquake struck, this area slid out toward the sea on a series of curving faults that connected in a slippery shale unit known as the Bootlegger shale. During the earthquake, this shale unit lost almost all cohesion, and the shaking of the soil and rock above it caused the entire neighborhood to slide toward the sea along the shale unit and be destroyed. Damage related to mass wasting and ground sliding

Figure 2.5a. Damage to Fourth Avenue, Anchorage, Alaska, from the 1964 m. 9.2 earthquake. Buildings have collapsed into graben formed by sliding and slumping associated with earthquake. Photo Credit: NOAA/NGDC

was extensive throughout the Anchorage area (Figure 2.5).

Liquefaction

Liquefaction is a process in which sudden shaking of certain types of water-saturated sands and muds turns these once-solid sediments into a slurry with a liquid-like consistency. The shaking causes individual grains to move apart and then water moves up in between the individual grains, making the whole water-sediment mixture behave like a fluid. Earthquakes often cause liquefaction of sands and muds, and any structures that are built on sediments that liquefy may suddenly sink into them as if they were resting on a thick fluid. It was the process of liquefaction that caused the Bootlegger shale in the 1964 Alaskan earthquake to suddenly become so weak, causing the destruction of Turnagain Heights. Liquefaction is also responsible for the sinking of sidewalks, telephone poles, building foundations, and other structures during earthquakes. One famous example of liquefaction occurred in the 1964 Japan earthquake, where entire rows of apartment buildings rolled onto their sides but were not severely damaged internally (Figure 2.6). Liquefaction also causes sand to bubble to the surface during an earthquake, forming mounds up to several meters high known as sand volcanoes or ridges of sand.

Figure 2.5b. A subsidence trough (or graben) formed at the head of the "L" Street landslide in Anchorage during the 1964 earthquake. The graben extends from the lower right corner of the picture to the upper left and passes beneath several houses. The slide block, which is the virtually unbroken ground to the left of the graben, moved to the left. The subsidence trough sank 7 to 10 feet (2.1 to 3.0 m) in response to 11 feet (3.3 m) of horizontal movement of the slide block. The volume of the trough is theoretically equal to the volume of the void created at the head of the slide by movement of the slide block. Note also the collapsed Four Seasons apartment building at the top center of the picture and the undamaged three-story reinforced concrete frame building behind it, which are on the stable block beyond the graben. Like the Turnagain Heights area, this neighborhood was built on the Bootlegger shale, a relatively weak unit under a firmer surface layer. During the earthquake this shale unit became almost cohesionless, and the shaking of the soil and rock above it caused the entire area to slide towards the sea along the shale unit and be destroyed. Photo Credit: NOAA/NGDC

Changes in Ground Level

During earthquakes, blocks of earth shift relative to one another. This may result in changes in the ground level, base level, water table, and high tide marks. Particularly large shifts have been recorded from some of the historically large earthquakes, such as the 1964 Alaskan one. An area more than 600 miles (1,000 km) long in south-central Alaska recorded significant changes in ground level, including uplifts of up to twelve yards (eleven meters), downdrops of more than two yards, and lateral shifts from several yards to tens of yards. Areas along the coastline that were uplifted experienced dramatic changes in the marine ecosystem: Clam banks were suddenly uplifted out of the water and remained high and dry. Towns built around docks were suddenly located many yards above the convenience of being at the shoreline. Areas that were downdropped experienced different effects: Forests that relied on fresh water for their root systems suddenly were inundated by salt water and were effectively "drowned." Populated areas located at previously safe distances from the high tide (and storm) line became prone to flooding and storm surges, and had to be relocated.

Areas that were far inland also suffered from changes in ground level: When some areas were uplifted by many meters, the water table recovered to a lower level relative to the land's surface and soon became out of reach of many water wells that had to be redrilled. Thus, changes in ground level, although seemingly a minor hazard associated with earthquakes, are significant and cause a large amount of damage that may cost millions of dollars to mitigate.

Tsunami and Seiche Waves

There are several types of large waves associated with earthquakes, including tsunami and seiche waves (Chapter 4). Tsunami, also known as seismic sea waves, form most usually from submarine landslides that displace a large volume of rock and sediment on the sea floor, which in turn displaces a large amount of water. Tsunami may be particularly destructive as they travel very rapidly (hundreds of miles per hour) and may reach many tens of yards above normal high tide levels. Two particularly devastating examples include a tsunami generated by a magnitude 8.7 earthquake in the Atlantic Ocean in 1775 that is estimated to have killed more than 60,000 people in Portugal; this number is from Lisbon alone, although the tsunami struck a large section of coastline and other tsunami were reported from North Africa, the British Isles, and the Netherlands. The other tsunami, generated in the Aleutian Islands of Alaska in 1946, traveled across the Pacific Ocean at 500 miles per hour (800 km/hour) and hit Hilo, Hawaii, with a crest eighteen yards higher than the normal high tide mark, killing 159 people, destroying approximately 500 homes, and damaging 1000 more structures.

Seiche waves may be generated by the back-and-forth motion associated with earthquakes, causing a body of water (usually lakes or bays) to rock back and forth, gaining amplitude and splashing up to higher levels than normal. The effect is analogous to shaking a glass of water back and forth, and watching the ripples in the water suddenly turn into large waves that splash out of the glass. Other seiche waves may be formed when landslides or rockfalls drop large volumes of earth into bodies of water. The largest recorded seiche wave of this type formed suddenly on July 9, 1958, when a large earthquake-initiated rockfall generated a seiche wave 1,700 feet (518 meters) tall. The seiche wave raced across Lituya Bay in Alaska, destroying the forest and killing several people, including a geologist who had warned authorities that such a wave could be generated by a large landslide in Lituya Bay.

Figure 2.6. One famous example of liquefaction occurred in the 1964 Japan earthquake, where entire rows of apartment buildings rolled onto their sides, but were not severely damaged internally. Photo shows aerial view of leaning apartment houses in Niigata produced by soil liquefaction and the behavior of poor foundations. Most of the damage was caused by cracking and unequal settlement of the ground such as is shown here. About ⅓ of the city subsided by as much as 2 meters as a result of sand compaction. Photo Credit: NOAA/NGDC

Damage to Utilities

Much of the damage and many of the casualties associated with earthquakes are associated with damage to the infrastructure and system of public utilities. For instance, much of the damage associated with the 1906 San Francisco earthquake came not from the earthquake itself, but from the huge fire that resulted from the numerous broken gas lines, from overturned wood and coal stoves, and even from fires set intentionally to collect insurance money on partially damaged buildings. Likewise, in the 1995 Kobe, Japan, earthquake, a large percentage of the damage was from fires that raged uncontrolled, with fire and rescue teams unable to reach the areas worst affected. Water lines were broken so that even in places that were accessible, firefighters were unable to put out the flames.

One of the lessons from these examples is that evacuation routes need to be set up in earthquake hazard zones in anticipation of post-earthquake hazards such as fires, aftershocks, and famine. These routes should ideally be clear of obstacles such as overpasses and buildings that may block access, and efforts should be made to clear these routes soon after earthquake disasters both for evacuation purposes and for emergency access to the areas worst affected.

WHAT CAN BE DONE TO REDUCE EARTHQUAKE HAZARDS?

Predicting Earthquakes

Knowing when and where an earthquake will occur and how strong it will be could save innumerable lives. Being able to predict earthquakes has therefore been the goal of many geologists and national planners for some time. Some observations have been used to predict in a very general manner which areas are most likely to have earthquakes in specific intervals of time. These include observing the recurrence interval of earthquakes along individual segments of faults, mapping gaps in minor seismic activity along faults (reflecting places where the fault is stuck, or locked, and may slip causing an earthquake), and observing minor changes in the physical properties of rocks. Other earthquake precursors are also being studied, including release of gases before earthquakes, releases of electrical energy before earthquakes, and changes in animal behavior in the few hours immediately preceding large earthquakes.

Predicting earthquakes based on recurrence interval is based on the statistics of how frequently earthquakes of specific magnitude occur along individual segments of faults. In many places, the historical records of past earthquakes do not go back very far compared to the typical interval between earthquakes. For instance, most of the seismically active parts of the western United States only have historical records going back a couple of hundred years, and in parts of the eastern United States these records go back somewhat less than 400 years. In other parts of the world, such as Japan and China, historical records of earthquakes go back more than a thousand years, enabling better documentation of earthquake intervals.

Earthquake forecasting based on recurrence interval of past events is based on statistics and yields only probabilities that earthquake events will happen in certain time intervals. For instance, if historical records show that earthquakes of magnitude 7 occur along a segment of a fault roughly every 150 years (perhaps with a twenty-year error margin, so that they really occur every 130 to 170 years), and if it has been 149 years since the last magnitude 7 earthquake, does it mean that a magnitude 7 earthquake will definitely occur along that segment of the fault in the next year? The answer is no, but the probability of a magnitude 7 earthquake occurring within the next ten years is high.

Seismic gaps are places along large fault zones that have little or no seismic activity compared to adjacent parts of the same fault. Seismic gaps are generally interpreted as places where the fault zone is stuck and where adjacent parts of the fault are gradually creeping or slipping along, slowly releasing seismic energy and strains associated with relative motion of opposing sides of the fault. Since the areas of the seismic gaps are not slipping, the energy gradually builds up in these sections until it is released in a relatively large earthquake. If the size of the seismic gaps and the amount of unslipped relative motion on either side of the fault can both be measured, then the size of the impending earthquake in the seismic gap can be predicted. Predicting when the earthquakes may occur in seismic gaps is another matter, and estimates must be based on recurrence intervals from past earthquakes or predicted from estimates of when the strength of the rock in the fault zone will be exceeded and cause rupture (associated with the earthquake) to occur.

Some of the physical properties of rocks in fault zones actually change in measurable ways prior to some earthquakes, and monitoring of these changes may also ultimately help predict impending earthquake events. One measurable change in rocks is called *dilation,* where rocks expand because of the development of numerous minor cracks or fractures that form in response to the stresses concentrated along the fault zone. The amount of dilation may be expressed by surface bulging along the fault zone. In theory, if the strength of rocks and the stress across the fault zone are both known, then the amount of dilation can be related to the decrease in rock strength and scientists can estimate when the earthquake may occur. Other physical properties of the rocks in fault zones may also change prior to earthquakes. For instance, the velocity of seismic waves is known to change in fault zones prior to some earthquakes and is thought to reflect changes in the number of small cracks in the rock as the strain accumulates before rupture events. Likewise, the electrical conductivity may also change and may also be related to changes in the physical properties of the rocks.

Sometimes large earthquakes are preceded by distinctive swarms of small earthquakes known as *foreshocks.* These are related to the formation of many small cracks and may be associated with the tilting of the land surface because of the built-up strain in the rocks adjacent to the fault. With the establishment of seismic monitoring stations near active faults, these small earthquake swarms can be observed and can ultimately help warn of impending large earthquake events.

There are several other poorly understood phenomena that may someday be used to help predict earthquakes. For instance, it has been documented that radon gas levels increase in some groundwater wells before some earthquakes. Also, groundwater levels in wells may drop before earthquakes, which may be caused by the water filling up the many microcracks that form in the fault zone's rocks before major earthquake events. Perhaps most peculiar among earthquake precursors is anomalous animal behavior. Some dogs and horses have exhibited unusual and erratic behavior before earthquakes, as have snakes, chickens, and fish that leap from the water of ponds. Why these animals behave so unusually before earthquakes is unknown, but may be related to their senses being able to identify changes in gases, light emitted from highly strained crystals before earthquakes, or changes in the local electromagnetic field immediately before large earthquakes. Perhaps these animals hold the key to short-term prediction of earthquakes, and we will be able to learn from observing them more closely and analyzing what it is in the environment that they sense changing before catastrophe strikes.

Seismic Hazard Zones and Risk Mapping

One of the ways to make the public and private sectors aware of the specific risks of earthquakes in individual areas is known as seismic risk zone mapping. In this technique, geologists assess the likelihood of having earthquakes of specific magnitudes in an area, then integrate knowledge of how the earth materials in a region or small area will respond to seismic shaking to make a map of likely effects throughout the region or area. These maps incorporate information such as the nature of the underlying rock, because different rock types respond to the passage of seismic waves in different ways (for instance, thick clay and soil tend to amplify seismic waves, creating longer, more intense shaking than on solid bedrock). The information in seismic risk maps is used by planners for designing further construction and may also be used with knowledge of population distributions, types of existing structures, and utilities to formulate emergency plans before large earthquakes occur. This type of planning can greatly reduce casualties from earthquakes.

Architecture and Building Codes

Places that are particularly prone to earthquakes should have strict building codes with buildings and other structures designed to withstand the shaking and jolting that accompany earthquakes. Skyscrapers can be constructed to sway, and smaller structures can be built with expansion joints and reinforced concrete so that they can better withstand the passage of the seismic waves. Bridges, tunnels, and pipelines should have the ability to expand and contract and to vibrate back and forth so that they do not rupture or collapse in earthquakes. Many of these expensive yet important construction features are commonly implemented in western countries such as the United States, where earthquakes are common. Unfortunately, in many other parts of the world, there are no enforced building codes in earthquake zones, and there is accordingly a much greater loss of life associated with property destruction during earthquakes. This was dramatically illustrated by the 1999 m 7.4 earthquake in Izmit, Turkey, during which much of the old parts of the cities in western Turkey survived the earthquake with little damage and loss of life. In contrast, recent, hastily built suburbs and apartment blocks that were built in violation of existing codes collapsed, causing tremendous destruction and loss of life.

There is great variation in the types of materials and styles of construction that must be used to build earthquake-resistant structures in different conditions. For instance, if a building is erected on solid bedrock close to likely epicenters, it should be made to withstand high-frequency shaking. The building should be made of flexible material and designed with a tall frame so that the natural frequency of the short-period shaking does not match that of the building height and thus amplify the effect. If the building is constructed on thick unconsolidated sediments, then relatively short stiff buildings may be the safest structures. In addition to these considerations, many variations in the design of floors, roofs, trusses, shear walls, and frames can greatly improve the chances for a building to survive an earthquake and for lives to be saved. Other design features have recently been implemented in some cases, including methods to effectively isolate the building from its foundation by adding layers of rubber, ball bearings, and even wheels!

Utilities Infrastructure and Emergency Response Readiness

Experience in disaster mitigation has shown that many lives are lost because emergency response teams have not been ready for large destructive earthquakes. Although it is virtually impossible to have disaster teams ready to handle the largest of earthquakes, municipalities should have plans in place to deal with massive building collapses, large numbers of casualties, and large numbers of displaced people. They also need to give special consideration to nuclear and other power plants, high-voltage transmission lines, large pipelines, and aqueducts. All this needs to be done in times when many roads may be impassable, electricity and gas lines may be shut off, fresh water and food may be unobtainable, and communication could be difficult. Many countries have such plans in place, and several international teams of rapid-response earthquake relief crews have been formed, which have helped immensely following several recent earthquakes. These plans need to be continuously revised, tested, and improved, and they serve as examples for less-developed countries that are just forming such plans.

WHAT IF YOU ARE IN AN EARTHQUAKE?

If you live in an area prone to earthquakes, or in an area identified by the U.S. Geological Survey or other agency as an area of high seismic risk, then the best way to insure your safety is to be prepared. You don't need to plan your daily activities in fear of or in preparation for an earthquake. However, it is wise to plan your home and office so that large heavy objects will not tumble on beds or heads or into other places where you may be, and it is wise to take precautions such as attaching heavy furniture to walls. Avoid purchasing properties that do not have adequate construction, and look out for areas along steep hillsides that may fail or slump during an earthquake. If you have a choice, you may decide to locate your home in an area that does not have thick unconsolidated sediments, and you certainly do not want a home built on loose fill. Have a battery-operated radio, flashlight, first aid kit, water, and other emergency supplies on hand in your home. It is also wise to make plans with your family for meeting up after an earthquake, should you be separated when it happens.

If you are in an earthquake, you should try to get outdoors into an open area after the quake is over. Stay away from structures that may collapse and away from coastlines that may experience tsunami or seiche waves. If you are in a crowded city with tall office buildings that may collapse or send glass or other material flying into the street, you will have to use extreme caution when leaving the building and seeking an open area. Stay away from steep slopes that could collapse. If you find yourself inside during the quake, stay inside for its duration but try to anchor yourself under a strong desk, in a supported doorway, in a corner, or at least against a wall. Avoid being next to shelves and heavy furniture that could collapse.

If you are in a major earthquake and your village, town, or city experiences major damage, it will become critical for your survival to know what to do after the earthquake is over. First of all, the wounded and trapped must be assisted if possible, and you should then check your house for broken gas and water mains, electrical connections, and fires. Remember that most large earthquakes are associated with numerous aftershocks that may be nearly as severe (or, in rare cases, more severe) than the initial earthquake, and buildings that are damaged may be prone to further damage or collapse. Try to find battery-operated radios so that you may be able to tune to local broadcasts and emergency messages.

Be prepared to save fresh water as it may quickly become difficult to find. Emergency supplies of water may be found in your home's water heater, toilet's cistern, melted ice cubes from the freezer, and in cans of food you may have in the pantry. In many cases, it is wise to set up camp outside away from buildings that may collapse, and to sleep and cook outside. Be careful of further building collapse if you are trying to remove items from your home, but do not re-enter seriously damaged buildings as they could collapse without warning.

Try to organize your block or community so that everyone may help each other. In severe events, it may be days or even a week before organized help arrives, and it becomes essential that neighbors work together to help each other and to search for missing or injured people. Know where nonpotable water supplies may be located so that these supplies may be used to put out fires, and prepare lists of people and their conditions to help authorities when they do arrive.

WHO'S IN CHARGE

Immediately after a disaster, you may be in charge of your own safety and well-being and for assisting your neighbors. Soon afterwards, local police, fire, and rescue teams will be in charge of rescue operations, search and recovery, evacuations, and establishing centers for obtaining fresh water and food. In the case of particularly large or disastrous events, the National Guard or other military units may participate in the rescue operations, establishing communication and hospital services. Hotlines and bulletin boards may be set up to reunite families who were separated when the earthquake struck. These teams will be in constant contact with scientific teams who will be monitoring aftershocks. In most cases, the Federal Emergency Management Agency (FEMA) will step in at this stage and begin to take orderly charge of the rescue and cleanup operations. They will work closely with the Red Cross and also help to organize the massive assessment of damage costs by the insurance companies and banks.

Some extremely sophisticated earthquake warning systems are being developed for places like southern California. These warning systems may be able to alert residents or occupants of part of the region that a severe earthquake has just occurred in another part of the region and that they have several seconds or perhaps a minute to take cover. The thought is that if structures are adequately constructed, and if people have an earthquake readiness plan already implemented, they will know how and where to take immediate cover when the warning whistles are sounded and this type of system may be able to save numerous lives.

EARTHQUAKE STATISTICS

Every year, there are hundreds of thousands of earthquakes on the planet, but there are only a few large earthquake events each year. Most earthquakes are relatively small events that do not release large amounts of seismic energy, or they are located far from populated regions and so do not affect many people. For instance, in November 2002 there was a magnitude 7.9 earthquake in central Alaska. The epicenter was located so far from densely populated areas that nobody was killed, and only minor damage was reported, even though shaking and seiche waves were reported from as far away as New Orleans. When large earthquakes do strike populated areas, the amount of damage is strongly dependent on the types of buildings and materials used in construction, the population density, and the preparedness of local officials to deal with such catastrophes.

In recorded history, there have only been seventeen earthquakes that have caused more than 50,000 deaths (Table 2.1) and only four earthquakes have had magnitudes exceeding nine on the Richter scale (Table 2.3). The earthquake that cost the highest toll in human lives occurred in Shaanxi, China, in 1556. Eight hundred-thirty thousand people died when the earthquake hit, causing their cliff-dwelling-style homes that were cut into windblown silt (loess) to collapse. The huge loss of life was thus directly related to the building materials and population density. Likewise, in 1976, in the worst earthquake disaster of the twentieth century, 240,000 people died in T'ang Shan, China, when dual magnitude 7.8 and 7.1 earthquakes leveled the city. In this case, the building materials were mostly unreinforced bricks, again explaining the huge loss of life.

WHAT AGENCIES DEAL WITH EARTHQUAKES?

The U.S. Geological Survey coordinates monitoring of more than 2,500 seismograph stations in the United States, forming a network known as the United States National Seismograph Network (USNSN). These are grouped in Regional Seismograph Networks (RSNs) operated by local institutions, and information collected from these is sent to the National Earthquake Information Center (NEIC) in Colorado, where it is made available for distribution. After an earthquake, the NEIC rapidly reports and updates federal, state, and local emergency coordinators, utilities, the media, and the public. The U.S. Geological Survey works with the NEIC and RSNs to rapidly interpret earthquake magnitude, and it assists in disaster response in coordination with local authorities. This can include identifying areas that have experienced the most intense seismic waves and areas that may be prone to tsunami, and warning utilities of nuclear power plants and other facilities that may require special attention. These efforts are coordinated with local authorities including fire and police, as well as national teams including FEMA, the National Guard, and the Red Cross.

The U.S. Geological Survey also works with local RSNs and with government and academic geologists to make seismic risk maps, which are used during earthquake disaster relief to quickly identify and assist areas that may be the worst affected and require the most rapid attention.

EXAMPLES OF NATURAL EARTHQUAKE DISASTERS

The amount of destruction from earthquakes depends on the earthquake's magnitude, as well as where and when it occurs, the types of building materials used for construction, population density, preparedness of emergency rescue teams, and the underlying geology. Table 2.1 highlights the huge numbers of casualties associated with some historical earthquakes, whereas Table 2.3 lists earthquakes with magnitudes > 8.5 that have occurred since 1990. In the following section, we provide some brief descriptions of specific earthquakes to illustrate the variability in their destructiveness.

DESCRIPTIONS OF SPECIFIC EARTHQUAKES

San Francisco, California, 1906 (m. 7.8)

Perhaps the most infamous earthquake of all time is the magnitude 7.8 temblor that shook San Francisco at 5:12 A.M. on April 18, 1906, virtually destroying the city, crushing 315 people to death in the city, and killing 700 people throughout the region. Many of the unreinforced masonry buildings that were common in San Francisco at that time immediately collapsed, but most steel and wooden frame structures remained upright. Ground shaking and destruction were most intense where structures were built on old fill and least intense where the buildings were anchored in solid bedrock. Most of the destruction from the 1906 earthquake came not from ground shaking, but from the intense firestorm that followed. Gas lines and water lines were ruptured, and fires started near the waterfront and worked their way into the city. Other fires were inadvertently started by people cooking in residential neighborhoods and by others dynamiting buildings in order to avoid collapses and stop the spread of the huge fire. As mentioned earlier, it has even been reported that some fires were started by individuals in attempts to collect insurance money on their slightly damaged homes. In all, 490 city blocks were burned.

The problems did not cease after the earthquake and fires, but continued because of the poor sanitary and health conditions that followed due to the poor infrastructure. Hundreds of cases of bu-

Table 2.3
The Ten Worst Earthquakes (1900–2000) in Terms of Magnitude

Place	Year	Richter Magnitude
Concepción, Chile	1960	9.5
Valdez, Alaska	1964	9.2
Alaska	1957	9.1
Kamchatka, Russia	1952	9.0
Ecuador	1906	8.8
Alaska	1965	8.7
Assam, northeast India	1950	8.6
Gansu, China	1920	8.6
Banda Sea	1938	8.5
Kuril Islands, Russia	1963	8.5

bonic plague claimed lives, and other diseases combined to bring the death total to as high as 5,000.

One of the lessons that could have been learned from the 1906 San Francisco earthquake was not appreciated until many years later. San Franciscans noted that some of the areas that shook the most and had the most destruction were those built on unconsolidated fill. After the 1906 earthquake, much of the rubble was bulldozed into the Bay, and later construction on this fill became the Marina district, which saw some of the worst damage during the 1989 Loma Prieta earthquake.

Valdez, Alaska, 1964 (m. 9.2)

One of the largest earthquakes ever recorded is the "Good Friday" earthquake that struck southern Alaska at 5:36 P.M. on Friday, March 27, 1964. Although the energy released during this earthquake was more than that of the world's largest nuclear explosion, and was greater than the earth's total average annual release of seismic energy, only 131 people died during this event. Damage is estimated at $240 million (1964 dollars), a remarkably small figure for an earthquake this size. During the initial shock and several other shocks that followed in the next one to two minutes, a 600-mile-long by 250-mile-wide (1,000 km long × 400 km wide) slab of subducting oceanic crust slipped further beneath the North American crust of southern Alaska. Ground displacements above the area that slipped were remarkable—much of the Prince William Sound and Kenai Peninsula area moved horizontally almost sixty-five feet (twenty meters), and moved upward by more than thirty-five feet (11.5 meters). Other areas more landward of the uplifted zone were downdropped by several to ten feet. Overall, almost 125,000 square miles (200,000 square kilometers) of land saw significant movements upward, downward, and laterally during this huge earthquake.

The ground shook in most places for three to four minutes during the Good Friday earthquake, causing widespread destruction in southern Alaska, damage as far away as southern California, and noticeable effects across the planet. Entire neighborhoods and towns slipped into the sea during this earthquake, and ground breaks, landslides, and slumps were reported across the entire region. Tsunami swept across many towns that had just seen widespread damage by building collapses, and other tsunami swept across the Pacific, destroying marinas as far away as southern California.

Southern Chile, 1960 (m. 9.5)

The largest earthquake ever recorded struck southern Chile on Sunday, May 22, 1960. This was a subduction zone earthquake, and a huge section of the downgoing oceanic slab moved during this and related precursors and aftershocks spanning a few days. The main shock was preceded by a large foreshock at 2:45 P.M., which was fortunate because it scared most people into the streets and away from buildings that were soon to collapse. Thirty minutes later, at 3:15 P.M., the magnitude 9.5 event struck and affected a huge area of southern Chile. An approximately 600-mile-long by 190-mile-wide (1,000 km long by 300 km wide) section of the fault separating the downgoing oceanic slab from the overriding plate slipped, allowing the oceanic plate to sink further into the mantle. The area that slipped during this event is roughly the size of California.

Loma Prieta, California, 1989 (m. 7.1)

The Santa Cruz and San Francisco areas were hit by a moderate-sized earthquake (magnitude 7.1) at 5:04 P.M. on Tuesday, October 17, 1989, during a World Series baseball game being played in San Francisco. Sixty-seven people died, 3,757 people were injured, and 12,000 were left homeless. Tens of millions of people watched on television as the earthquake struck just before the beginning of game three, and the news coverage that followed was unprecedented.

The earthquake was caused by a rupture along a twenty-six-mile-long (forty-two km) segment of the San Andreas Fault near the Loma Prieta peak in the Santa Cruz Mountains south of San Francisco. The segment of the fault that had ruptured was the southern part of the same segment that ruptured in the 1906 earthquake, but the 1989 rupture occurred at greater depths and involved some vertical motion as well as horizontal motion. The actual rupturing lasted only eleven seconds, during which time the western (Pacific) plate slid almost six feet (1.9 meters) to the northwest and parts of the Santa Cruz Mountains were uplifted by up to four feet (1.3 meters). The rupture propagated at 1.24 miles per second (2 km/sec) and was a relatively short-duration earthquake for one of this magnitude. Had it been much longer, the damage would have been much more extensive. As it was, the damage totaled more than $6 billion.

The actual fault plane did not rupture the surface, although many cracks appeared and slumps formed along steep slopes. The Loma Prieta earthquake had been predicted by seismologists because the segment of the fault that slipped had a noticeable paucity of seismic events since the 1906 earthquake, and was identified as a seismic gap with a high potential for slipping and causing a significant earthquake. The magnitude 7.1 event and the numerous aftershocks filled in this seismic gap, and the potential for large earthquakes along this segment of the San Andreas Fault is now significantly lower. There are, however, other seismic gaps along the San Andreas Fault in heavily populated areas that should be monitored closely.

Kobe, Japan, 1995 (m. 6.9)

The industrial port city of Kobe, Japan, was hit by history's costliest earthquake ($100 billion in property damage) at 5:46 A.M. on January 17, 1995. The thirty-mile-long (fifty km) fault rupture passed directly through the world's third-busiest port city and home to 1.5 million people. Six thousand three-hundred eight people died in Kobe before sunrise on that cold January morning. The rupturing event took fifteen seconds and moved each side of the fault more than six feet (1.7 meters) horizontally relative to the other side, and uplifted the land by three feet (one meter). There are many areas of unconsolidated sediment in and around Kobe; these areas saw some of the worst damage and shook for as long as 100 seconds because of the natural amplification of the seismic waves. Liquefaction was widespread and caused much of the damage, including the collapse of buildings and port structures and the destruction of large parts of the transportation network. Water, sewer, gas, and electrical systems were rendered useless. More than 150,000 buildings were destroyed, and a huge fire that ensued consumed the equivalent of seventy U.S. blocks.

Turkey, 1999 (m. 7.8)

On August 17, 1999, a devastating earthquake measuring 7.4 on the Richter scale hit heavily populated areas in northwestern Turkey at 3:02 A.M. local time. The epicenter of the earthquake was near the industrial city of Izmit, about sixty miles (100 km) east of Istanbul and near the western segment of the notorious North Anatolian strike-slip fault. The earthquake formed a > 75-mile-long (120 km) surface rupture, along which offsets were measured between four and fifteen feet (–1.5–5 meters). This was the deadliest and most destructive earthquake in the region in more than sixty years, causing more than 30,000 deaths and the largest property losses in Turkey's recorded history. The World Bank estimated that direct losses from the earthquake were approximately $6.5 billion, with perhaps another $20 billion in economic impact from secondary and related losses.

The losses from this moderate-sized earthquake were so high because the region in which it occurred contains approximately 25 percent of Turkey's population, hosts much of the country's industrial activity, and has seen a recent construction boom in which building codes were ignored. Large numbers of high-rise apartment buildings were constructed with substandard materials including extra-coarse and sandy cement, too few reinforcement bars in concrete structures, and a general lack of support structures.

"THE BIG ONE": WHAT CAN SOUTHERN CALIFORNIA EXPECT?

For many years, there has been much speculation about the effects of the anticipated "big one" in southern California. This speculation and fear are not unfounded. The unrepentant forces of plate tectonics continue to slide the Pacific plate northward relative to North America along the San Andreas Fault, and the stick-slip type of behavior that characterizes segments of this fault system generates many earthquakes. Several segments of the fault have seismic gaps, where the tectonic stresses may be particularly built up. One of these gaps released its stress during the Loma Prieta earthquake of 1989, but other gaps remain, including ones in the area generally east of Los Angeles. Some models predict that the seismic energy may be released in this area by a series of moderate earthquakes (approximately m. 7), but other, more sinister predictions also remain plausible. The last major rupture along the southern California segment was in 1680, and studies of prehistoric earthquakes in this region show that major, catastrophic events recur roughly every 250 years.

RESOURCES AND ORGANIZATIONS

Print Resources

Bolt, B. A. *Earthquakes.* 4th ed. New York: W. H. Freeman, 1999, 366 pp.

Boraiko, A. A. "Earthquake in Mexico." *National Geographic* 169 (1986), 654–75.

California Division of Mines and Geology. *The Loma Prieta (Santa Cruz Mountains) Earthquake of October 17, 1989.* Special Publication 104. Sacramento, Calif.: Author, 1990.

Coburn, A., and Spence, R. *Earthquake Protection.* Chichester, England: Wiley, 1992.

Earthquakes and Volcanoes, a bimonthly publication of the U.S. Geological Survey, provides current information on earthquakes and seismology, volcanoes, and related natural hazards of interest to both generalized and specialized readers. Available from *Earthquakes and Volcanoes,* U.S. Geological Survey, Mail Stop 967, Federal Center, Denver, CO, 80225; 303–273–8408; (fax) 303–273–8450.

Kendrick, T. D. *The Lisbon Earthquake.* London: Methuen, 1956.

Logorio, H. *Earthquakes: An Architect's Guide to Non-Structural Seismic Hazards.* New York: Wiley, 1991.

Reilinger, R., Toksoz, N., McClusky, S., and Barka, A. "1999 Izmit, Turkey, Earthquake Was No Surprise." *GSA Today* 10, no. 1 (2000), 1–6.

Reiter, L. *Earthquake Hazard Analysis.* New York: Columbia University Press, 1990.

Richter, C. F. *Elementary Seismology.* San Francisco: W. H. Freeman, 1958, 137–38.

U.S. Geological Survey. *The San Francisco Earthquake and Fire of April 18, 1906.* U.S. Geological Survey Bulletin 324, 1907, 170 pp. A rare book that provides a thorough description of the damage resulting from the earthquake and fire of 1906, including many black-and-white photographs and firsthand descriptions.

———. *The Alaska Earthquake, March 27, 1964.* 1966.

———. Geological Survey Professional Papers on the Alaska Earthquake: 542-B (Effects on Communities—Whittier); 542-D (Effects on Communities—Homer), 542-E (Effects on Communities—Seward); 542-G (Effects on Communities—Various Communities); 543-A (Regional Effects; Slide-induced Waves, Seiching and Ground Fracturing at Kenai Lake), 543-1 (Regional Effects—Tectonics); 543-B, Regional Effects—Martin-Bering Rivers Area), 543-F (Regional Effects—Ground Breakage in the Cook Inlet Area);

543-H (Regional Effects—Erosion and Deposition on a Raised Beach, Montague Island); 543-J (Regional Effects—Shore Processes and Beach Morphology); 544-C (Effects on Hydrologic Regime—Outside Alaska); 544-D (Effects on Hydrologic Regime—Glaciers); 544-E (Effects on Hydrologic Regime—Seismic Seiches); and 545-A (Effects on Transportation and Utilities—Eklutna Power Project).

———. "Lesson Learned from the Loma Prieta Earthquake of October 17, 1989." Circular 1045, 1989.

Verney, P., *The Earthquake Handbook.* New York: Paddington Press, 1979, 224 pp.

Wallace, R. E., ed. *The San Andreas Fault System, California.* U.S. Geological Survey Professional Paper 1515, 1990.

Non-Print Resources

Videos

The Day the Earth Shook, 1995, NOVA, 55 mins.

The Earth Revealed—Earthquakes, 1992, Annenberg CPB Project, 30 mins.

Hidden Fury: The New Madrid Earthquake Zone, 1993, Bullfrog Films, 27 mins.

Killer Quake, 1994, NOVA/KCET-TV, 60 mins.

Loma Prieta Earthquake, 1992, U.S. Geological Survey, 53 mins.

When the Earth Quakes, 1990, National Geographic, 28 mins.

Web Sites

Council of National Seismic Networks Web site:
http://www.cnss.org/
Site describes the activities of the nationwide seismograph networks.

Current Seismic Activity Web sites:
http://quake.wr.usgs.gov/recenteqs
http://wwwneic.cr.usgs.gov/
These two sites show current earthquake activity as the earthquakes occur.

FAQ Web site:
http://quake.wr.usgs.gov/more/eqfaq.html
Site answers frequently asked questions about earthquakes.

Incorporated Research Institutions for Seismology (IRIS) Consortium Data Center Web site:
http://www.iris.washington.edu
The IRIS Consortium Data Center provides facts about recent earthquakes.

Map Web site:
http://fingerrecent@eqinfo.seis.utah.edu/HTML/SeismicityMaps.html
Shows maps of recent seismicity.

Map Web site:
http://www.geo.ed.ac.uk/scratch/quake_all.html
Offers maps of global distribution of earthquakes.

National Center for Earthquake Engineering Research Web site:
http://nceer.eng.buffalo.edu

St. Louis University Earthquake Center Web site:
http://www.eas.slu.edu/Earthquake_Center/earthquakecenter.html
The St. Louis University Earthquake Center specializes in earthquakes of the central U.S.

San Francisco Bay Area Earthquake Hazard Information Web site:
http://www.abag.ca.gov/bayarea/eqmaps/eqhouse.html

Organizations

Boston College
Weston Geophysical Observatory
Weston, Massachusetts

California Institute of Technology
Pasadena, California

Saint Louis University
Department of Earth and Atmospheric Sciences
Saint Louis, MO 63103

University of Utah Seismograph Stations
Salt Lake City, Utah

U.S. Geological Survey
National Earthquake Information Center
Federal Center
Box 25046, MS 967
Denver, CO 80225–0046
303–273–8500

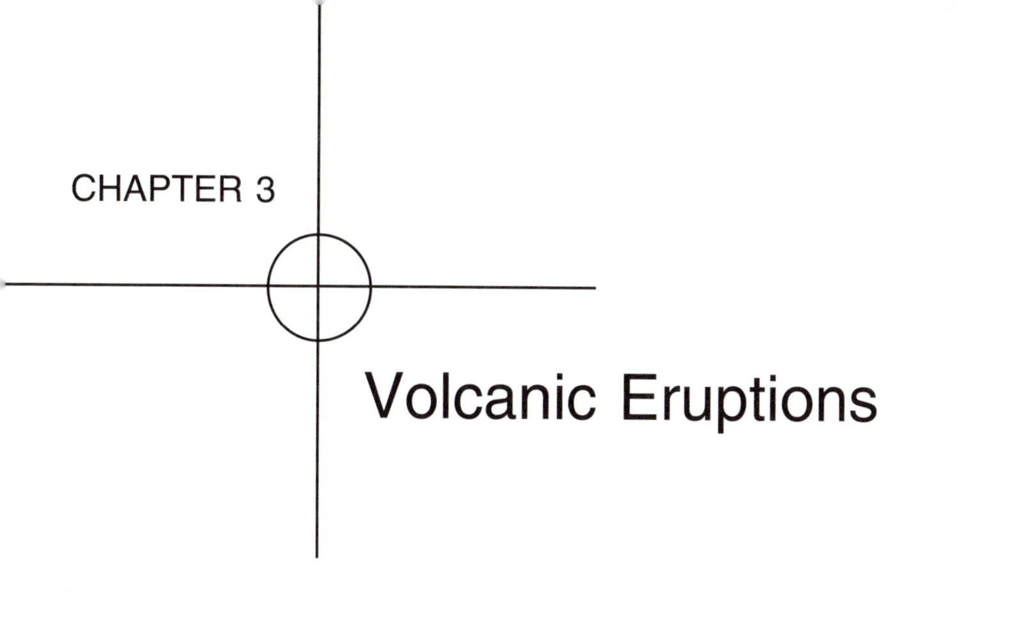

CHAPTER 3

Volcanic Eruptions

INTRODUCTION

Volcanic eruptions provide one of the most spectacular of all natural phenomena yet they also rank among the most dangerous of geological hazards. Eruptions may send blocks of rock, ash, and gas tens of thousands of feet into the atmosphere in beautiful eruption plumes, yet individual eruptions have also killed tens of thousands of people. The hazards associated with volcanic eruptions are not limited to the immediate threat from the flowing lava and ash, but include longer-term atmospheric and climate effects, and changes to land use patterns and the livelihood of populations.

People have been awed by the power and fury of volcanoes for thousands of years, as evidenced by biblical passages referring to eruptions, and more recently by the destruction of Pompeii and Herculaneum by the eruption of Italy's Mount Vesuvius in A.D. 79. Sixteen thousand people died in Pompeii alone, buried by a fast-moving hot incandescent ash flow known as a *nuee ardent.* This famous eruption buried Pompeii in thick ash that quickly solidified and preserved the city and its inhabitants remarkably well. In the sixteenth century, Pompeii was rediscovered and has been the focus of archeological investigations since then. Mount Vesuvius is still active, looming over the present-day city of Naples.

This chapter examines the different types of volcanic eruptions, starting with a fairly detailed description of the various characteristics of magma that lead to differences in eruption style. Following that is a discussion of different types of volcanic rocks, landforms, eruptions, and the tectonic settings that typically host each type of eruption. Then we will cover the different hazards associated with volcanism, including the eruption cloud, poisonous gases, volcanic flows, mudflows, floods and debris avalanches, tsunami, and earthquakes, and the chapter ends with an analysis of the long-range consequences of volcanic eruptions in terms of changes in climate. In-depth discussions of historical eruptions will illustrate each of these hazards.

WHERE ARE VOLCANIC ERUPTIONS MOST AND LEAST LIKELY TO OCCUR?

Most volcanic eruptions on the planet are associated with the boundaries of tectonic plates (see Chapter 1). Extensional plate boundaries where plates are being pulled apart, such as along the mid-ocean ridges, experience the greatest volume of erupted magma each year, yet these eruptions are not generally hazardous, especially since most of these eruptions occur many kilometers below sea level. A few exceptions to this rule are noted where the mid-ocean ridges rise above sea level, such as in Iceland. There, volcanoes including the famous Hekla Volcano have caused significant damage. Hekla has even erupted beneath a glacier, causing catastrophic melting and generating fast-moving catastrophic floods called *Jökulhlaups*. Icelanders have learned to benefit from living in a volcanically active area, tapping a large amount of heat in geothermal power generating systems.

Larger volcanoes are associated with extensional plate boundaries located in continents. For instance, the east African rift system is where the African continent is being ripped apart, and it hosts some spectacular volcanic cones, including Kilamanjaro, Oldu Lengai, Nyiragongo, and many others. The most hazardous volcanoes on the planet are associated with convergent plate boundaries, where one plate is sliding beneath another in a subduction zone (see Chapter 1). The famous "Ring of Fire" rims the Pacific Ocean (Figure 1.6), and its name refers to the ring of numerous volcanoes located above subduction zones around the Pacific Ocean. The Ring of Fire extends through the western Americas, Alaska, Kamchatka, Japan, Southeast Asia, and Indonesia. Volcanoes in this belt, including Mount Saint Helens in Washington State, typically have violent explosive eruptions that have killed many people and altered the landscape over wide regions.

A third style of volcanic activity is not associated with plate boundaries, but is characterized by large broad volcanic shields that typically have spectacular but not explosive volcanic eruptions. The most famous of these "intraplate" volcanoes is Mauna Loa on Hawaii. Hazards associated with these volcanoes include lava flows and traffic jams from tourists trying to see the flows.

Why is there such a variation in the style of volcanic eruptions, and what causes specific types of eruptions to be associated with different types of plate boundaries? The next section examines the different types of magmas and eruptions in some detail to answer these questions, and these answers are essential for planning how to respond to volcanic hazards in different parts of the world.

ORIGINS OF VOLCANIC ERUPTIONS

The wide variety of eruption styles and hazards associated with volcanoes around the world can be linked to different types of *magma*, or molten rock within the earth. Different types of magma form in different tectonic settings, and when the magma reaches the surface, it is known as *lava*, which may flow or explosively erupt from volcanoes.

Kinds of Igneous Rocks

Most magma solidifies below the surface, forming certain types of igneous rocks ("igneous" is the Latin word for fire). Igneous rocks that form below the surface are called *intrusive* or *plutonic rocks,* whereas those that crystallize on the surface are called *extrusive* or *volcanic rocks.* Rocks that crystallize at a very shallow depth are called *hypabyssal rocks.* Intrusive igneous rocks crystallize slowly, giving crystals an extended time to grow, thus forming rocks called *phanerites* that have large mineral grains that are clearly distinguishable with the naked eye. In contrast, magma that cools rapidly forms fine-grained igneous rocks called *aphanites,* in which the component grains can not be distinguished readily without a microscope; aphanites are formed when magma from a volcano falls or flows across the surface and cools quickly. Some igneous rocks, known as *porphyries,* have two populations of grain size. One very large group of crystals (called phenocrysts) is mixed with a uniform groundmass or matrix filling the space between the large crystals. This indicates two stages of cooling, as when magma has resided for a long time beneath a volcano, growing big crystals. When the volcano erupts, it spews out a mixture of the large crystals and liquid magma that then cools quickly.

Intrusive Igneous Bodies

Once magmas are formed from melting rocks in the earth, they intrude the crust and may take several forms (Figure 3.1). A *pluton* is a general name for a large, cooled, igneous intrusive body in the earth. Plutons are classified based on their geometry, size, and relations to the older rocks surrounding them, which are known as country rock. Concordant plutons have boundaries parallel to layering in the country rock, whereas discordant plutons have boundaries that cut across layering in the country rock. Dikes are tabular but discordant intrusions and sills are tabular and concordant intrusives. Volcanic necks are conduits connecting a volcano with its underlying magma chamber. A famous example of a volcanic neck is Devils Tower in Wyoming. Some plutons are so large that they have special names. Batholiths are plutons with a surface area of more than sixty square miles (100 square kilometers).

Batholith Emplacement Mechanisms

Scientists have long speculated on how such large volumes of magma intrude the crust and what relationships these magmas have on the style of volcanic eruption. One mechanism that may operate is *assimilation,* by which the hot magma melts surrounding rocks as it rises, causing it to become part of the magma. In doing this, the magma becomes cooler and its composition changes to reflect the added melted country rock. It is widely thought that magmas may rise only a very limited distance by the process of assimilation. Some magmas may forcefully push their way into the crust if there are high pressures in the magma. One variation of this forceful emplacement style is *diapirism,* where the weight of surrounding rocks pushes down

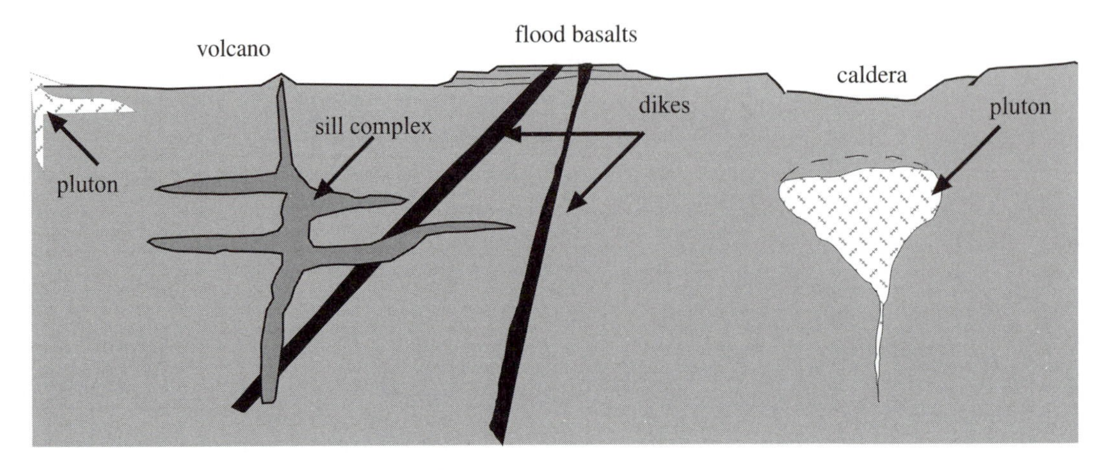

Figure 3.1. Cross-section of a part of the continental crust showing some common forms of plutons and intrusive igneous rocks.

on the melt layer, which squeezes its way up through cracks that can expand and extend, forming volcanic vents at the surface. *Stoping* is a mechanism whereby big blocks get thermally shattered, drop off the top of the magma chamber, and fall into the chamber, much like a glass ceiling breaking and falling into the space below.

Naming Igneous Rocks

Determining whether an igneous rock is phaneritic or aphanitic is just the first stage in giving it a name. The second stage is determining its mineral constituents. The chemical composition of a magma is closely related to how explosive and hazardous a volcanic eruption will be. The variation in the amount of silica (SiO_2) in igneous rocks is used to describe the variation in their composition—and the magmas that formed them. Rocks with low amounts of silica (basalt, gabbro) are known as mafic rocks, whereas rocks with high concentrations of silica (rhyolite, granite) are known as silicic or felsic rocks. Intermediate rocks have intermediate silica contents, between 52–65%. Table 3.1 shows how the different kinds of igneous rocks are related.

THE ORIGIN OF MAGMA

Some of the variation in the nature of different types of volcanic eruptions can be understood by understanding the origin of magma and examining what causes magmas to have such a wide range in composition. Magmas come from deep within the earth, but what conditions lead to the generation of melts in the interior of the Earth?

The *geothermal gradient* is a measure of how temperature increases with depth in the earth, and it provides information about the depths at which melting occurs and the depths at which magmas form. The differences in the composition of the oceanic and continental crusts lead to differing abilities to conduct the heat from the interior of the earth, and thus different geo-

Table 3.1
Magma Types

	SiO2 (%)	Volcanic rock	Plutonic rock
mafic	45–52%	basalt	gabbro
intermediate	53–65%	andesite	diorite
felsic/silicic	>65%	rhyolite	granite

thermal gradients. The geothermal gradients show that temperatures within the earth quickly exceed 1000°C, so why are these rocks not molten? The answer is that pressures are very high, and pressure influences the ability of a rock to melt: *As the pressure rises, the temperature at which the rock melts also rises.* This effect of pressure on melting is modified greatly by the presence of water, because wet minerals melt at lower temperatures than dry minerals. As the pressure rises, the amount of water that can be dissolved in a melt increases. Therefore, *increasing the pressure on a wet mineral has the opposite effect as increasing the pressure on a dry mineral: It decreases the melting temperature.*

Partial Melting

If a rock melts completely, the magma has the same composition as the rock. However, rocks are made of many different minerals, *all of which melt at different temperatures.* So if a rock is slowly heated, the resulting melt or magma will first have the composition of the first mineral that melts, and then the first two minerals that melt, and so on. If the rock continues to completely melt, the magma will eventually end up with the same composition as the starting rock, but this does not always happen. What often happens is that the rock only partially melts so that the minerals with low melting temperatures contribute to the magma, whereas the minerals with high melting temperatures didn't melt and are left as a residue (or restite). In this way, the end magma can have a composition different than the rock it came from.

The phrase *magmatic differentiation by partial melting* refers to the process of forming magmas with differing compositions through the incomplete melting or rocks. For magmas formed in this way, the composition of the magma depends on both the composition of the parent rock and the percentage of melt.

Basaltic Magma Partial melting in the mantle leads to the production of *basaltic magma,* which forms most of the oceanic crust. By looking at the mineralogy of the oceanic crust, which is dominated by olivine, pyroxene, and feldspar, we conclude that very little water is involved in the production of the oceanic crust. These minerals are all *anhydrous,* or without water in their structure. Thus, dry partial melting of the upper mantle must lead to the formation of oceanic crust. By collecting samples of the mantle that have been erupted through volcanoes, we know that it has a composition of *garnet peridotite* (olivine + garnet + orthopyroxene). By taking samples of this back to the laboratory and raising its temperature and pressure so that it is equal to that found 100 km (sixty-two miles) below the sea floor, we find that 10 percent to 15 percent partial melt of this garnet peridotite yields a basaltic magma.

Magma that forms at fifty miles' depth is *less dense than the surrounding solid rock, so it rises,* sometimes quite rapidly (at rates of half a mile per day, measured by earthquakes under Hawaii). In fact, it may rise so fast that it does not cool off appreciably, erupting at the surface at more than 1000°C. That is where basalt comes from.

Granitic Magma Granitic magmas are very different than basaltic magmas. They have about 20 percent more silica, and the minerals in granite (mica, amphibole) have a lot of water in their crystal structures. Also, granitic magmas are found almost exclusively in regions of continental crust. From these observations, we infer that the source of granitic magmas is within the continental crust. Laboratory experiments suggest that when rocks with the composition of continental crust start to melt at temperature and pressure conditions found in the lower crust, a granitic liquid is formed with 30 percent partial melting. These rocks can begin to melt by either the addition of a heat source such as basalt intruding the lower continental crust, or by burying water-bearing minerals and rocks to these depths.

These granitic magmas rise slowly because of their high SiO_2 and high viscosities, until they reach the level in the crust where the temperature and pressure conditions are consistent with freezing or solidifying a magma with this composition. This is about 3–6 miles (5–10 km) beneath the surface, which explains why large portions of the continental crust are not molten lava lakes. There are many regions with crust above large magma bodies (called batholiths) that is heated by the cooling magma. An example is the Yellowstone National Park, where there are hot springs, geysers, and many features indicating that there is a large hot magma body underneath. Much of Yellowstone Park is a giant valley called a caldera, formed when an ancient volcanic eruption emptied an older batholith of its magma and the overlying crust collapsed into the empty hole formed by the eruption.

Andesitic Magma The average composition of the continental crust is *andesitic,* or somewhere between the composition of basalt and rhyolite. Laboratory experiments show that *partial melting of wet oceanic crust yields an andesitic magma.* Remember that oceanic crust is dry, but after it forms it interacts with *seawater,* which fills cracks to several kilometers' depth. Also, the sediments on top of the oceanic crust are full of water, but these are for the large part nonsubductable. Figure 3.2 shows water escaping from the crust, forming andesitic magmas.

Solidification of Magma

Just as rocks partially melt to form different liquid compositions, magmas may solidify to different minerals at different times to form different solids (rocks). This process also results in the continuous change in the composition of the magma: If one mineral is removed, the resulting composition is different. If some process removes these solidified crystals from the

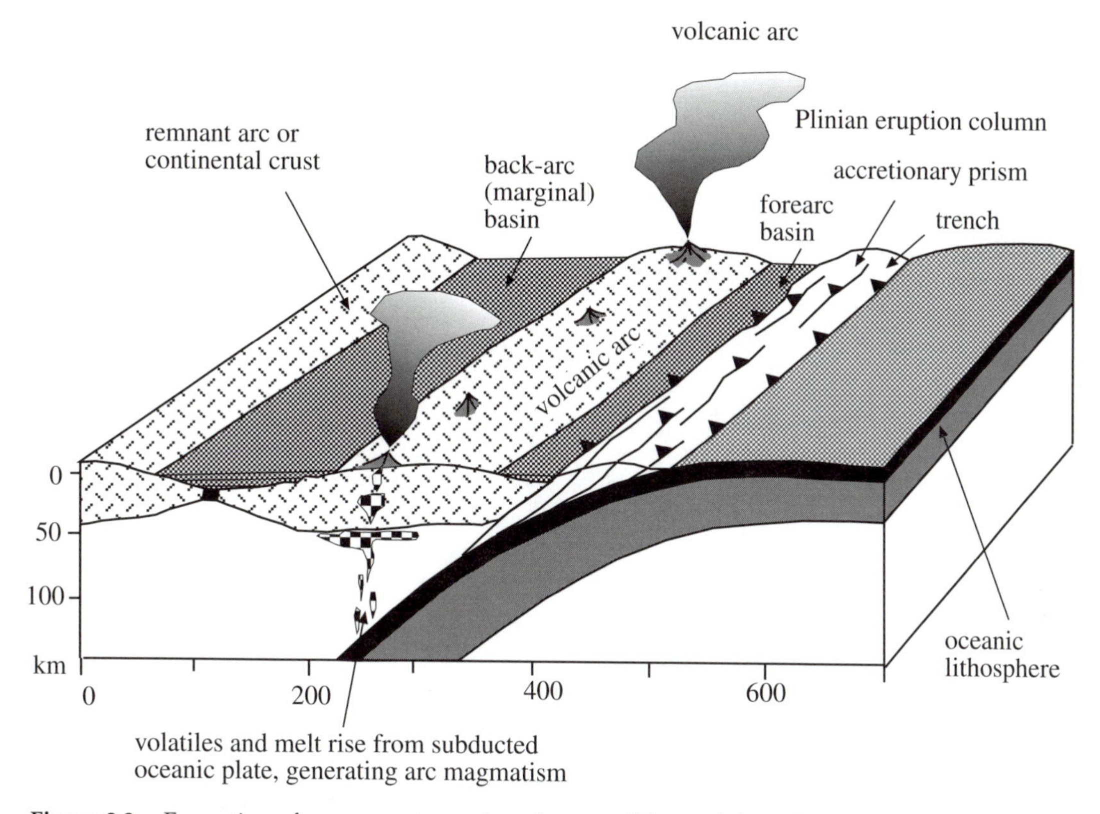

Figure 3.2. Formation of convergent margin volcanoes. When subducted oceanic crust reaches a depth of about 100 kilometers, water is released causing melting of the overlying rock in the mantle. These magmas rise and form the explosive, water-rich volcanoes that characterize convergent margins such as the "Ring of Fire."

system of melts, a new magma composition results.

The removal of crystals from the melt system may occur by several processes, including the squeezing of melt away from the crystals or the sinking of dense crystals to the bottom of a magma chamber. These processes lead to *magmatic differentiation by fractional crystallization,* which was first described by Norman L. Bowen. Bowen systematically documented how crystallization of the first minerals changes the composition of the magma and results with decreasing temperatures in the formation of progressively more silicic rocks.

TYPES OF VOLCANIC ERUPTIONS AND LANDFORMS

There is tremendous variety in the style of volcanic eruptions, both between volcanoes and from a single volcano during the course of an eruptive phase. This variety is related to the different types of magma produced by the different mechanisms described above. Geologists have found it useful to classify volcanic eruptions based on how explosive the eruption may be, which materials erupted, and the type of landform produced by the volcanic eruption.

Tephra is material that comes out of a volcano during an eruption, and it may be thrown through the air or transported over the land as part of a hot moving flow. Tephra includes both new magma from the volcano and older broken rock fragments that got caught in the eruption. It includes ash and *pyroclasts,* rocks ejected by the volcano. Large pyroclasts are called volcanic bombs, smaller fragments are lapilli, and the smallest grade is ash.

Although the most famous volcanic eruptions produce huge explosions, many eruptions are relatively quiet and nonexplosive. Nonexplosive eruptions have magma types that have low amounts of dissolved gases, and they tend to be basaltic in composition. Basalt flows easily and for long distances and tends not to have difficulty flowing out of volcanic necks. Nonexplosive eruptions may still be spectacular, as any visitor to Hawaii lucky enough to witness the fury of Pele, the Hawaiian goddess of the volcano, can testify. Mauna Loa, Kilauea, and other nonexplosive volcanoes produce a variety of eruption styles, including fast-moving lava flows, liquid rivers of lava, lava fountains that spew fingers of lava trailing streamers of light hundreds of feet into the air, and thick sticky lava flows that gradually creep downhill. The Hawaiians devised clever names for these flows, including *aa* for a blocky rubble flow because walking across these flows in bare feet makes one exclaim "Aa!" in pain. *Pahoehoes* are ropy textured flows, after the Hawaiian term for rope.

Explosive volcanic eruptions are among the most dramatic natural events on Earth (Figures 3.3 and 3.4). With little warning, long-dormant volcanoes can explode with the force of hundreds of atomic bombs, pulverizing whole mountains and sending the existing material, together with millions of tons of ash, into the stratosphere. Explosive volcanic eruptions tend to be associated with volcanoes that produce andesitic or rhyolitic magma and have high contents of dissolved gases. These are mostly associated with convergent plate boundaries. Volcanoes that erupt magma with high contents of dissolved gases often produce a distinctive type of volcanic rock known as pumice, which is full of bubble holes that in some cases make the rock light enough to float on water.

When the most explosive volcanoes erupt, they produce huge eruption columns known as Plinian columns (named after Pliny the Elder, the Roman statesman who died in A.D. 79 taking samples of volcanic gas during an eruption of Mount Vesuvius). These eruption columns can reach forty-five km (thirty miles) in height, and they spew hot turbulent mixtures of ash, gas, and tephra into the atmosphere where winds may disperse them around the planet (Figure 3.4). Large ashfalls and tephra deposits may be spread across thousands of square kilometers. These explosive volcanoes also produce one of the scariest and most dangerous clouds on the planet. *Nuee ardents* are hot glowing clouds of dense gas and ash that may reach temperatures of nearly 1000°C, rush down vol-

Figure 3.3. Photograph of explosive eruption from Mount Augustine, Alaska. This stratovolcano island with a summit lava dome is located in the Cook Inlet, southern Alaska. It has erupted explosively six times since 1812. Three of these eruptions are among the largest in Alaska's recorded history, and the one in 1883 produced a 9 m tsunami at Port Graham. Dangers from major ash falls or volcanically-generated tsunamis worry the local residents. The photo depicts the volcano on the first day of an eruption cycle. This west-southwest view of the summit was taken on March 27, 1986. Photo Credit: M.E. Yount, U.S. Geological Survey

canic flanks at 450 miles per hour (700 km/hour), and travel more than sixty miles (100 km) from the volcanic vent. Nuee ardents have been the nemesis of many a volcanologist and curious observer, as well as thousands upon thousands of unsuspecting or trusting villagers. Nuee ardents are but one type of *pyroclastic flow,* which includes a variety of mixtures of volcanic blocks, ash, gas, and lapilli that produce volcanic rocks called *ignimbrites.*

Most volcanic eruptions emanate from the central vents at the top of volcanic cones. However, many flank eruptions have been recorded in which eruptions blast out of fissures on the side of the volcano. Occasionally volcanoes blow out their sides, forming a lateral blast like the one that initiated the 1980 eruption of Mount St. Helens in Washington State (Figure 3.4). This blast was so forceful that it began at the speed of sound, killing everything in the initial blast zone.

Volcanic landforms and landscapes are wonderful, dreadful, beautiful, and barren. They are as varied as the volcanic rocks

Figure 3.4. Photograph of May 18, 1980 eruption of Mt. St. Helens. Looking more like a smoke cloud from an A-bomb blast than steam and ash from a volcano eruption, Mt. St. Helens sent this plume of steam and ash some 60,000 feet in the air, as the volcano awoke from a six-week nap. ©Bettmann/CORBIS

and eruptions that produce them. Shield volcanoes include the largest and broadest mountains on the planet (for example, Mauna Loa is more than 100 times as large as Mount Everest). These have slopes of only a few degrees, produced by basaltic lavas that flow long distances before cooling and solidifying. Stratovolcanoes, in contrast, are the familiar steep-sided cones like Mount Fuji, are made of stickier lavas such as andesites and rhyolites, and may have slopes of 30°. Other volcanic constructs include cinder or tephra cones, including the San Francisco Peaks in Arizona, which are loose piles of cinder and tephra. Calderas, like Crater Lake in Oregon, are huge circular depressions, often many kilometers in diameter, that are produced when a deep magma chamber under a volcano empties out during an eruption and the overlying land collapses inward, producing a topographic depression. Yellowstone Valley occupies one of the largest calderas in the United States. Many geysers, hot springs, and fumaroles in the valley are related to groundwater circulating to certain depths, being heated by shallow magma, and mixing with volcanic gases that escape through minor cracks in the crust of the earth.

VOLCANISM IN RELATIONSHIP TO PLATE TECTONIC SETTING

The types of volcanism and associated volcanic hazards differ in various tectonic settings because each setting produces a different type of magma through the processes described above. Mid-ocean ridges and intraplate "hot spot" types of volcanoes typically produce nonexplosive eruptions, whereas convergent tectonic margin volcanoes may produce tremendously explosive and destructive eruptions. Much of the variability in the eruption style may be related to the different types of magma produced in these different settings and also to the amount of dissolved gases, or *volatiles,* in these magmas. Magmas with large amounts of volatiles tend to be highly explosive, whereas magmas with low contents of dissolved volatiles tend to be nonexplosive.

Eruptions from mid-ocean ridges are mainly basaltic flows with low amounts of dissolved gases. These eruptions are relatively quiet, with basaltic magma flowing in underwater tubes and breaking off in bulbous shapes called pillow lavas. The eruption style in these underwater volcanoes is analogous to toothpaste being squeezed out of a tube. Eruptions from mid-ocean ridges may be observed in the few rare places where the ridges emerge above sea level, such as Iceland. Eruptions there include lava fountaining, where basaltic cinders are thrown a few hundred feet in the air and accumulate as cones of black glassy fragments, and they also include long stream-like flows of basalt.

Hot spot volcanism tends to be much like that at mid-ocean ridges, particularly where the hot spots are located in the middle of oceanic plates. The Hawaiian Islands have the most famous hot spot type of volcanoes in the world, the active volcanoes on the island of Hawaii known as Kilauea and Mauna Loa. Mauna Loa is a huge shield volcano, characterized by a very gentle slope of a few degrees from the base to the top. This gentle slope is produced by lava flows that have a very low viscosity and can flow and thin out over large distances before they solidify. Magmas with high viscosity would be much stickier and would solidify in short distances, producing volcanoes with steep slopes. Measured from its base on the Pacific Ocean seafloor to its summit, Mauna Loa is the tallest mountain in the world, a fact attributed to the large distances that its low-viscosity lavas flow and to the large volume of magma produced by this hot spot volcano.

Volcanoes associated with convergent plate boundaries produce by far the most violent and destructive eruptions. Recent convergent margin eruptions include Mount Saint Helens, two volcanoes in the Philippines, Mount Pinatubo, Mount Etna in Italy, and a Mayan volcano. The magmas from these volcanoes tend to be much more viscous and higher in silica content, and they have the highest concentration of dissolved gasses. Many of the dissolved gasses and volatiles, such as water, are released from the subducting oceanic plate as high mantle temperatures heat it up as it slides beneath the convergent margin volcanoes.

Table 3.2
Examples of Volcanic Disasters

Volcano and location	Year	Number of deaths
Tambora, Indonesia	1815	92,000
Krakatau, Indonesia	1883	36,500
Mt. Peleé, Martinique	1902	32,000
Nevada del Ruiz, Colombia	1985	24,000
Santa Maria, Guatemala	1902	6,000
Galunggung, Indonesia	1822	5,500
Awu, Indonesia	1826	3,000
Lamington, Papua New Guinea	1951	2,950
Agung, Indonesia	1963	1,900
El Chichon, Mexico	1982	1,700

VOLCANIC HAZARDS

Volcanic eruptions are responsible for the deaths of thousands of people, and they directly affect large portions of the world's population, land-use patterns, and climate. Table 3.2 lists examples of the ten worst volcanic disasters since 1800, measured in the numbers of lives lost. The following sections will demonstrate that different phenomena associated with volcanic eruptions have been responsible for the greatest loss of life in these various examples. Understanding the specific hazards associated with volcanoes is important for reducing losses from future eruptions, especially considering that millions of people live close to active volcanoes.

Hazards of Lava Flows

Lava flows are most common around volcanoes that are characterized by eruptions of basalt with low contents of dissolved gasses. These places include Hawaii, Iceland, and other places characterized by nonexplosive eruptions. In January 2002, massive lava flows erupted from Nyiragongo volcano in Congo, devastating the town of Goma and forcing 300,000 people to flee their homes as the lava advanced through the town (Figure 3.5). Lava flows generally follow topography, flowing from the volcanic vents downslope into valleys, much as streams or water from a flood would travel. Some lava flows may move as fast as water, up to almost forty miles per hour (sixty-five km/hour) on steep slopes, but most lava flows move considerably slower. More typical rates of movement are about ten feet per hour (several meters per hour), or ten feet per day for slower flows. These rates of lava movement allow most people to move out of danger to higher ground, but lava flows are responsible for significant amounts of property damage in places like Hawaii. Basaltic lava is extremely hot (typ-

Figure 3.5. Lava flow hazards. In 2001, Nyiragongo volcano erupted, sending thick rivers of molten lava into the city of Goma. A river of lava from the erupting volcano of Mount Nyiragongo burns the houses of Goma January 18, 2002. Tens of thousands of residents of Goma fled their homes and vast areas of the town were set ablaze by lava flows from the volcano. ©AFP/CORBIS

ically about 1000°C) when it flows across the surface, so when it encounters buildings, trees, and other flammable objects, these objects typically burst into flame and are destroyed. Lava flows also are known to bury roads, farmlands, and other low-lying areas (Figure 3.5). It must be kept in mind, however, that the entire Hawaiian island chain was built by lava flows, and the real estate that is being damaged would not even be there if it were not for the lava flows.

It is significantly easier to avoid hazardous lava flows than some other volcanic hazards. It is generally unwise to build or buy homes in low-lying areas adjacent to volcanoes. Lava flows have been successfully diverted in a few examples in Hawaii, Iceland, and elsewhere in the world. One of the better strategies is to build large barriers of rock and soil to divert the flow from its natural course to a place where it will not damage property. More creative methods have also had limited success: Lava flows have been bombed in Hawaii and Sicily, and spraying large amounts of water on active flows in Hawaii and Iceland has chilled these flows enough to stop their advance into harbors and populated areas.

Hazards of Pyroclastic Flows

Violent pyroclastic flows present one of the most severe hazards associated with volcanism. Unlike slow-moving lava flows, pyroclastic flows may move at hundreds of kilometers per hour by riding on a cushion of air, burying entire villages or cities before anyone has a chance to escape. Nuee ardents are particularly hazardous varieties of pyroclastic flows, with temperatures that may exceed 800°C and down-slope velocities of many hundreds of miles per hour. On an otherwise quiet day in 1902, the city of St. Pierre on the beautiful Caribbean island of Martinique was quickly buried by a nuee ardent from Mt. Pelée, killing more than 29,000 people.

In A.D. 79, Mount Vesuvius erupted and buried the towns of Pompeii and Herculaneum under pyroclastic ash clouds, permanently encapsulating these cities and their inhabitants in volcanic ash. Crater Lake in Oregon provides a thought-provoking example of the volume of volcanic ash that may be produced during a pyroclastic eruption. The eruption that caused this caldera to collapse covered an area of 13,000 square kilometers with more than six inches (fifteen cm) of ash and blanketed most of present-day Washington, Oregon, and Idaho, and large parts of Montana, Nevada, California, British Columbia, and Alberta, with ash. Larger eruptions can occur: For instance, a thick rock unit known as the Bishop tuff covers a large part of the southwestern United States. If any of the active volcanoes in the western United States were to produce such an eruption, it would be devastating for the economy and huge numbers of lives would be lost.

Hazards of Mudflows, Floods, and Debris Avalanches

When pyroclastic flows and nuee ardents move into large rivers, they quickly cool and mix with water, becoming fast-moving mudflows known as *lahars*. Lahars may also result from the extremely rapid melting of icecaps on volcanoes. A type of lahar in which ash, blocks of rock, trees, and other material are chaotically mixed together is known as a *debris flow*. Such lahars and mudflows were responsible for much of the burial of buildings and deaths from the trapping of automobiles during the 1980 eruption of Mount Saint Helens. Big river valleys were also filled by lahars and mudflows during the 1991 eruption of Mount Pinatubo in the Philippines, resulting in extensive property damage and loss of life. One of the greatest volcanic disasters of the twentieth century resulted from the generation of a huge mudflow when the icecap on the volcano Nevada del Ruiz in Colombia catastrophically melted during the 1985 eruption. When the mudflow moved downhill, it buried the towns along the rivers leading to the volcano, killing 23,000 people and causing more than $200 million U.S. dollars of damage to property (Figure 3.6).

Pyroclastic flows may leave thick unstable masses of unconsolidated volcanic ash that can be easily remobilized long after an eruption during heavy rains or earthquakes. These thick piles of ash may remain for many years after an eruption and may not be remobilized until a hurricane or other heavy rainstorm (itself a catastrophe) fluidizes the ash, initiating destructive mudflows.

Figure 3.6. *Top*: Aerial view of mudflow from Nevada del Ruiz, covering the former site of the town of Armero. *Bottom*: Detail of the destruction of downtown Armero by mudflows from Nevada del Ruiz. Photo Credit: U.S. Geological Survey

Hazards of Poisonous Gases

One of the lesser-known hazards of active volcanoes stems from their emission of poisonous gases. These gases normally escape through geysers, fumaroles, and fractures in the rock. In some instances, however, volcanoes emit poisonous gases, including carbon monoxide, carbon dioxide, and sulfurous gases. These may also mix with water to produce acidic pools of hydrochloric, hydrofluoric, and sulfuric acid. It is generally not advisable to swim in strange colored or unusual smelling ponds on active volcanoes.

Some of the more devastating emissions of poisonous gases from volcanoes occurred in 1984 and 1986 in Africa. In the larger of the emissions, in 1986, approximately 1,700 people and thousands of cattle were killed when a huge cloud of carbon dioxide bubbled out of the volcanic crater lakes Nyos and Monoun in Cameroon, quickly suffocating the people and animals downwind from the vent lakes. More than 300 million cubic feet (100 million cubic meters) of gas emissions escaped without warning and spread out over the area in less than two hours, highlighting the dangers of living on and near active volcanoes. Lakes similar to Lakes Nyos and Manoun are found in many other active volcanic areas, including heavily populated parts of Japan, Zaire, and Indonesia.

Scientists have recently taken steps to reduce the hazards of additional gas emissions from these lakes. In 2001, a team of scientists from Cameroon, France, and the United States installed the first of a series of degassing pipes into the depths of Lake Nyos, in an attempt to release the gases from depths of the lake gradually before they erupt catastrophically. The first pipe extends to 672 feet in the lake and causes a pillar of gas-rich water to squirt up the pipe and form a fountain on the surface, slowly releasing the gas from depth. The scientific team estimates that they need five additional pipes to keep the gas levels at a safe level, which will cost an additional $2 million.

Hazards of Volcanic-induced Earthquakes and Tsunamis

Minor earthquakes generally accompany volcanic eruptions. The typical scenes from old disaster movies where the ground shakes and the volcano erupts are actually not far from the true experience, though the earthquakes tend to be *slightly* exaggerated in the movies. The earthquakes are generated by magma forcing its way upward from the depths of the magma chamber, through cracks and fissures, and into the volcano. These earthquakes have become one of the more reliable methods of predicting exactly when an eruption is imminent, as geologists can trace the movement of the magma by very detailed seismic monitoring. The earthquakes typically form a continuous, low frequency rhythmic ground shaking known as a harmonic tremor. If you are near a volcano and you feel these harmonic tremors, it is a good warning that it is time to leave.

Tsunami, or giant seismic sea waves, may be generated by volcanic eruptions, particularly if the eruptions occur underwater. These giant waves may inundate coastlines with little warning. In 1883, more than 36,500 people in Indonesia were killed by a tsunami generated by the eruption of Krakatau volcano. Tsunami are discussed in detail in Chapter 4.

Hazards from Changes in Climate

Some of the larger, more explosive volcanic eruptions spew vast amounts of ash and finer particles called *aerosols* into the atmosphere and stratosphere, and it may take years for these particles to settle back down to earth. They get distributed around the planet by high-level winds, and they have the effect of blocking out some of the sun's rays, which lowers global temperatures. A side effect is that the extra particles in the atmosphere also produce more spectacular sunsets and sunrises, as does extra pollution in the atmosphere. These effects were readily observed after the 1991 eruption of Mount Pinatubo, which spewed more than five cubic miles (eight cubic kilometers) of ash and aerosols into the atmosphere, causing global cooling for two years after the eruption. Even more spectacularly, the 1815 eruption of Tambora in Indonesia, in which 92,000 people died, caused three days of total darkness for approximately 300 miles (500 kilometers) around the volcano, and it initiated the famous "year without a summer" in Europe because the ash from this eruption lowered global temperatures by more than a degree. Even these amounts of gases and small airborne particles are dwarfed by the amount of material placed into the atmosphere during some of Earth's largest eruptions, known as flood basalts. Flood basalts and the environmental consequences of their eruption are discussed in Chapter 12.

Other Hazards and Long-Term Effects of Volcanic Eruptions

Dispersed ash may leave thin layers on agricultural fields, which may be beneficial or detrimental depending on the composition of the ash. Some ash basically fertilizes soil, whereas other ash is toxic to livestock. Much of the richest farmland on the planet is on volcanic ash layers near volcanoes. Volcanic ash consists of tiny but jagged and rough particles that may conduct electricity when wet. Therefore, ash also poses severe hazards to electronics and machinery that may last many months past an eruption, and ash has been known to disrupt power generation and telecommunications. The particles are abrasive, and if inhaled may cause serious heart and lung ailments.

Ash clouds have some unexpected and long-term consequences to the planet and its inhabitants. Airplane pilots have sometimes mistakenly flown into ash clouds, thinking they are normal clouds, which has caused engines to fail. For instance, KLM Flight 867 with 231 people aboard flew through the ash cloud produced during the 1989 eruption of Mount Redoubt in Alaska, causing engine failure. The plane suddenly dropped two miles in altitude before the pilots were able to restart the engines, narrowly averting disaster. This event led the U.S. Geological Survey to formulate a series of warning codes for eruptions in Alaska and their level of danger to aircraft.

After volcanic eruptions, large populations of people may be displaced from their homes and livelihoods for extended or permanent time periods. In many cases, these peoples are placed into temporary refugee camps, which all too often become permanent shanty villages riddled with disease, poverty, and famine. Many of the casualties from volcanic eruptions come from these long-term effects and not from the initial eruption. More needs to be done to insure that populations displaced by volcanic disasters are relocated into safe settings.

WHAT CAN BE DONE TO REDUCE VOLCANIC HAZARDS?

Predicting Volcanic Eruptions

One of the best ways to understand what to anticipate from an active volcano is to study its past history. Historical records may be examined to learn about geologically recent eruptions. Geological mapping and analysis can reveal what types of material the volcano has spewed forth in the more distant past. A geologist who studies volcanic deposits can tell through examination of these deposits whether the volcano is characterized by explosive or nonexplosive eruptions, whether it has nuee ardents or mudflows, and how frequently the volcano has erupted over long time intervals. This type of information is crucial for estimating what the risks are for any individual volcano, and programs of risk assessment and volcanic risk mapping need to be done around all of the nearly 600 active volcanoes on continents of the world. Which areas are prone to ash falls? Which areas have been repeatedly hit by mudflows? Are any areas characterized by periodic emissions of poisonous gas? Approximately sixty eruptions occur globally every year, so these data would prove immediately useful when eruptions appear imminent.

Precursors to Eruptions Volcanic eruptions are sometimes preceded by a number of precursory phenomena, or warnings that an eruption may be imminent. Many of these involve subtle changes in the shape or other physical characteristics of the volcano. Many volcanoes develop bulges, swells, or domes on their flanks when magma rises in the volcano before an eruption. These shape changes can be measured using sensitive devices called tiltmeters, which measure tilting of the ground surfaces, or devices that precisely measure distances between points such as geodolites and laser measuring devices. Bulges were measured on the flanks of Mount St. Helens before its 1980 eruption.

Eruptions may also be preceded by other more subtle precursory events, such as increase in the temperature or heat flow from the volcano, measurable both on the surface and in crater lakes, hot springs, fumaroles, and hot springs on the volcano. There may also be detectable changes in the composition of gases emitted by the volcano, such as increases in the hydrochloric acid and sulfur dioxide gases, before an eruption.

One of the most reliable precursors to an eruption is the initiation of the harmonic seismic tremors that reflect the movement of magma into the volcano. These tremors typically begin days or weeks before an eruption and steadily change their characteristics, enabling successively more accurate predictions of how imminent the eruption is before it actually happens. Careful analysis of precursor phenomena including the harmonic tremors, change in the shape of the volcano, and emission of gases has enabled accurate prediction of volcanic eruptions, including those of Mount St. Helens and Mount Pinatubo. These predictions saved innumerable lives.

Volcano Monitoring Precursory phenomena may only be observed if volcanoes are carefully and routinely monitored. Volcanic monitoring is aimed at detecting the precursory phenomena described above and tracking the movement of magma beneath volcanoes. In the United States, the U.S. Geological Survey is in charge of comprehensive volcano monitoring programs in the Pacific Northwest, Alaska, and Hawaii.

One of the most powerful methods of determining the position and movement of magma in volcanoes is by using seismology, or the study of the passage of seismic waves through the volcano. These can be natural seismic waves generated by earthquakes beneath the volcano or seismic energy released by geologists who set off explosions and monitor how the energy propagates through the volcano. As discussed in Chapter 2, certain types of seismic waves (P-waves) travel through fluids like magma, whereas other types of seismic waves (S-waves) do not. Therefore, the position of the magma beneath a volcano can be determined by detonating an explosion on one side of the volcano and having seismic receivers placed around the volcano to determine the position of a "shadow zone" where P-waves are received but S-waves are not. The body of magma that creates the shadow zone can be mapped out in three dimensions by using data from the numerous seismic receiver stations. Repeated experiments over time can track the movement of the magma.

Other precursory phenomena are also monitored to track their changes with time, which can further refine estimates of impending eruptions. Changes in the temperature of the surface can be monitored by thermal infrared satellite imagery, and other changes, such as shifts in the composition of emitted gases, are monitored. Other promising precursors may be found in changes to physical properties, such as the electrical and magnetic fields around volcanoes prior to eruptions.

WHAT IF YOU ARE IN A VOLCANIC ERUPTION?

What you should do if you are in a volcanic eruption depends on what kind of eruption you are experiencing. If you are near an impending explosive eruption and eruption warnings have been issued, you should try to evacuate the area as quickly as possible. Avoid low-lying areas that may be prone to mudslides, lahars, or debris avalanches, and especially steer clear of possible routes of hot glowing avalanches. If you live in or frequently visit areas near active volcanoes, you should make yourself aware of areas where these hazards are particularly high. Information on hazards of specific volcanoes is available from the U.S. Geological Survey and local government offices. Some useful Web

Table 3.3
The Ten Worst Eruptions in Terms of Deaths and Destruction

Location	Year	Deaths
Tambora, Indonesia	1815	92,000
Krakatau, Indonesia	1883	36,500
Mt. Pelée, Martinique	1902	32,000
Nevada del Ruiz, Colombia	1985	24,000
Santa Maria, Guatemala	1902	6,000
Galunggung, Indonesia	1822	5,500
Awu, Indonesia	1826	3,000
Lamington, Papua New Guinea	1951	2,950
Agung, Indonesia	1963	1,900
El Chichon, Mexico	1982	1,700

sites for U.S. volcanoes are listed at the end of this chapter.

Even the relatively passive eruptions of Hawaii pose many risks and hazards to those who view eruptions. Specific hazards include collapse of lava benches and tubes, especially where lava flows meet the sea. It is not uncommon for large benches of recently formed lava in Hawaii to suddenly collapse into the sea, setting off a series of explosions and sending waves of scalding hot water onto the land. Tephra jets are hazardous explosions of lava and steam that form where waves from the ocean splash against lava pouring into the sea. Sometimes these explosions can send volcanic bombs that are as large as a person many tens of feet through the air, so it is best to stay at least a few hundred feet from where lava is entering the sea. Lava can also make lava haze and volcanic smog that can contain poisonous and irritating sulfur dioxide and other hazardous gases, and also severely limit visibility. It is best to stay away from suspicious white clouds near active lava flows.

VOLCANIC ERUPTION STATISTICS

Nearly a quarter million people have died in volcanic eruptions in the past 400 years, with a couple of dozen volcanic eruptions killing more than a thousand people each. Table 3.3 lists the ten worst volcanic eruptions in terms of the number of deaths. Eruptions in remote areas have little consequence except for global climate change. However, eruptions in populated areas can cause billions of dollars in damage and can cause entire towns and cities to be relocated. The eruption of Mount St. Helens in 1980 created $1 billion worth of damage.

The U.S. Geological Survey is the main organization in charge of monitoring volcanoes and eruptions in the United States, and it takes this responsibility for many other places around the world. In cases of severe eruptions or eruptions that threaten populated areas, other agencies such as the Federal Emergency and Management Association (FEMA) will join the U.S. Geological Survey and help in disseminating information and evacuating the population.

EXAMPLES OF VOLCANIC DISASTERS

Vesuvius, Italy, 79 A.D.

The most famous volcanic eruption is probably that of Vesuvius in A.D. 79, which buried the towns of Pompeii, Herculaneum, and Stabiae in present-day Italy and killed tens of thousands of people. Before the eruption, Pompeii was a well-known center of commerce and home to approximately 20,000 people. On August 24, A.D. 79, Vesuvius erupted after several years of earthquakes. The initial blast launched two and a half cubic miles (four cubic kilometers) of pumice, ash, and other volcanic material into the air, which quickly buried Pompeii under ten feet of volcanic debris. Successive pyroclastic flows, together killing about 4,000 people in Vesuvius and neighboring towns, followed the initial blast. Vesuvius immediately moved into a second phase of its massive eruption, during which huge quantities of ash were blown up to twenty miles upward into the atmosphere, alternately surging upward and dropping tons of ash onto the surrounding region, killing most of the people who were not killed in the initial eruption. Daylight was quickly turned into a dark impenetrable night, and the town of Pompeii was buried under another six to seven feet of ash. Thick ash also accumulated on the slopes of the volcano and was quickly saturated with water from rains created by the volcanic eruption. Water-saturated mudflows called lahars moved swiftly down the slopes of Vesuvius, burying the town of Herculaneum and killing its remaining inhabitants. This area has since been rebuilt, and the town of Ercolano now lies on top of the twenty feet of ash that buried Herculaneum. Vesuvius is still active and has experienced many eruptions since the famous eruption in A.D. 79. It is not recommended to visit Ercolano during an eruption of Vesuvius.

Mt. Pelée, Martinique, 1902

Martinique was a quiet, West Indies island whose French colonizers grew sugar for export since the mid-1600s. In the spring of 1902, the volcano on the north end of the island began to show some activity, with boiling lakes and intermittent minor pyroclastic flows and eruptions. By April, most people were becoming worried about the increasingly intense activity, and they were congregating in the city of St. Pierre, which was located only six miles from the volcano, to catch boats to leave the island. A larger eruption occurred on May 5 and killed forty people in a pyroclastic flow that raced down the Riviere Blanche. However, elections were five days away, and Governor Mouttet did not want the island's people to leave for fear that he might lose the election in a bad turnout. He ordered the military to halt the exodus from the island. At 7:50 A.M. on May 8, a huge nuee ardent or glowing avalanche erupted from Mt. Pelée at 2,200°F and moved over the six miles to St. Pierre at 115 miles per hour, killing all but two of the city's 30,000 residents (including Governor Mouttet) in a matter of minutes. One of these was an imprisoned murderer who spent the rest of his life being paraded as "The Prisoner of St. Pierre" in circuses to show his badly burned body. About 2,000 people from surrounding towns were killed in later eruptions. Today the whole region is resettled and densely inhabited.

Nevada del Ruiz, Colombia, 1985

The Nevada del Ruiz volcano in Colombia entered an active phase in November 1984 and began to show harmonic tremors on November 10, 1985. At 9:37 P.M. that night, a large Plinian eruption sent an ash cloud several miles into the atmosphere, and this ash settled onto the ice cap on top of the mountain. This ash, together with volcanic steam, quickly melted large amounts of the ice, which mixed with the ash and formed giant lahars (mudflows) that slid down the east side of the mountain into the village of Chinchina, killing 1,800 people. The eruption continued and melted more ice that mixed with more ash and sent additional larger lahars westward. Some of these lahars moved nearly thirty miles at nearly thirty miles per hour, and under a thunderous roar they buried the town of Armero under twenty-six feet of mud (Figure 3.6). 22,000 people died in Armero that night.

Mount St. Helens, Washington, 1980

The most significant eruption in the conterminous United States in the past fifty years is that of Mount St. Helens in 1980. On March 27, 1980, many small eruptions on Mount St. Helens began when magma rose high enough to meet groundwater, which caused steam explosions. The volcano gradually bulged by about 300 feet, and harmonic tremors indicated an impending large eruption. The U.S. Geological Survey issued eruption warnings, and the area was evacuated. On May 18, 1980, the upper 1,313 feet of the volcano were blown away in an unusual lateral blast. A magnitude 5.1 earthquake initiated the lateral blast, and the side of the mountain exploded outward at 150 miles per hour. The preexisting rock mixed with magma and rose to temperatures of 200°F. This mass flew into Spirit Lake on the northern flank of the volcano and formed a wave 650 feet high that destroyed much of the landscape. Lahars filled the north and south forks of the Toutle River, Pine Creek, and Muddy River. A hot pyroclastic flow blasted out of the hole on the side of the volcano and moved at 250 miles per hour, knocking over and burying trees and everything else in its path for hundreds of square miles. A huge Plinian ash cloud, which erupted to heights of twelve miles and carried ash that dropped over much of the western United States, complemented these local eruption features (Figure 3.4). Pyroclastic flows continued to move down the volcano at sixty miles per hour and at temperatures of 550–700°F. Sixty-two people died in this relatively minor eruption, and damage to property is estimated at $1 billion.

Nyiragongo, Congo, 2002

Nyiragongo volcano is located in Congo (formerly Zaire) in the East African rift valley, along an extensional plate boundary. Nyiragongo is an 11,380-foot-tall steep stratovolcano cone that has experienced significant eruptions, including one in 1977 that killed seventy people. It is one of the most active volcanoes along the East African rift valley, and it contained a lava lake in its crater from 1894 through 1977 and another lava lake formed in 1982 before the 2002 eruption. Nyiragongo entered a new active eruption period in January 2002, sending blocky lava flows into the town of Goma on the shores of Lake Kivu (Figure 3.5). Lava flows six feet deep covered much of the town, burning buildings and crops, blocking streets, and sending 300,000 people fleeing across the border into Rwanda. In some places, the lava moved so quickly, at speeds of up to forty miles per hour (sixty km/hr), that people could barely outrun it. The lava ignited fuel tanks and gasoline stations, causing many explosions and fires in Goma. People who were illegally trying to siphon gas out of an aboveground tank for their personal use initiated one particularly large explosion. Somebody accidentally dropped a bottle of gasoline on the hot lava (estimated to be 1000°C just below a thin surface crust), setting off a massive fireball in which several dozen people died. Nearly three-quarters of Goma was destroyed in the eruption.

RESOURCES AND ORGANIZATIONS

Print Resources

Fisher, R. V. *Out of the Crater: Chronicles of a Volcanologist.* Princeton, N.J.: Princeton University Press, 2000, 180 pp.

Fisher, R. V., Heiken, G., and Hulen, J. B. *Volcanoes: Crucibles of Change.* Princeton, N.J.: Princeton University Press, 1998, 334 pp.

Holloway, M. "Trying to Tame the Roar of Deadly Lakes." *New York Times,* February 27, 2001, sec. D3.

Lacey, M. "Tens of Thousands Flee a Devastating Volcano in Congo." *New York Times,* January 19, 2002, sec. A3.

Simkin, T., and Fiske, R. S. *Krakatau 1883: The Volcanic Eruption and Its Effects.* Washington, D.C.: Smithsonian Institution Press, 1993.

Non-Print Resources

Web Articles

"Benefits of Volcano Monitoring Far Outweigh Costs—The Case of Mount Pinatubo" by Chris Newhall, James Hendley II, and Peter Stauffer
http://geopubs.wr.usgs.gov/fact-sheet/fs030-97/
Analysis of the methods, costs, and benefits of volcano monitoring at Mount Pinatubo.

"The Cataclysmic 1991 Eruption of Mount Pinatubo, Philippines" by Chris Newhall, James Hendley II, and Peter Stauffer
http://geopubs.wr.usgs.gov/fact-sheet/fs113-97/
Detailed descriptions of the huge eruption of Mount Pinatubo in 1991. Also includes assessment of continuing hazards.

"Invisible CO_2 Gas Killing Trees at Mammoth Mountain, California" by Michael Sorey, Chris Farrar, William Evans, David Hill, Roy Bailey, James Hendley II, and Peter Stauffer
http://geopubs.wr.usgs.gov/fact-sheet/fs074-97/
Fascinating account of how large volumes of CO_2 gas are seeping out of the caldera beneath Long Valley and Mammoth Mountain, California, and killing many trees in the area. This may indicate an impending eruption.

"Lahars of Mount Pinatubo, Philippines" by Chris Newhall, Peter Stauffer, and James Hendley II
http://geopubs.wr.usgs.gov/fact-sheet/fs114-97/
Descriptions of the huge lahars and mudflows of Mount Pinatubo and what the warning signs of impending lahars were.

"Living on Active Volcanoes—The Island of Hawai'i"

http://geopubs.wr.usgs.gov/fact-sheet/fs074-97/
Discussion of the volcanic activity and hazards on the island of Hawaii. Includes maps showing areas of low through high risk for volcanic hazards.

"Living with a Restless Caldera—Long Valley, California" by David Hill, Roy Bailey, Michael Sorey, James Hendley II, and Peter Stauffer
http://geopubs.wr.usgs.gov/fact-sheet/fs108-96/
Description of the rising magma beneath Long Valley, California, and the volcanic hazards this poses to the western United States. Includes accounts of historical eruptions and systems of warnings that are in place in the event of an eruption.

"Living with Volcanic Risk in the Cascades" by Dan Dzurisin, Christina Heliker, Peter Stauffer, and James Hendley II
http://geopubs.wr.usgs.gov/fact-sheet/fs114-97/
General discussion of overall volcanic hazards in the Cascades and links to other sites with more specific information.

"Mount Baker—Living with an Active Volcano" by Kevin Scott, Wes Hildreth, and Cynthia Gardner
http://geopubs.wr.usgs.gov/fact-sheet/fs059-00/ (May 2000)
Discussion of hazards associated with Mount Baker and discussions of why certain areas have been closed to tourists.

"Mount Hood—History and Hazards of Oregon's Most Recently Active Volcano" by Cynthia Gardner, William Scott, Jon Major, and Thomas Pierson
http://geopubs.wr.usgs.gov/fact-sheet/fs060-00/
Discussion of historical activity of Mount Hood, and possible hazards it may pose to Portland, Oregon, and vicinity. Includes maps of proximal and distal hazard areas.

"Mount St. Helens—From the 1980 Eruption to 2000" by Steve Brantley and Bobbie Myers
http://geopubs.wr.usgs.gov/fact-sheet/fs036-00/
History of recent Mount St. Helens eruptions.

"Viewing Lava Safely—Common Sense Is Not Enough" by Jenda Johnson
http://geopubs.wr.usgs.gov/fact-sheet/fs152-00/
A short guide to some of the volcanic hazards associated with viewing lava flows on Kilauea Volcano, Hawaii.

"Volcanic Air Pollution—A Hazard in Hawai'i" by Jeff Sutton, Tamar Elias, James Hendley II, and Peter Stauffer
http://geopubs.wr.usgs.gov/fact-sheet/fs169-97/
Discussion of the various gases and pollution produced by Kilauea Volcano, Hawaii.

Web Sites

Hawaiian Volcano Observatory Web site:
http://hvo.wr.usgs.gov/
Site includes updates of current volcanic activity.

U.S. Geological Survey Volcano Hazards Program Web site:
http://volcanoes.usgs.gov/
Contains updates of U.S. and worldwide volcanic activity, and has feature articles on recent research. Also has links to sites on volcanic hazards, historical eruptions, monitoring programs, emergency planning, and warning schemes. Has resources including photos, fact sheets, videos, and an education page. Also offers grants to college students doing volcano research.

Volcanoworld Web site:
http://volcano.und.nodak.edu/vwdocs
Volcanoworld presents updated information about eruptions and volcanoes, and has many interactive pages designed for different grade levels from kindergarten through the college and professional levels.

Organizations

Alaska Volcano Observatory
U.S. Geological Survey
4200 University Drive
Anchorage, AK 99508

Cascades Volcano Observatory
U.S. Geological Survey
5400 MacArthur Blvd.
Vancouver, WA 98661
360–993–8900

Smithsonian Institution
Museum of Natural History E-421
Washington, D.C. 20560-0119
202–357–2476
The Smithsonian publishes the Bulletin of the Global Volcanism Network.

Volcano Disaster Assistance Program
This is a cooperative effort between the U.S. Agency for International Development (Office of Foreign Disaster Assistance) and the U.S. Geological Survey. These organizations head a mobile response team that has been mobilized many times since its initiation in 1986, saving numerous lives from areas at risk. Go to http://geopubs.wr.usgs.gov/fact-sheet/fs064–97/ for more information.

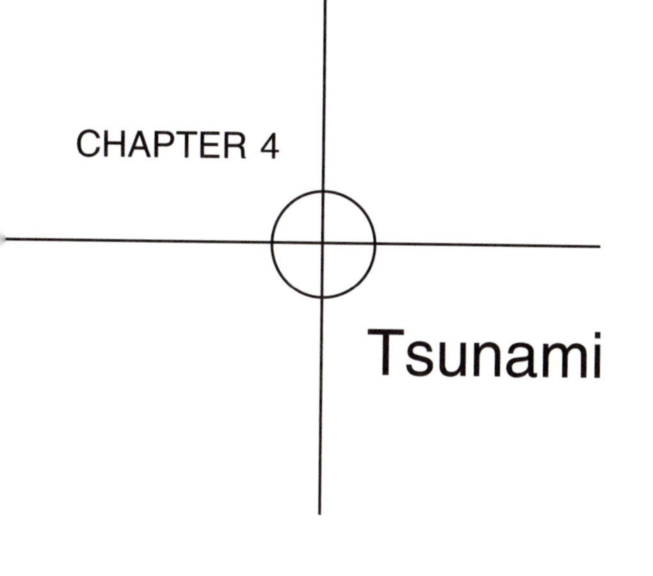

CHAPTER 4

Tsunami

INTRODUCTION

Every few years, giant sea waves rise unexpectedly out of the ocean and sweep over coastal communities, killing hundreds of people and causing millions of dollars in damage. Such events occurred in 1946, 1960, 1964, 1992, 1993, and 1998 in coastal Pacific areas. In 1998, a catastrophic fifty-foot-high wave unexpectedly struck Papua New Guinea, killing more than 2,000 people and leaving more than 10,000 homeless. What are these giant waves, and what causes them to strike without warning? Tsunami, spelled the same in its singular and plural forms, are seismic sea waves that are generated by the sudden displacement of the sea floor. The name is Japanese for "harbor wave." Tsunami are also commonly called tidal waves, although this is improper because they have nothing to do with tides.

Tsunami are generated most often by thrust earthquakes along deep-ocean trenches and convergent plate boundaries. Tsunami therefore occur most frequently along the margins of the Pacific Ocean, a region characterized by numerous thrust-type earthquakes. About 80 percent of all tsunami strike Pacific Rim shorelines, with the most by far being generated in and striking southern Alaska. Volcanic eruptions, giant submarine landslides, and the sudden release of gases from sediments on the sea floor may also generate them. Tsunami are common on Pacific islands, including Hawaii and Japan, which now have extensive warning systems in place to alert residents when they are likely to occur. Before these warning systems were in place, the tsunami would occasionally strike coastal areas, in some cases reaching fifty feet or more in height.

Some historical tsunami have been absolutely devastating to coastal communities, wiping out entire populations with little warning. One of the most devastating tsunami in recent history was generated by the eruption of the Indonesian volcano Krakatau in 1883. When Krakatau erupted, it blasted a large part of the center of the volcano out, and seawater rushed in to fill the hole. This seawater was immediately heated and it exploded outward in a steam eruption and a huge wave of hot water. The tsunami generated by this eruption reached more than 120 feet in height and killed more than 36,500 people in nearby coastal regions. Another famous tsunami was also generated by a volcanic eruption: Santorini (now called Thíra), Greece, on the Mediterranean island of Crete. In 1600 B.C., this volcano was the site of the most powerful eruption in recorded history, and it generated a tsunami that destroyed many Mediterranean coastal areas and probably led to the eventual downfall of the Minoan civilization on Crete. The tsunami deposited volcanic debris at elevations of up to 800 feet above the mean ocean level on the nearby island of Anaphi, and the wave was still more than twenty feet high when it ran up the shorelines on the far side of the Mediterranean in Israel.

MOVEMENT OF TSUNAMI

Tsunami are waves with exceptionally large distances between individual crests, and they move like other waves across the ocean (waves are described using the terms shown in Figure 7.2a). Wavelength is the distance between crests, wave height is the vertical distance from the crest to the bottom of the trough, and the amplitude is one-half of the wave height. Most ocean waves have wavelengths of 300 feet (100 meters) or less. Tsunami are exceptional in that they have wavelengths that can be 120 miles (200 kilometers) or greater. When tsunami are traveling across deep ocean water, their amplitudes are typically less than three feet. You would probably not even notice even the largest of tsunami if you were on a boat in the deep ocean (so the giant wave that sunk the leisure ship in the 1970s film *The Poseidon Adventure* must have been an entirely different creature, such as a *rogue wave*). Circular or elliptical paths that decrease in size with depth describe the motion of water in waves. All motion from the waves stops at a depth equal to one-half the distance of the wavelength. Tsunami therefore are felt at much greater depths than ordinary waves, and this effect may be used with deep ocean bottom tsunami detectors to help warn coastal communities when tsunami are approaching.

Waves with long wavelengths travel faster than waves with short wavelengths. Since the longer the wavelength, the faster the wave in deep open water, tsunami travel extremely fast across the ocean. Normal ocean waves travel at less than fifty-five miles per hour (ninety kilometers per hour), whereas many tsunami travel at 500 to 600 miles per hour (800 to 950 km/hour), faster than most commercial airliners!

When waves encounter shallow water, the friction of the sea floor along the base of the wave causes them to slow down dramatically, and the waves effectively pile up on themselves as successive waves move into shore. This causes the wave height or amplitude to increase dramatically, with tsunami sometimes rising fifteen to 150 feet above the normal still water line.

When tsunami strike the coastal environment, the first effect is sometimes a significant retreat or *drawdown* of the water level, whereas in other cases the water just starts to rise quickly. Since tsunami have long wavelengths, it typically takes several minutes for the water to rise to its full height. Also, since there is no trough right behind the crest of the wave, on account of the very long wavelength of tsunami, the water does not recede until a considerable time after the initial crest rises onto land. The rate of rise of the water in a tsunami depends in part on the shape of the sea floor and coastline. If the sea floor rises slowly, the tsunami may crest slowly, giving people time to outrun the rising water. In other cases, especially where the sea floor rises steeply or the shape of the bay causes the wave to be amplified, tsunami may come crashing in huge walls of water with breaking waves that pummel the coast with a thundering roar and wreak utmost destruction.

Because tsunami are waves, they travel in successive crests and troughs. Many deaths in tsunami events are related to people going to the shoreline to investigate the effects of the first wave, or to rescue those injured or killed in the initial crest, only to be drowned or swept away in a succeeding crest. Tsunami have long wavelengths, so successive waves have a long lag time between individual crests. The period of a wave is the time between the passage of individual crests, and for tsunami the period can be an hour or more. Thus, a tsunami may devastate a shoreline area, retreat, and then with another crest may strike an hour later, then another, and another in sequence.

Tsunami Run-up

Run-up is the height of the tsunami above sea level at the farthest point it reaches on the shore. This height may be considerably different from the height of the wave where it first hits the shore (Figure 4.1). Many things influence the run-up of tsunami, including the size of the wave, the shape of the shoreline, the profile of the water depth, and other irregularities particular to individual areas. Some bays and other places along some shorelines may amplify the effects of waves that come in from a certain direction, making run-ups higher than average. These areas are called wave traps, and in many cases the incoming waves form a moving crest of breaking water, called a bore. Tsunami magnitudes are commonly reported using the maximum run-up height along a particular coastline.

ORIGIN OF TSUNAMI

What causes tsunami? Tsunami may be generated by any event that suddenly displaces the sea floor, which in turn causes the sea water to move suddenly to compensate for the displacement. Most tsunami are earthquake-induced or caused by volcanic eruptions, although giant submarine landslides have initiated others. It is even possible that gases dissolved on the sea floor may suddenly be released, forming a huge bubble that erupts upward to the surface, generating a tsunami.

Earthquake-induced Tsunami

Earthquakes that strike offshore or near the coast have generated most of the world's tsunami. In general, the larger the earthquake, the larger the potential tsunami, but this is not always the case. Some earthquakes produce large tsunami, whereas others do not. Earthquakes that

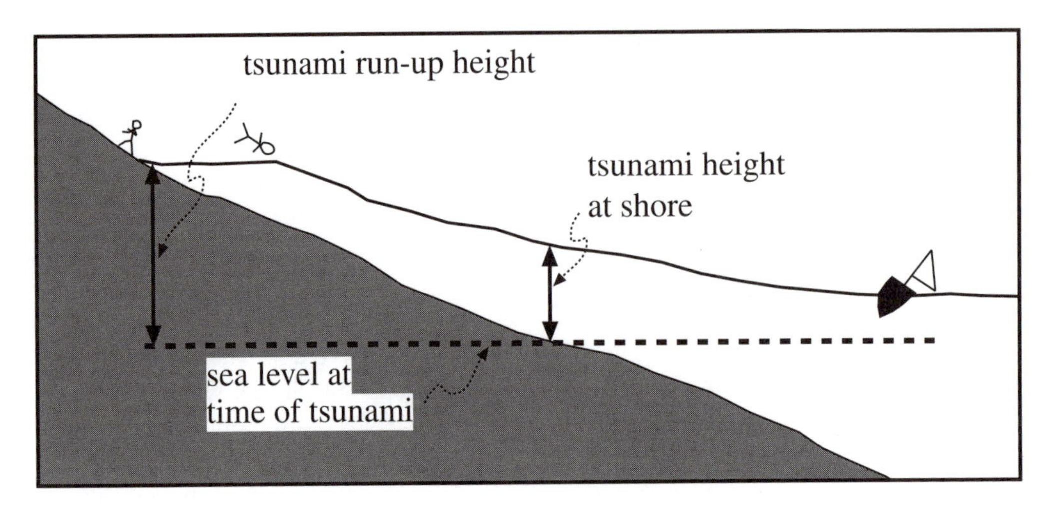

Figure 4.1. Diagram showing the definition of tsunami run-up height, which is the height above normal sea level that the tsunami reaches.

have large amounts of vertical displacement of the sea floor result in larger tsunami than earthquakes that have predominantly horizontal movements of the sea floor. This difference is approximately a factor of ten, probably because earthquakes with vertical displacements are much more effective at pushing large volumes of water upward or downward, generating tsunami. Another factor that influences the size of an earthquake-induced tsunami is the speed at which the sea floor breaks during the earthquake: Slower ruptures tend to produce larger tsunami.

Tsunami earthquakes are a special category of earthquakes that generate tsunami that are unusually large for the earthquake's magnitude. Tsunami earthquakes are generated by large displacements that occur along faults near the sea floor. Most are generated on steeply dipping faults that displace the sea floor vertically, displacing the maximum amount of water (Figure 4.2). These types of earthquakes also frequently cause large submarine (underwater) landslides or slumps, which also generate tsunami. In contrast to tsunami generated by vertical slip on vertical faults, which cause a small region to experience a large uplift, other tsunami are generated by movement on very shallowly dipping faults. These are capable of causing large regions to experience minor uplift, displacing large volumes of water and generating a tsunami (Figure 4.2). Some of the largest tsunami may have been generated by earthquake-induced slumps along convergent tectonic plate boundaries. For example, in 1896 a huge seventy-five-foot-high tsunami was generated by an earthquake-induced submarine slump in Sanriku, Japan, killing 26,000 people in the wave. Another famous tsunami generated by a slump from an earthquake is the 1946 wave that hit Hilo, Hawaii. This tsunami was fifty feet high, killed 150 people, and caused about $25 million in damage to Hilo and surrounding areas. The amazing thing about this tsunami is that it was generated by an earthquake-induced slump off Unimak Island in the Aleutian Islands of Alaska just 4.5 hours earlier! This tsunami traveled at 500 miles per hour (800 km/hour) across the Pacific, hitting Hawaii without warning.

Another potent kind of tsunami-generating earthquake occurs along subduction zones (see Chapter 1). Sometimes, when certain kinds of earthquakes strike in this environment, the entire forearc region above the subducting plate may snap upward by up to a few tens of feet, displacing a huge amount of water (Figure 4.2). The tsunami generated during the 1964 m 9.2 Alaskan earthquake formed a tsunami of this sort, and it caused numerous deaths and extensive destruction in places as far away as California. (This tsunami is discussed in detail below.)

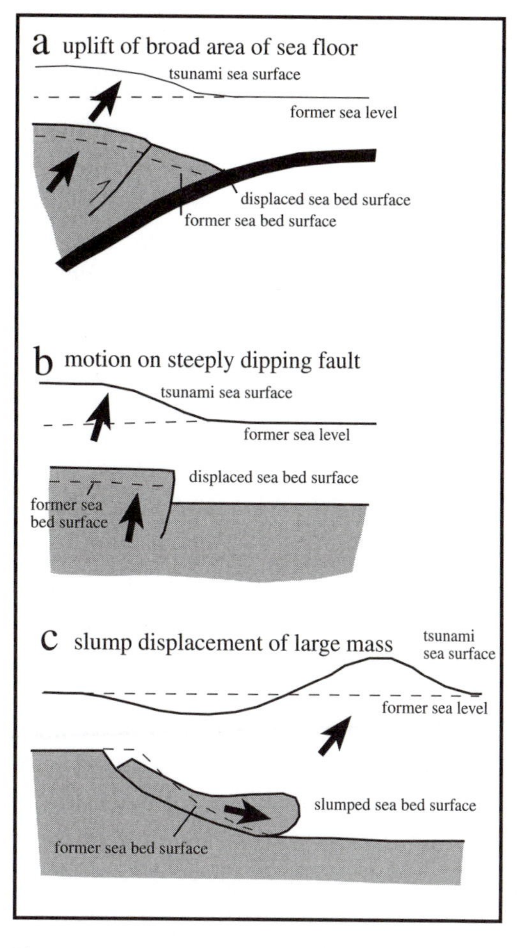

Figure 4.2. Tsunami generation by sea-floor faults of different types. (a) Shows tsunami generated by the displacement of a large area of the sea floor, such as an accretionary prism above a subduction zone. (b) Shows tsunami generation by vertical motion on a steeply dipping sea floor fault, and (c) shows tsunami generation by slumping of a large mass on the sea floor.

Volcanic Eruption–induced Tsunami

Some of the largest recorded tsunami have been generated by volcanic eruptions. These may be associated with the collapse of volcanic slopes, debris and ash flows (see Chapter 3) that displace large amounts of water, or submarine eruptions that explosively displace water above the volcano. The most famous volcanic eruption–induced tsunami include the series of huge waves generated by the eruption of Krakatau in 1883, which reached run-up heights of 120 feet and killed 36,500 people. The number of people that perished in the eruption of Santorini in 1600 B.C. is not known, but the toll must have been huge. The waves reached 800 feet in height on islands close to the volcanic vent of Santorini. Flood deposits have been found 300 feet above sea level in parts of the Mediterranean Sea and extend as far as 200 miles southward up the Nile River. Several geologists suggest that these were formed from a tsunami generated by the eruption of Santorini. As discussed earlier, the floods from this eruption may also, according to some scientists, account for some historical legends such as the great biblical flood, the parting of the Red Sea during the exodus of the Israelites from Egypt, and the destruction of the Minoan civilization of the island of Crete.

Landslide-induced Tsunami

Many tsunami are generated by landslides that displace large amounts of water. These may be from rock falls and other debris that falls off cliffs into the water, such as the huge avalanche that triggered a 1,720-foot-high tsunami in Lituya Bay, Alaska (Figure 4.3). Submarine landslides tend to be larger than avalanches that originate above the water line, and they have generated some of the largest tsunami on record. Many submarine landslides are earthquake-induced, but tsunami are thought to be landslide-induced when the earthquake is not large enough to produce the observed size of the associated tsunami. Sediments near deep-sea trenches are often saturated in water and resting unstably on slopes that are close to the point of failing and sliding downhill. When an earthquake strikes these areas, large parts of the submarine slopes may give out simultaneously, displacing water and generating a tsunami. The 1964 m 9.2 earthquake in Alaska generated more than twenty tsunami, and these were responsible for most of the damage and deaths from this earthquake.

Some steep submarine slopes that are not characterized by earthquakes may also be capable of generating huge tsunami. Recent studies off the coast of Atlantic City, New Jersey, along the east coast of North America have revealed significant tsunami hazards. A pile of unconsolidated sediments, thousands of feet thick, on the continental slope is so porous and saturated with water that it is on the verge of collapsing under its own weight. A storm or minor earthquake may be enough to trigger a giant submarine landslide in this area, possibly generating a tsunami that could sweep across the beaches of New York, New Jersey, Delaware, and much of the east coast of the United States.

In a recent interview on *Science* on the BBC, Simon Day of the U.K. describes some new computer models of the potential tsunami effects from giant landslides that may one day be generated from oversteepened coastal cliffs in the volcanic Cape Verde Islands. The Cape Verde Islands are located in the eastern Atlantic off the coast of west Africa, and they were constructed from hot spot style volcanism (i.e., they are not associated with the mid-ocean ridge or island arcs; see Chapter 1). The islands have very steep western slopes, and Simon Day has completed field work that suggests that these cliffs are unstable. He and his colleagues have suggested that if new magma enters the volcanic islands, it may heat the groundwater in the fractures in the rock, creating enough pressure to induce the giant cliffs to collapse. Any landslides generated from such an anticipated collapse would have the potential to

Figure 4.3. Lituya Bay (Alaska) landslide and tsunami aftermath. *Top*: Fourteen years later, on July 9, 1958, an 8.3 earthquake along the Fairweather Fault triggered a debris avalanche. The avalanche created the largest known tsunami in history, which in turn stripped trees up to an elevation of 1720 feet along Lituya Bay. *Bottom*: The 1958 Lituya Bay earthquake triggered a debris avalanche that stripped the trees from this mountainside. ©Lloyd Cluff/CORBIS

generate giant tsunami, several thousands of feet high, that could sweep the shores of the Atlantic. The wave height will diminish with distance from the Cape Verde Islands, but the effects on the shores of eastern North America, the Caribbean, and eastern South America are expected to be devastating. The waves will probably also wrap around the Cape Verde Islands, hitting the U.K. and the west coast of Africa with smaller but still damaging waves.

Gas Hydrate Eruption–induced Tsunami

Decaying organic matter on the sea floor releases large volumes of gas. Under some circumstances, including cold water at deep depths, these gases may coagulate, forming gels called gas hydrates. It has recently been recognized that these gas hydrates occasionally spontaneously release their trapped gases in giant bubbles that rapidly erupt to the surface. Such catastrophic degassing of gas hydrates poses a significant tsunami threat to regions of thick sediment accumulation such as along the east coast of the United States. This gas–hydrate eruption induced tsunami threat, along with the newly recognized slump–induced tsunami threat have significantly changed the recognized hazard potential in areas that were not previously thought to have a significant threat.

Other Tsunami

Consider the tsunami generated by the massive landslide in Lituya Bay, Alaska, in 1958, as described above. Now, consider what kind of tsunami may be generated by the impact of a giant asteroid with the earth. These types of events do not happen very often, but when they do they are cataclysmic. Geologists are beginning to recognize deposits of impact-generated tsunami and now estimate that they may reach several thousand feet in height. One such tsunami was generated about 66 million years ago by an impact that struck the shoreline of the Yucatán Peninsula, producing the Chicxulub impact structure. This impact produced a huge crater and sent a 3,000-foot-high tsunami around the Atlantic, devastating the Caribbean and the U.S. Gulf coast. Subsequent fires and atmospheric dust that blocked the sun for several years killed off much of the planet's species, including the dinosaurs (see Chapter 12). Imagine if such an impact occurred today.

SEICHE WAVES

Seiche waves are similar to tsunami, except that they are confined to enclosed bodies of water such as lakes. They are generally generated by similar phenomena as true tsunami and may also be initiated by the rocking motion of the ground associated with large earthquakes. Many seiche waves were generated on lakes in southern Alaska during the 1964 m. 9.2 earthquake, including some on Kenai Lake that washed away piers and other structures near the shore. Seiche-like waves sometimes resonate in bays and fiords during large earthquakes, but these are not truly seiche waves as they form in bodies of water connected to the sea.

HOW ARE TSUNAMI STUDIED, AND WHO STUDIES THEM?

Many countries around the Pacific cooperate in monitoring the generation and movement of tsunami. The *seismic sea wave warning system* was established and became operational after the great 1946 tsunami that devastated Hilo, Hawaii, parts of Japan, and many other Pacific Rim coastlines. The seismic sea wave warning system and other tsunami warning systems generally operate by monitoring seismograms to detect potentially seismogeneic earthquakes, then monitor tide gauges to determine if a tsunami has been generated. These systems issue warnings if a tsunami is detected and pay special attention to areas that have greater potential for being inundated by the waves.

It therefore takes several different specialists to be able to warn the public of impending tsunami danger. First, seismologists are needed to monitor and quickly interpret the earthquakes and determine which ones are potentially dangerous for tsunami generation. Second, oceanographers are needed to predict the travel characteristics of the tsunami. Coastal geomorphologists must interpret the shape of coastlines and submarine topography to determine which areas may be the most prone to being hit by tsunami, and geologists are needed to search for any possible ancient tsunami deposits to see what the history of tsunami run-up is along specific coastlines. Finally, engineers are needed to try to modify coastlines to reduce the risk from tsunami. Features such as sea walls and breakwaters can be built, and buildings can be sited in places that are outside of reasonable tsunami striking distance. Loss control engineers typically work with insurance underwriters to identify areas and buildings that are particularly prone to tsunami-related flooding.

Several detailed reports have described areas that are particularly prone to repeated tsunami hazards. The U.S. Army Engineer Waterways Experiment Station has produced several reports that are useful for city planners, the Federal Insurance Administration, and state and local governments.

National Ocean and Atmospheric Administration (NOAA)

The NOAA has been operating tsunami gauges in the deep ocean since 1986. These instruments must be placed on the deep sea floor (typically up to 1,500–2,000 feet [1000 meters] deep) and recovered and redeployed each year. The recordings from these instruments are sent back to shore by cables.

U.S. Geological Survey (USGS)

The USGS runs the Pacific Tsunami Warning Center in Honolulu, Hawaii. It also has been actively engaged in mapping tsunami hazard areas and establishing ancient tsunami run-up heights on coastlines prone to tsunami, to help in predicting future behavior in individual areas.

TSUNAMI HAZARDS

Tsunami generally present very minor hazards to those at sea because of the low amplitude of the waves in deep water (see above). Exceptions may occur when the sea traveler is located very near the source of the tsunami. If an earthquake-induced tsunami occurs in shallow water, the initial wave close to the epicenter and displaced sea floor may be quite large before it stabilizes into an organized, deep water wave. If the tsunami is generated by a volcanic eruption, the waves near the explosion may be quite severe as well. Striking accounts of devastating near-source hazards of tsunami at sea are found in historical records of the 1883 eruption of Krakatau. Sea captains reported many vessels that were run aground or tipped by tsunami in the Straits of Sunda near the eruption, and the waters were reportedly exceedingly rough and littered with pumice, corpses, and rubble for some time after the eruption.

The most serious tsunami hazards are associated with where the tsunami encounters shallow water and runs up onto land. In these coastal areas, the main hazard is from the wave itself, which can overtake and drown people and animals, and destroy most structures in its path. Tsunami also carry large amounts of debris that act as projectiles and can do serious damage when they crash into structures. Tsunami also are known to have ruptured fuel storage tanks, started electrical fires, and eroded foundations, seawalls, and other constructions during their retreat from the land.

WHAT CAN BE DONE TO REDUCE TSUNAMI HAZARDS?

Predicting Tsunami

Great progress has been made in predicting tsunami, both in the long term and in the short term following tsunami earthquakes. Much of the long-term progress reflects recognition of the association of tsunami with plate tectonic boundaries, particularly convergent margins. Certain areas along these convergent margins are susceptible to tsunami-generating earthquakes, either because of the types of earthquakes that characterize that region or because thick deposits of loose unconsolidated sediments characteristically slide into trenches in other areas. Progress in short-term prediction of tsunami stems from the recognition of the specific types of seismic wave signatures that are associated with tsunami-generating earthquakes. Seismologists are in many cases able to immediately recognize certain earthquakes as potentially tsunami-generating and issue an immediate warning for possibly affected areas.

Tsunami are generated mainly along convergent tectonic zones, mostly in subduction zones. The motion of the tsunami-generating faults in these areas is typically at right angles to the trench axis. After many years of study, it is now understood that there is a relationship between the direction of the motion of the fault block and the direction toward which most of the tsunami energy (expressed as wave height) is directed. Most earthquakes along subduction zones move at right angles to the trench, and the tsunami are also preferentially directed at right angles away from the trench. This relationship causes certain areas around the Pacific to be hit by more tsunami than others, because there is a preferential orientation of trenches around the Pacific. Most tsunami are generated in southern Alaska (the tsunami capital of the world) and are directed toward Hawaii while glancing the west coast of the United States, whereas earthquakes in South America direct most of their energy at Hawaii and Japan.

Tsunami Hazard Zones and Risk Mapping

The U.S. Geological Survey and other civil defense agencies have mapped many areas that are particularly prone to tsunamis. Recent tsunami, historical records, and deposits of ancient tsunami identify some of these areas. Many coastal communities, especially those in Hawaii, have posted coastal areas with tsunami warning systems, showing maps of specific areas prone to tsunami inundation. Tsunami warning signals are in place and residents are told what to do and where to go if the alarms are sounded.

Tsunami Warning Systems

Today's tsunami warning systems are capable of saving many lives by alerting residents of coastal areas that a tsunami is approaching their location. These systems are most effective for areas located more than 500 miles (750 km), or one hour, away from the source region of the tsunami, but may also prove effective at saving lives in closer areas. The tsunami warning system operating in the Pacific Ocean basin integrates data from several different sources and involves several different government agencies. As mentioned earlier, the National Oceanographic and Atmospheric Administration (NOAA) operates the Pacific Tsunami Warning Center in Honolulu, Hawaii. This center includes many seismic stations that record earthquakes, and it quickly sorts out those earthquakes that are likely to be tsunamogenic based on the earthquake's characteristics. A series of tidal gauges placed around the Pacific monitors the passage of any tsunamis past their location, and if these stations detect a tsunami, warnings are quickly issued for local and regional areas likely to be affected. Analyzing all of this information takes time, however, so this Pacific-wide system is most effective for areas located far from the earthquake source.

Tsunami warning systems designed for shorter-term, more local warnings are also in place in many communities, including Japan, Alaska, Hawaii, and on many other Pacific islands. These warnings are based mainly on quickly estimating the magnitude of nearby earthquakes and the ability of public authorities to rapidly issue the warning so that the population has time to respond. For local earthquakes, the time between the shock event and the tsunami hitting the shoreline may be only a few minutes. Therefore, if you are in a coastal area and feel a strong earthquake, you should take that as a natural warning that a tsunami may be imminent and leave low-lying coastal areas. This is especially important considering that approximately 99 percent of all tsunami-related fatalities have historically occurred within 150 miles

(250 km) of the tsunami's origin, or within thirty minutes of when the tsunami was generated.

How to See a Tsunami before It Hits You in the Face

If you are near the sea or in an area prone to tsunami (as indicated by warning signs in places like Hawaii), then you need to pay particular attention to some of the subtle and not-so-subtle warning signs that a tsunami may be imminent. First, there may be warning sirens if you are in an area that is equipped with a tsunami warning system. If you hear the sirens, do not waste time—run to high ground immediately! If you are in a more remote location, such as if you are camping on a beach in Alaska, you may need to pay attention to the natural warning signs. If you feel an earthquake, you should run for higher ground. You may have only minutes before the tsunami hits, or maybe an hour or two. And remember, tsunami travel in groups with periods between crests that can be an hour or more. So don't go back to the beach to investigate the damage after the first crest passes. If the tsunami-generating earthquake occurred far away and you don't feel the ground motion, you may not have any warning of the impending tsunami except for the thunderous crash of waves right before it hits you in the face. In other cases, the water may suddenly recede to unprecedented levels right before it quickly rises up again in the tsunami crest. In either case, if you are enjoying the beachfront, remain aware of the dangers. If you are camping, pick a sheltered spot where the waves might be refracted and not run up so far. In general, the heads of bays receive the highest run-ups and the sides and mouths record lower run-up heights. But this may vary considerably, depending on the submarine topography and other factors.

WHAT IF YOU ARE IN A TSUNAMI?

If you do find yourself in a tsunami emergency, follow the directions of emergency officials and stay away from the waterfront! You should quickly seek high ground, so climb a hill or go into a tall building designed to withstand a tsunami. Do not return to the seafront until you are instructed by authorities that it is safe to do so, because there may be more waves coming, spaced an hour or more apart. It is especially important to remember these hazards when considering purchasing or spending large amounts of time in seafront homes.

Many U.S. and foreign agencies have completed exhaustive studies of ways to mitigate or reduce the hazards of tsunami. For instance, the Japanese Disaster Control Research Center has worked with Japan's Ministry of Construction to evaluate the effects of tsunami on their coastal road network, in order to plan better for inundation by the next tsunami. They considered historical records of types of damage to the road systems (e.g., washouts, flooding, and blockage by debris from destroyed homes and cars) and detailed surveys of the region to devise a plan of alternate roads to use during tsunami. They have devised a mechanism of communicating the immediate danger and alternate plans to motorists. Their system includes plans for the installation of multiple wireless electronic bulletin boards at key locations, warning motorists to steer away from hazardous areas. Similar studies have been undertaken by other agencies that deal with the coastal area, such as Japan's Fisheries Agency. They have suggested building a series of levees, emergency gates, and cut-off facilities to maintain the fresh water supply to residents.

TSUNAMI STATISTICS

Tsunami have taken hundreds of thousands of lives in the past few hundred years, and some of the larger tsunami have caused up to billions of dollars in damage. Some of the worst tsunami disasters are listed in Table 4.1.

DESCRIPTIONS OF SPECIFIC TSUNAMI

Lisbon, Portugal, Earthquake and Tsunami, 1755

Tsunami do not regularly strike Atlantic regions, and only about 10 percent of all tsunami occur in the Atlantic Ocean. A few tsunami have been associated with earthquakes in the Caribbean region, such as in 1867 in the Virgin Islands, 1918 in Puerto Rico, and on June 6, 1692, when 3,000 people were killed by a tsunami that leveled Port Royal, Jamaica. The most destructive historical tsunami to hit the Atlantic region struck on November 1, 1755. Lisbon, Portugal, was the worst hit because it was near the epicenter of the earthquake that initiated the tsunami. At least three large waves, each from fifteen to forty feet high, struck Lisbon in quick succession, killing at least 60,000 people in Lisbon alone. England was hit by six- to ten-foot waves, and the tsunami even affected the Caribbean region, hitting Antigua with twelve-foot waves and Saba and St. Martin with waves more than twenty feet in height.

Flores, Indonesia, Tsunami, 1992

One of the most deadly tsunami in recent history hit the island of Flores, Indonesia, located several hundred miles from the coast of northern Australia near the popular resort island of Bali. The tsunami hit on December 12, 1992 and was triggered by a magnitude 7.9 earthquake, with the earthquake faulting event lasting for a long seventy seconds. The tsunami had run-up heights of fifteen to ninety feet along the northeastern part of Flores Island, where more than 2,080 people were killed and at least another 2,000 injured. It is believed that large amounts of sediment slumped underwater on the north side of the island during the earthquake, generating the unusually large and destructive tsunami.

There was very little warning time for the residents of Flores Island, because the epicenter of the earthquake that generated the tsunami was only thirty miles (fifty km) off the northern coast of the island. The first waves hit less than five minutes after the initial earthquake shock, with five or six individual waves being recorded by residents in different places. The residents of Flores were unaware that they lived in a tsunami hazard area, and they did not connect the ground shaking with possible sea hazards. Therefore, no warnings were

Table 4.1
The Ten Worst Tsunami in Terms of Deaths and Destruction

Location	Date	Deaths
Santorini, Greece	1500 B.C.	Devastation of Mediterranean
Eastern Atlantic	November 1, 1755	60,000 in Lisbon, Portugal
Peru and Chile	August 13, 1868	
Krakatau, Indonesia	August 27, 1883	36,500
Honshu, Japan	June 15, 1896	26,000
Honshu, Japan	March 2, 1933	3,000
Indonesia	December 12, 1992	2,000
Chile	May 23, 1960	1850 dead or missing
Nicaragua	September 2, 1992	170 dead
Aleutian Islands	April 1, 1946	150 in Hawaii, $25 million damage
Indonesia	December 2, 1992	137 dead
Alaska	March 28, 1964	119 dead in California, $104 million damage

issued and nobody fled the coastal areas after the quake. The villagers had built their homes and villages right along the coastline not far above the high tide line, further compounding the threat. Many of the homes were made of bricks, but even these relatively strong structures were washed away by the tsunami (Figure 4.4). In some cases, the concrete foundations of basements moved many tens of feet inland as coherent blocks as the tsunami rushed across the area with its associated strong currents. In this way, the tsunami made a less-than-subtle suggestion about where the safe building zone should begin. When the tsunami receded, 28,118 homes were washed away and thousands of other structures were leveled. In many cases, only white wave-washed beaches remained where there were once villages with populations of hundreds of people. Houses, furniture, clothing, and animal and human remains were scattered through the forests behind the villages.

The Flores Island tsunami also caused severe coastal erosion, including cliff collapse and the scouring of coral reef complexes. Forests, brush, and grasses were destroyed, leaving coastal hills with vegetation only at their tops. Thick deposits of loose sediment were moved inland and also redeposited as sheets of sediment in deeper water by the tsunami backwash, causing an overall sudden erosion of the land. Some large coral reef boulders, up to four feet in diameter, were moved inland many hundreds of feet from the shoreline.

Nicaragua Tsunami, 1992

On September 1, 1992, a relatively small (m. 7) earthquake centered thirty miles off the coast of Nicaragua generated a huge tsunami that swept across 200 miles of

Figure 4.4. Photos of Flores Island (Indonesia) tsunami damage, from 1992. A sandy beach is all that remains after the waves removed all trace of Riangkroko. An extremely large tsunami run-up (26 m) was measured at this small rural village on Flores Island, and 137 people lost their lives to the earthquake and several tsunami. The village was located at the mouth of the Nipah River, a small river with its northwest side facing the Flores Sea. The inundation distance from the shoreline along the river is approximately 600 m. Photo Credit: Harry Yeh, University of Colorado and NOAA/NGDC

coastline, killing 150 people and making 14,500 more homeless. This tsunami was generated by a tsunamogenic earthquake that ruptured near the surface in an area where thick sediments have been subducted and that are now plastered along the interface between the downgoing oceanic plate and the overriding continental plate of Central America. It has been suggested that these sediments lubricated the fault plane and caused the earthquake to rupture slowly and last a particularly long time, generating an unusually large tsunami. The resulting tsunami was up to forty feet high at nearby beachfronts, and it swept homes, vehicles, and unsuspecting people to sea within minutes of the earthquake. But the hazards were not over, because the tsunami had ripped through water storage and sewage treatment plants, and the resulting contamination caused an outbreak of cholera that took even more lives.

Alaska Earthquake-Related Tsunami, 1964

Southern Alaska is hit by a significant tsunami every ten to thirty years. During the m. 9.2 earthquake in Alaska on March 27, 1964, approximately 35,000 square miles of seafloor and adjacent land were suddenly thrust upward by up to fifty feet. This mass movement generated a huge tsunami (Figure 4.5) as well as many related seiche-like waves in surrounding bays. The tsunami generated from this earthquake caused widespread destruction in Alaska, especially in Seward, Valdez, and Whittier. These towns all experienced large earthquake-induced submarine landslides, some of which tore away parts of the towns' waterfronts. The submarine landslides also generated large tsunami that cascaded over these towns barely after the ground stopped shaking from the earthquake.

The town of Seward on the Kenai Peninsula experienced strong ground shaking for about four minutes during the earthquake, and this initiated a series of particularly large submarine landslides and seiche-like waves that removed much of the waterfront docks, trainyards, and streets. The movement of such significant quantities of material underwater generated a tsunami with a thirty-foot run-up that washed over Seward about twenty minutes after the earthquake. The earthquake and landslides tore apart many seafront oil storage facilities (Figures 4.6 and 4.7), and the oil exploded, sending flames 200 feet into the air. The oil on the waves caught fire and rolled inland on the thirty- to forty-foot-high tsunami crests. These flaming waves crashed into a train loaded with oil, causing each of forty tankers to successively explode as it was torn from the tracks. The waves moved inland, carrying boats, train cars, houses, and other debris, much of it in flames. The mixture moved overland at 50–60 miles per hour and raced up the airport runway, blocking the narrow valley marking the exit from the town into the mountains. The tsunami had several large crests with many reports of the third being the largest. Twelve people died in Seward, and the town of Seward was declared a total loss after the earthquake and ensuing tsunami. The town was then moved to a new location a few miles away, on ground thought to be more stable from submarine landslides. Steep mountains mark the southwest side of the current town.

As the tsunami reached into the many bays and fiords in southern Alaska, it caused the water in many of these bays to oscillate back and forth in series of standing waves that caused widespread and repeated destruction. In Whittier, wave run-ups were reported to be up to 104 feet, and the town suffered submarine landslides, which destroyed oil and train facilities. Submarine landslides, seiche-like waves, twenty-three-foot-high tsunami, and exploding oil tanks similarly affected Valdez. The town of Kodiak, 500 miles from the epicenter, also experienced extensive damage, and eighteen people were killed. The land at Kodiak subsided 5.6 feet during the earthquake, and then the town was hit by a series of at least ten tsunami crests up to twenty feet high that destroyed the port and dock facilities as well as more than 200 other buildings.

When the tsunami reached the state of Washington four hours after the earthquake, it washed up as a five-foot-tall wall

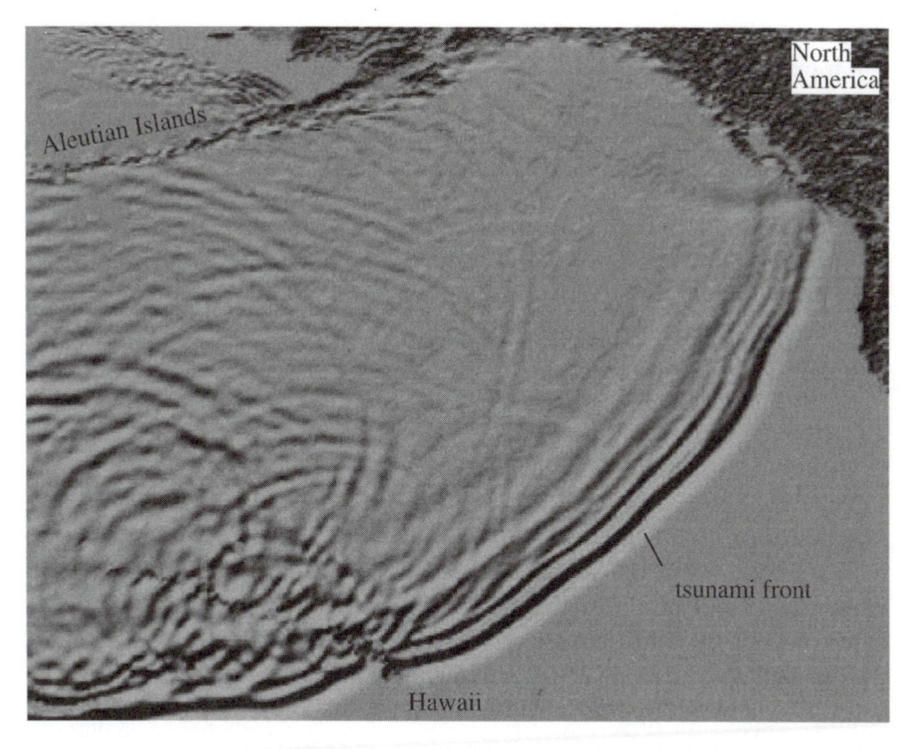

Figure 4.5. Computer-generated graphic illustration (modified after Titov et al., 2002, of NOAA's Pacific Disaster Center) showing a tsunami generated by an earthquake in Alaska, moving across the Pacific and striking the Hawaiian Islands. Note wave interference patterns generated by waves reflecting off islands and seamounts.

of water and it decreased in amplitude to two feet by the time it got to Astoria, Oregon. However, when the wave headed south to Crescent City, California, the shape of the sea floor and bay was able to focus the energy from this wave into a series of five tsunami crests. The fifth was a twenty-one-foot-high crest that swept into downtown, washing away much of the waterfront district and killing eleven people after they had returned to assess the damage from the earlier wave crests.

Lituya Bay, Alaska, Tsunami, 1958

One of the largest-known landslide-induced tsunami struck Lituya Bay of southeastern Alaska on July 9, 1958. Lituya Bay is located about 150 miles southeast of Juneau and is a steep sided, seven-mile-long, glacially carved fiord (Figure 4.3a). Forest-covered mountains rise 6,000 feet out of the water. At 10:15 P.M., an earthquake struck the region and a huge mass of rock was released from about 3,000 feet

Figure 4.6. Photograph of Seward Alaska after 1964 tsunami, where houses and debris were tossed about like toys. The Alaskan Railroad, one of the lifelines of our largest state, was cut here as well as in many other areas by the March 27th tsunami stemming from magnitude 9.2 earthquake. At least eight were killed here as a result of the wave. ©Bettmann/CORBIS

up the cliffs near the head of the bay, and this material plunged into the water below. The wave generated by this massive collapse was enormous. The first wave (really a splash) soared up to 1,720 feet on the opposite side of the bay, removing trees and soil with the force of the wave and the backwash (Figure 4.3b). It washed over an island in the middle of the bay, destroying a government research station and killing two geologists stationed there who happened to be investigating the possibilities of tsunami hazards in the bay. The collapse also generated a 170-foot-high tsunami that moved at forty miles per hour out toward the mouth of the bay, erasing shoreline features along its path and shooting a fishing troller out of the bay into open water. Lituya Bay and others like it have experienced numerous tsunami as shown by distinctive scour marks and debris deposits now known in the bay. This phenomenon was also well known to the native Tlingit, who had legends of spirits who lived in the bay and who would send huge waves out to punish those who angered them.

Figure 4.7. The tsunami caused much damage to the railroad facilities at Seward Port. Rails were stripped from the railroad ties by the tsunami. Most of the Alaska Railroad dock was washed away by the waves. The railroad also lost two cranes and its waterfront trackage. Note also the fire-damaged oil storage tanks. Damage at Resurrection Bay totaled $14.6 million dollars. Photo credit: NOAA/NGDC

Hilo, Hawaii, Tsunami, 1946

On April 1, 1946, an earthquake generated near the Unimak Islands in Alaska devastated vast regions of the Pacific Ocean. This tsunami was one of the largest and most widespread of tsunami to spread across the Pacific Ocean this century.

A lighthouse at Scotch Cap in the Aleutians was the first to feel the effect of the tsunami. At about 1:30 P.M., the crew of five at the lighthouse recorded feeling an earthquake lasting 30–40 seconds, but no serious damage occurred. A second quake was felt nearly half an hour later, also with no damage. Fifty minutes after the earthquake, the crew of a nearby ship recorded a "terrific roaring of the sea, followed by huge seas." They reported a tsunami that rose over the top of the Scotch Cap lighthouse and over the cliffs behind the station, totally destroying the lighthouse and a coast guard station. The lighthouse was built of steel-reinforced concrete and sat on a bluff forty-six feet above sea level, but the wave is estimated to have been 90–100 feet high. A rescue crew sent to the Scotch Cap lighthouse five hours after the disaster reported that the station was gone, and debris (including human organs) was strewn all over the place. There were no survivors.

Many hours later, the tsunami had traveled halfway across the Pacific and was encroaching on Hawaii. Residents of Hilo, Hawaii, first noticed that Hilo Bay drained, and springs of water sprouted from the dry sea floor that was littered with dying fish. As the residents wondered at the cause of the water suddenly draining from their bay, a series of huge waves came crashing in from the ocean and quickly moved into the downtown district. Buildings were ripped from their foundations and thrown into adjacent structures, bridges were pushed hundreds of feet upstream from their crossings, and boats and railroad cars were tossed about like toys. After the tsunami receded, Hilo was devastated, with a third of the city destroyed and ninety-six people dead. Outside Hilo, entire villages disappeared and were washed into the sea along with their residents.

Kamaishi (Sanriku), Japan, Tsunami, 1896

Twenty-seven thousand people died in a huge tsunami that swept over the seaport of Kamaishi, Japan, on June 15, 1896. A local earthquake caused mild shaking of the port city, which was not unusual in this tectonically active area. However, twenty minutes later the bay began to recede, then forty-five minutes after the earthquake, the port city was inundated with a ninety-foot-high wall of water that came in with a tremendous roar. The town was nearly completely obliterated, and 27,000 people perished in a few short moments. Kamaishi was a fishing port, and when the tsunami struck the fishing fleet was at sea and did not notice the tsunami, since it had a very small amplitude in the deep ocean. When the fishing fleet returned the next morning, they sailed through many miles of debris and thousands of bodies, and reached their homes only to find a few smoldering fires among a totally devastated community.

Chile Earthquake and Tsunami, 1960

The great m. 9.5 Chilean earthquake of May 22, 1960, generated a huge tsunami that killed more than 1,000 people near the earthquake epicenter and killed almost 1,000 more as the wave propagated across the Pacific Ocean. Many of the deaths in Chile were reportedly related to the fact that after the first fifteen-foot-high tsunami crest passed, many Chileans assumed the danger was over and returned to the shoreline. Fifty minutes later, a second larger tsunami crest crashed ashore at 125 miles per hour with a run-up of twenty-six feet, and it was followed by a third crest with a height of thirty-five feet that finished off much of the coastline.

The tsunami was accurately predicted to hit Hawaii and arrived within a minute of the predicted time. This should have saved lives, but about sixty people were killed, including many who heard the warnings but rushed to the coast to watch the waves strike. Approximately twenty-three hours after the earthquake, another 185 people perished when the tsunami swept over coastal regions of Japan.

TSUNAMI IN THE TWENTY-FIRST CENTURY

Areas that have been repeatedly hit by tsunami in the past are likely to be hit again during future earthquakes. Therefore, if you live or travel to Pacific Rim beaches, you need to maintain an awareness of the potential hazards. Local officials need to plan for potential tsunami disasters, even if they only occur every several tens of years in individual places. However, other areas that have not seen many historical tsunami may also be prone to less frequent, albeit destructive, tsunami. Some potential tsunami hazards have been identified along the east coast of the United States. In one recent study by geologists from the Woods Hole Oceanographic Institute, Lamont-Doherty Earth Observatory, and the University of Texas, some giant submarine scarps have been identified on the continental shelf off the coast of Virginia and North Carolina. Geologists Neal Driscoll, Jeff Weissel, and John Goff believe that these scarps may represent the early stages of a large-scale submarine slope failure, and that such a huge failure could generate large tsunami that could sweep up Chesapeake Bay and along the Virginia–North Carolina coastline. What could trigger the submarine slope off the east coast to suddenly fail? Earthquakes and storms are likely triggers, as is the sudden release of gases from decaying organic matter. These scientists estimate that collapse of the submarine slope off Virginia and North Carolina could easily produce a tsunami with a run-up of greater than ten feet. A similar slope-failure off the coast of Newfoundland in 1929 formed a 30–40 foot high tsunami, killing fifty-one people along the coast.

RESOURCES AND ORGANIZATIONS

Print Resources

Bernard, E. N., ed. *Tsunami Hazard: A Practical Guide for Tsunami Hazard Reduction.* Dordrecht, The Netherlands: Kluwer Academic Publishers, 1991.

Booth, J. S., O'Leary, D. W., Popencoe, P., and Danforth, W. W. "US Atlantic Continental Slope Landslides: Their Distribution, General Attributes, and Implications." *US Geological Survey Bulletin* 2002 (1993), 14–22.

Dawson, A. G., and Shi, S. "Tsunami Deposits." *Pure and Applied Geophysics* 157 (2000), 493–511.

Driscoll, N. W., Weissel, J. K., and Goff, J. A. "Potential for Large-scale Submarine Slope Failure and Tsunami Generation along the U.S. Mid-Atlantic Coast." *Geology* 28 (2000), 407–10.

Dvorak, J., and Peek, T. "Swept Away." *Earth* 2, no. 4 (1993), 52–59.

Latter, J. H. "Tsunami of Volcanic Origin, Summary of Causes, with Particular Reference to Krakatau, 1883." *Journal of Volcanology* 44 (1981), 467–90.

McCoy, F., and Heiken, G. "Tsunami Generated by the Late Bronze Age Eruption of Thera (Santorini), Greece." *Pure and Applied Geophysics* 157 (2000), 1227–56.

Minoura, K., Inamura, F., Nakamura, T., Papadopoulos, A., Takahashi, T., and Yalciner, A. "Discovery of Minoan Tsunami Deposits." *Geology* 28 (2000), 59–62.

Minoura, K., Imamura, F., Takahashi, T., and Shuto, N. "Sequence of Sedimentation Processes Caused by the 1992 Flores Tsunami, Evidence from Babi Island." *Geology* 25 (1997), 523–26.

Okazaki, S., Shibata, K., and Shuto, N. "A Road Management Approach for Tsunami Disaster Planning," in Y. Tsuchiya and N. Shuto, eds., *Tsunami: Progress in Prediction Disaster Prevention and Warning.* Boston: Kluwer Academic Publishers, 1995, pp. 223–34.

Revkin, A. C. "Tidal Waves Called Threat to

East Coast." *New York Times,* July 14, 2000, sec. A18.

Satake, K. "Tsunamis." *Encyclopedia of Earth System Science* 4, 1992, 389–97.

Steinbrugge, K. V. *Earthquakes, Volcanoes, and Tsunamis: An Anatomy of Hazards.* New York: Skandia America Group, 1982.

Tsuchiya, Y., and Shuto, N., eds. *Tsunami: Progress in Prediction Disaster Prevention and Warning.* Boston: Kluwer Academic Publishers, 1995, 336 pp.

U.S. Geological Survey. "Surviving a Tsunami—Lesson from Chile, Hawaii, and Japan." U.S. Geological Survey Circular 1187, 1987.

Yeh, H., Imamura, F., Syndakis, C., Tsuji, Y., Liu, P., and Shi, S. "The Flores Island Tsunami." *EOS: Transactions of the American Geophysical Union* 73 (1993), N33.

Non-Print Resources

Web Addresses

http://ngdc.noaa.gov/seg/hazard/tsu.html

http://walrus.wr.usgs.gov/tsunami

http://www.pmel.noaa.gov/tsunami

Organizations

National Tsunami Hazard Mitigation Program
NOAA/Pacific Marine Environmental Laboratory
7600 Sand Point Way N.E.
Seattle, WA 98115
202–761–1683
This partnership between the states of Hawaii, Alaska, California, Oregon, and Washington, and the Federal Emergency Management Agency, National Oceanic and Atmospheric Administration, and U.S. Geological Survey is preparing maps showing tsunami inundation areas and implementing mitigation plans for the states in the program. The NTHMP is also developing an early warning system, including seismic stations and deep ocean tsunami detectors.

U.S. Geological Survey
345 Middlefield Road
Menlo Park, CA 94025
650–329–5042

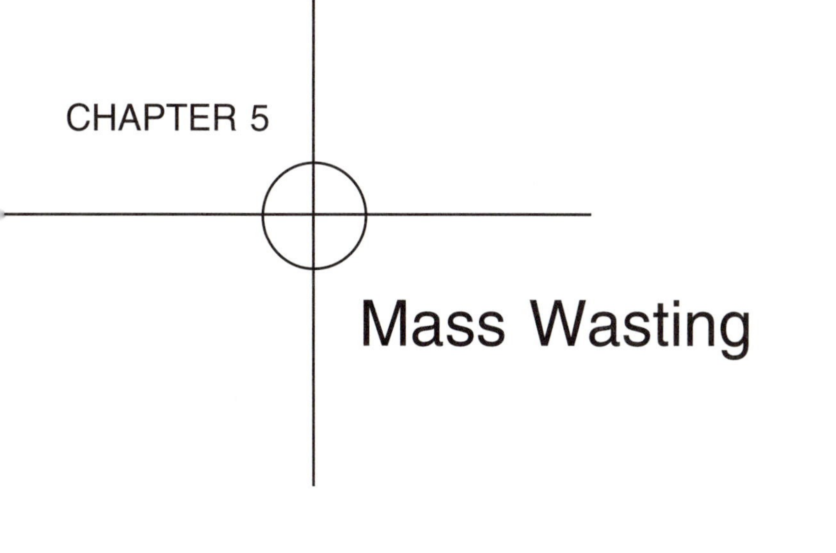

CHAPTER 5

Mass Wasting

INTRODUCTION

Mass wasting is the downslope movement of soil, rock, and other earth materials (together called *regolith*) by gravity without the direct aid of a transporting medium such as ice, water, or wind. It is estimated that more than 2 million mass movements occur each year in the United States alone. Mass movements occur at various rates, from a few inches per year to sudden catastrophic rock falls and avalanches that can bury entire towns under tons of rock and debris. In general, the faster the mass movement, the more hazardous it is to humans, although even slow movements of soil down hill slopes can be extremely destructive to buildings, pipelines, and other constructions. In the United States alone, mass movements kill tens of people and cost more than $1.5 billion a year. Other mass movement events overseas have killed tens to hundreds of thousands of people in a matter of seconds (Table 5.1). Mass wasting occurs under a wide variety of environmental conditions and forms a continuum with weathering, as periods of intense rain reduce friction between regolith and bedrock, making movement easier. Mass movements also occur underwater, such as the giant submarine landslides associated with the 1964 Alaskan earthquake discussed in Chapters 2 and 4.

Mass movements are a serious concern and problem in hilly or mountainous terrain, especially for buildings, roadways, and other features engineered into hillsides. Mass movements are also a problem along riverbanks and in places with large submarine escarpments, such as along deltas like the Mississippi Delta in Louisiana. The problems are further compounded in areas prone to seismic shaking or severe storm-related flooding. Imagine building a million-dollar mansion on a scenic hillside, only to find it tilting and sliding down the hill at a few inches per year. Less spectacular but common effects of slow downhill mass movements are the slow tilting of telephone poles along hillsides and the slumping of soil from oversteepened embankments onto roadways during storms.

Mass wasting is becoming more of a

Table 5.1
The Ten Worst Downslope Flows in Terms of Deaths and Destruction

Location	Date	Deaths
Shaanxi Province, China	1556	830,000
Shaanxi Province, China	1920	200,000
Nevados Huascaran, Peru	1970	70,000
Honduras, Nicaragua	1998	10,000
Venezuela	1999	10,000
Nevados Huascaran, Peru	1962	4,000
Vaiont, Italy	1963	3,000
Rio de Janeiro, Brazil	1966–1967	2,700
Mount Coto, Switzerland	1618	2,430
Kure, Japan	1945	1,145

problem as populations move from the overpopulated flat land to new developments in hilly terrain. In the past, small landslides in mountains, hills, and canyons were not a serious threat to people, but now, with large numbers of people living in landslide-prone areas, landslide hazards and damage are rapidly increasing.

WEATHERING AND SOILS: INGREDIENTS OF MASS WASTING

Mass wasting often involves the downslope movement of products of weathering, and there is a continuum between the processes of weathering, erosion, and mass wasting. *Weathering* is a process of mechanical and chemical alteration marked by the interaction of the lithosphere, atmosphere, hydrosphere, and biosphere. The resistance to weathering varies with climate, composition, texture, and how much a rock is exposed to the elements of weather. Weathering processes occur at the lithosphere/atmosphere interface. This is actually a zone that extends down into the ground to the depth that air and water can penetrate—in some regions this is a few meters, in others it is a kilometer or more. In this zone, the rocks make up a porous network with air and water migrating through cracks, fractures, and pore space. The effects of weathering can often be seen in outcrops on the side of the roads, where they cut through the zone of alteration into underlying bedrock. These roadcuts and weathered outcroppings of rock show some similar properties. The upper zone near the surface is made of soil or regolith in which the texture of the fresh rock is not apparent, the middle zone is where the rock is altered but retains some of its organized appearance, and a lower zone is composed of fresh unaltered bedrock.

Processes of Weathering

There are three main types of weathering. *Chemical weathering* is the decomposition of rocks through the alteration of individual mineral grains and is a common process in the soil profile. *Mechanical weathering* is the disintegration of rocks, generally by abrasion. Mechanical weathering is common in the talus slopes at the bottom of the mountains, along beaches, and along river bottoms. *Biological weathering* involves biological agents breaking down rocks and minerals. Some organisms attack rocks for nutritional purposes; for instance, chitons bore holes through limestone along the seashore, extracting their nutrients from the rock.

Generally, mechanical and chemical weathering are the most important, and they work hand-in-hand to break down rocks into the regolith. The combination of chemical, mechanical, and biological weathering produces soils, or a *weathering profile.*

Mechanical Weathering There are several different types of mechanical weathering that may act separately or together to break down rocks. The most common process of mechanical weathering is abrasion, by which the movement of rock particles in streams, along beaches, in deserts, or along the bases of slopes causes fragments to knock into each other. These collisions cause small pieces of each rock particle to break off, gradually rounding the particles, making them smaller, and creating more surface area for the processes of chemical weathering to act upon.

Some rocks develop *joints* or parallel sets of fractures from differential cooling, from the pressures exerted by overlying rocks, or from tectonic forces. Joints are fractures along which no observable movement has occurred. Joints promote weathering in two ways: First, they are planes of weakness across which the rock can break easily, and second, they act as passageways for fluids to percolate along and promote chemical weathering.

Crystal growth may aid mechanical weathering. When water percolates through joints or fractures, it can precipitate minerals such as salts, which grow larger and exert large pressures on the rock along the joint planes. If the blocks of rock are close enough to a free surface such as a cliff, large pieces of rock may be forced off in a rock fall that the gradual growth of small crystals along joints initiates.

When water freezes to form ice, its volume increases by 9 percent. Water is constantly seeping into the open spaces provided by joints in rocks. When water filling the space in a joint freezes, it exerts large pressures on the surrounding rock. These forces are very effective agents of mechanical weathering, especially in areas with freeze-thaw cycles. They are responsible for most rock debris on talus slopes of mountains.

Heat may also aid mechanical weathering, especially in desert regions where the daily temperature range may be extreme. Rapid heating and cooling of rocks sometimes exerts enough pressures on the rocks to shatter them to pieces, thus breaking large rocks into smaller fragments.

Plants and animals may also aid mechanical weathering. Plants grow in cracks and push rocks apart. This process may be accelerated if plants such as trees become uprooted or blown over by wind, exposing more of the underlying rock to erosion. Burrowing animals, worms, and other organisms bring an enormous amount of chemically weathered soil to the surface and continually turn the soils over and over, greatly assisting the weathering process.

Chemical Weathering Minerals that form in igneous and metamorphic rocks at high temperatures and pressures may be unstable at temperatures and pressures at the Earth's surface, so they react with the water and atmosphere to produce new minerals. This process is known as *chemical weathering*. The most effective chemical agents are weakly acidic solutions in water. Therefore, chemical weathering is most effective in hot and wet climates.

Rain water mixes with CO_2 from the atmosphere and from decaying organic matter, including smog, to produce carbonic acid according to the following reaction:

$$H_2O + CO_2 \rightarrow H_2CO_3$$

Water + carbon dioxide → carbonic acid

Carbonic acid ionizes to produce the hydrogen ion (H^+), which readily combines with rock-forming minerals to produce alteration products. These alteration products may then rest in place and become soils or may be eroded and accumulate somewhere else.

Hydrolysis is a processes that occurs when the hydrogen ion from carbonic acid combines with K-feldspar to produce kaolinite, a clay mineral, according to the reaction:

$$2\ KAlSi_3O_8 + 2\ H_2CO_3 + H_2O \rightarrow Al_2Si_2O_5(OH)_4 + 4\ SiO_2 + 2\ K^{+1} + 2\ HCO_3$$

feldspar + carbonic acid + water → kaolinite + silica + potassium + bicarbonate ion

This reaction is one of the most important ones in chemical weathering. The product, kaolinite, is common in soils and is virtually insoluble in water. The other products, silica, potassium, and bicarbonate, are typically dissolved in water and carried away during weathering.

Much of the material produced during chemical weathering is carried away in solution and deposited elsewhere, such as in the sea. The highest-temperature minerals are leached the easiest. Many minerals combine with oxygen in the atmosphere to form another mineral by oxidation. Iron is very easily oxidized from the Fe+2 state to the Fe+3 state, forming goethite or, with the release of water, hematite.

$$2FeO + OH \rightarrow Fe_2O_3 + H_20$$

iron oxide + hydroxyl ion → hematite + water

Different types of rock weather in different ways. For instance, granite contains K-feldspar and weathers to clays. Building stones are selected to resist weathering in different climates, but now increasing acidic pollution is destroying many old landmarks. Chemical weathering results in the removal of unstable minerals and a consequent concentration of stable minerals. Included in the remains are quartz, clay, and other rare minerals such as gold and diamonds, which may be physically concentrated in placer deposits.

On many boulders, weathering only penetrates a fraction of the diameter of the boulder, resulting in a rind of the altered products of the core. The thickness of the rind itself is useful for knowing the age of the boulder, if rates of weathering are known. These types of weathering rinds are useful for determining the age of rock slides and falls, and the time interval between rock falls in any specific area.

Exfoliation is a weathering process by which rocks break off in successive shells like the skin of an onion. Exfoliation is caused by differential stresses within a rock formed during chemical weathering processes. For instance, feldspar weathers to clay minerals, which take up a larger volume than the original feldspar. When the feldspar minerals turn to clay, they exert considerable outward stress on the surrounding rock, which is able to form fractures parallel to the rock's surface. This need for increased space is accommodated by the minerals through the formation of these fractures, and the rocks on the hillslope or mountain are then detached from their base and more susceptible to sliding or falling in a mass wasting event.

If weathering proceeds along two or more sets of joints in the subsurface, it may result in shells of weathered rock that surround unaltered rocks, looking like boulders. This is known as spheroidal weathering. The presence of the several sets of joint surfaces increases the effectiveness of chemical weathering, because the joints increase the available surface area to be acted on by chemical processes. The more subdivisions within a given volume, the greater the surface area.

Factors That Influence Weathering

The effectiveness of weathering processes is dependent upon several different factors, explaining why some rocks weather one way at one location and a different way in another location. Rock type is an important factor in determining the weathering characteristics of a hillslope, because different minerals react differently to the same weathering conditions. For instance, quartz is resistant to weathering, and quartz-rich rocks typically form large mountain ridges. Conversely, shales readily weather to clay minerals, which are easily washed away by water, so shale-rich rocks often occupy the bottoms of valleys. Examples of topography being closely related to underlying geology in this manner are abundant in the Appalachian Mountains, Rocky Mountains, and most other mountain belts of the world.

Rock texture and structure is important in determining the weathering characteristics of a rock mass. Joints and other weaknesses promote weathering by increasing the surface area for chemical reactions to take place on, as described above. They also allow water, roots, and mineral precipitates to penetrate deeply into a rock mass, exerting outward pressures that can break off pieces of the rock mass in catastrophic rock falls and slides.

The slope of a hillside is important for determining what types of weathering and mass wasting processes occur on that slope. Steep slopes let the products of weathering get washed away, whereas gentle slopes promote stagnation and the formation of deep weathered horizons.

Climate is one of the most important factors in determining how a site weathers. Moisture and heat promote chemical reactions, so chemical weathering processes are strong and fast, and they dominate over mechanical processes in hot wet climates. In cold climates, chemical weathering is much less important: Mechanical weathering is very active during freezing and thawing, so mechanical processes such as ice wedging tend to dominate over chemical processes in cold climates. These differences are exemplified by two examples of weathering. In much of New England, a hike over mountain ridges will reveal fine, millimeter-thick striations that were formed by glaciers moving over the region more than 10,000 years ago. Chemical weathering has not removed even these one-millimeter-thick marks in 10,000 years. In contrast, new construction sites in the tropics, such as roads cut through mountains, often expose fresh bedrock. In a matter of ten years, these road cuts will be so weathered to a red soil-like material called gruse that the original rock will not be recognizable.

As in most things, time is important. It takes tens of thousands of years to wash away glacial grooves in cold climates, but in tropics, weathered horizons including soils that extend to hundreds of meters may form over a few million years. These variations are important when considering the variations in mass wasting processes, the subject of the remaining parts of this chapter.

FORMATION OF SOILS

Differences in soil profile and types result from differences in climate, the rock type it started from, the types of vegetation and organisms, topography, and time. Normal weathering produces a characteristic soil profile that is marked by a succession of distinctive horizons in a soil from the surface downward. The A-horizon is closest to the surface and typically has a gray or black color because of high concentrations of humus (decomposed plant and animal tissues). The A-horizon has typically lost some substances through downward leaching. The B-horizon is typically brown or reddish, enriched in clay produced in place, and transported downward from the A-horizon. The C-horizon of a typical soil consists of slightly weathered parent material. Young soils typically lack a B-horizon, and the B-horizon grows in thickness with increasing age.

Some unusual soils form under unusual climate conditions. Polar climates are typically cold and dry, and the soils produced in polar regions are typically well-drained without an A-horizon, sometimes underlying layers of frost-heaved stones. In wetter polar climates, tundra may overlie permafrost, which prevents the downward draining of water. These soils are saturated in water and rich in organic matter. These polar soils are very important for the global environment and global warming. They have so much organic material in them that is effectively isolated from the atmosphere, that they may be thought of as locking up much of the carbon dioxide on the planet. Cutting down northern forests, as is going on in Siberia, may affect the global carbon dioxide budget and possibly contribute to climate change and global warming (see Chapter 9).

Dry climates limit the leaching of unstable minerals such as carbonate from the A-horizon, which may also be enhanced by evaporation of groundwater. Extensive evaporation of groundwater over prolonged times leads to the formation of caliche crusts. These are hard, generally white carbonate minerals and salts that were dissolved in the groundwater but got precipitated when the groundwater moved up through the surface and evaporated, leaving the initially dissolved minerals behind.

In warm wet climates, most elements except for aluminum and iron are leached from the soil profile, forming laterite and bauxite. Laterites are typically deep red in color and are found in many tropical regions. Some of these soils are so hard that they are used for bricks.

Rates of Soil Formation

Soils form at various rates in different climates and other conditions, ranging from about fifty years in moderate temperatures and wet climates to about 10,000–100,000 years for a good soil profile to develop in dry climates such as the desert southwest of the United States. Some soils, such as those in the tropics, have been forming for several million years and are quite mature. Deforestation causes erosion of soils that can not be reproduced quickly. In many places, such as parts of Madagascar, South America, and Indonesia, deforestation has led to accelerated rates of soil erosion, removing thick soils that have been forming for millions of years. These soils supported a rich diversity of life, and it is unlikely that they will ever be restored in these regions.

DRIVING FORCES OF MASS WASTING

Gravity is the main driving force behind mass wasting processes, as it is constantly pulling on material and attempting to force it downhill. This is because, on a slope, gravity can be resolved into two components: one perpendicular to the slope and one parallel to the slope (Figure 5.1). The steeper the angle of the slope, the greater the influence of gravity. The effect of gravity reaches a maximum along vertical or overhanging cliffs.

It is the tangential component of gravity that tends to pull material downhill and results in mass wasting. When the tangential component of gravity is great enough to overcome the force of friction at the base of the boulder, the boulder falls downhill, with grave consequences to those in the house below. The friction is really a measure of the resistance to gravity: the greater the friction, the greater the resistance to gravity's pull. Friction can be greatly reduced by lubrication of surfaces in contact, allowing the two materials to slide past one another more easily. Water is a common lubricating agent, so mass wasting events tend to occur more frequently during times of heavy or prolonged rain. For a mass wasting event or a mass movement to occur (in this case, the crashing of the boulder into the house below), the lubricating forces must be strong enough to overcome the resisting forces that tend to hold the boulder in place against the wishes of gravity. Lubricating forces include the cohesion between similar particles (like one clay molecule to another) and the adhesion be-

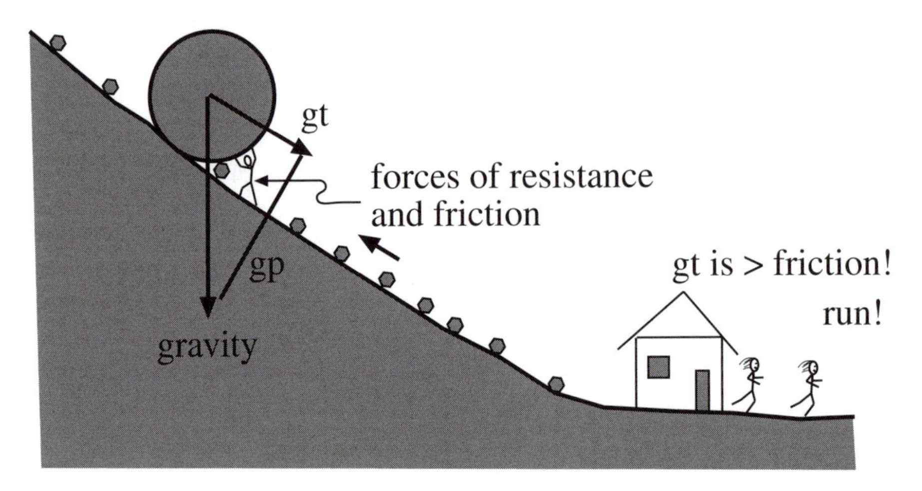

Figure 5.1. Driving forces of mass wasting. The force of gravity acting on a boulder on a hill can be resolved into a component parallel to the hill surface (gt) and a component perpendicular to the hill surface (gp). Resistive forces such as friction counteract the component parallel to the surface. When gt > friction, mass wasting may occur, and the boulder may suddenly roll downhill.

tween different or unlike particles (like the boulder to the clay beneath it). When the resisting forces are greater than the driving force (a tangential component of gravity), the slope is steady and the boulder stays in place. When lubricating components reduce the resisting forces so much that the driving forces are greater than the resisting forces, slope failure occurs. In our example above, the boulder comes crashing down into the house below, landing squarely on the cauliflower soufflé being served for supper.

The process of the downslope (or underwater) movement of regolith may occur rapidly, as in this case, or it may proceed slowly. In any case, slopes on mountain sides typically evolve toward steady state angles known as the *angle of repose*, balanced by material moving in from upslope and out from downslope. This angle of repose is also a function of the grain size of the regolith.

Human activity also increases driving forces for mass wasting. Excavation for buildings, roads, or other cultural features along the lower portions of slopes may actually remove parts of the slopes, causing them to become steeper than they were before construction and to exceed the angle of repose. This will cause the slopes to be unstable or metastable and thus susceptible to collapse. Building structures on the tops of slopes will also make them unstable, as the extra weight of the building adds extra stresses to the slope that may be enough to initiate the collapse of the slope.

Physical Conditions That Control Mass Wasting

Whether or not mass wasting occurs and the type of resulting mass wasting are controlled by many factors. These include characteristics of the regolith and bedrock and the presence or absence of water, overburden, the slope angle, the stable angle of repose for the material, and the way that the particles are packed together.

Mass wasting in solid bedrock terrain is strongly influenced by preexisting weaknesses in the rock that make movement along them easier than if the weaknesses were not present. For instance, bedding planes, joints, and fractures, if favorably oriented, may act as planes of weakness along which giant slabs of rock may slide downslope. If the rock or regolith has many pores, or open spaces between grains, it will be weaker than a rock without pores. This is because there is no material in the pores, whereas if the open spaces were filled, the material in the pore space could hold the rock together. Furthermore, pore spaces allow fluids to pass through the rock or regolith, and the fluids may further dissolve the rock, creating more pore space and further weakening the material. Water in open pore space may also exert pressure on the surrounding rocks, pushing individual grains apart and making the rock weaker.

Water may act to either enhance or inhibit movement of regolith and rock downhill. Water inhibits downslope movement when the pore spaces are only partly filled with water, and the surface tension (bonding of water molecules along the surface) acts as an additional force holding grains together. This surface tension is able to bond water grains to each other, water grains to rock particles, and rock particles to each other. An everyday example of how effectively surface tension may hold particles together is found in sand castles at the beach: When the sand is wet, tall towers can be constructed, but when the sand is dry, only simple piles of sand can be made.

Water more typically acts to reduce the adhesion between grains, promoting downslope movements. When the pore spaces are filled, the water acts as a lubricant and may actually exert forces that push individual grains apart. The weight of the water in pore spaces also exerts additional pressure on underlying rocks and soils, and this is known as loading. The loading from water in pore spaces is in many cases enough so that the strength of the underlying rocks and soil is exceeded, and the slope fails, resulting in a downslope movement.

Another important effect of water in pore spaces occurs when the water freezes; when water freezes, it expands by a few percent, and this expansion exerts enormous pressures on surrounding rocks, in many cases pushing them apart. The freeze-thaw cycles found in many climates are responsible for many of the downslope movements.

Steep slopes are less stable than shallow slopes. Loose unconsolidated material tends to form slopes at specific angles that, depending on the specific characteristics of the material, range from about 33–37°. The way that the particles are arranged or packed in the slope is also a factor, and the denser the packing, the more stable the slope.

PROCESSES OF MASS WASTING

There are three basic types of mass movements, which are distinguished from each other by the way that the rock, soil, water, and debris move. *Slides* move over and in contact with the underlying surface. *Flows* include movements of regolith, rock, water, and air, in which the moving mass breaks into many pieces that flow in a chaotic mass movement. *Falls* move freely through the air and land at the base of the slope or escarpment. There is a continuum between different processes of mass wasting, but many differ in terms of the velocity of downslope movement and also in the relative concentrations of sediment, water, and air. A *landslide* is a general name for any downslope movement of a mass of bedrock, regolith, or a mixture of rock and soil, and it is used here to indicate a mass wasting process. Since all mass wasting processes occur on slopes, we will first discuss one of the most common events: the failure of the slope.

Slumps

A *slump* is a type of sliding slope failure in which a downward and outward rotational movement of rock or regolith occurs along a concave upslip surface. This produces either a singular or a series of rotated blocks, each with the original ground surface tilted in the same direction. Slumps are especially common after earthquakes and heavy rainfalls, and are common along roadsides and other slopes that have been artificially steepened to make room for buildings or other structures. Slump blocks may continue to move after the initial sliding event, and in some cases this added slippage is enhanced by rainwater that falls on the back-tilted surfaces and infiltrates along the fault, acting as a lubricant for added fault slippage.

The 1964 magnitude 9.2 earthquake in Alaska triggered one of the more spectacular examples of slumping in recent history. The Turnagain Heights neighborhood that overlooked the scenic Cook Inlet, with the Alaska Range and Aleutian volcanoes in the background, was built on a series of layered rocks. One unit, known as the Bootlegger shale, is a weak unit that acts like quicksand when it is shaken (a process known as liquefaction). When the earthquake struck, the Bootlegger shale was unable to support overlying rock layers and they all moved toward Cook Inlet on a series of tilted slump blocks. The entire neighborhood of Turnagain Heights sat on top of the slumped blocks, and houses tilted on the surfaces of the slump blocks, twisted, and slid toward the inlet. Seventy homes in the Turnagain Heights neighborhood were destroyed, killing three people in the twisted buildings. The entire neighborhood was condemned, bulldozed, and is now a park known as Earthquake Park.

A *translational slide* is a variation of a slump in which the sliding mass moves not on a curved surface, but downslope on a preexisting plane such as a weak bedding plane or a joint. Translational slides may remain relatively coherent or break into small blocks, forming a debris slide.

One of the worst translational slides ever occurred in Vaiont, Italy, in 1963. In 1960, a large dam was built in a deep valley in northern Italy. The valley occupies the core of a synclinal fold in which the limestone rock layers dip inward toward the valley center and form steeply dipping bedding surfaces along the valley walls. The bottom of the valley was oversteepened by downcutting from streams, forming a steep V-shaped valley in the middle of a larger U-shaped glacial valley. The rocks on the sides of the V-shaped valley are highly fractured and broken into many individual blocks, and there was an extensive cave network carved in the limestones. The reservoir behind the dam held approximately 500 million cubic feet of water.

After the dam was constructed, the pores and caves in the limestone filled with water, exerting extra, unanticipated pressures on the valley walls and dam. Heavy rains in the fall of 1963 made the problem worse, and authorities predicted that sections of the valley might experience landslides. The rocks on the slopes surrounding the reservoir began creeping downhill, first at a quarter inch per day (3–6.5 mm/day), then accelerating to one-and-a-half inches per day (forty mm per day) by October 6, 1963. Even though authorities were expecting landslides, they had no idea of the scale of what was about to unfold.

At 10:41 P.M. on October 9, a 1.1-mile-long and one-mile-wide section of the mountain on the south wall of the reservoir suddenly failed and slid into the reservoir at more than sixty miles per hour. Approximately 750 million cubic feet of debris fell into the reservoir, creating an earthquake shock and creating an air blast that shattered windows and blew roofs off nearby houses. The debris that fell into the reservoir displaced a huge amount of water, generating a series of monstrous waves that raced out of the reservoir, devastating nearby towns (Figure 5.2). A 780-foot-tall wave moved out of the north side of the reservoir, followed by a 328-foot-tall second wave. The waves combined and formed a 230-foot-tall wall of water that moved down the Vaiont Valley, inundating the town of Longarone, where more than 2,000 people were killed by the fast-moving floodwaters. Other waves bounced off the walls of the reservoir and emerged out the upper end of the reservoir, smashing into the town of San Martino where another 1,000 people were killed in the raging waters.

Sediment Flows

When mixtures of rock debris, water, and air begin to move under the force of gravity, they are said to be a *sediment flow*. This is a type of deformation that is continuous and irreversible. The way in which this mixture flows depends on the relative amounts of solid, liquid, and air; the grain size distribution of the solid fraction; and the physical and chemical properties of the sediment. Mass wasting processes that involve flow are transitional within themselves and to stream-type flows in the amounts of sediment/water and in velocity. There are many names for the different types of sediment flows, including slurry flows, mudflows, debris flows, debris avalanches, earthflows, and loess flows. Many mass movements begin as one type of flow and evolve into another during the course of the mass wasting event. For instance, it is common for flows to begin as rock falls

Figure 5.2. The Variont Dam (Italy) Disaster, October 9, 1963. Aerial view taken October 10 shows aftermath of flooding of the 873-foot Variont Dam the day before. A 300-foot wall of water, unleashed by a landslide, spilled over the dam, killing thousands of people and causing untold destruction. ©Bettmann/CORBIS

or debris avalanches and evolve into debris avalanches, debris flows, and mudflows along their length as they pick up water and debris and flow over differing slopes.

Creep

Creep is the imperceptibly slow downslope-flowing movement of regolith. It involves the very slow plastic deformation of the regolith as well as repeated microfracturing of bedrock at nearly imperceptible rates. Creep occurs throughout the upper parts of the regolith, and there is no single surface along which slip has occurred. Creep rates range from a fraction of an inch per year, up to about two inches per year on steep slopes. Creep accounts for leaning telephone poles, fences, and many of the cracks in sidewalks and roads. Although creep is slow and not very spectacular, it is one of the most important mechanisms of mass wasting and it accounts for the greatest total volume of material moved downhill in any given year. One of the most common creep mechanisms is through frost heaving (Figure 5.3). Creep through frost heaving is extremely effective at moving rocks, soil, and regolith downhill because when the ground freezes, ice crystals form and grow, pushing rocks upward perpendicular to the surface. As the ice melts in the freeze-thaw cycle, gravity takes over and the pebble or rock moves vertically downward, ending up a fraction of an inch downhill from where it started. Other mechanisms of surface expansion and contraction, such as warming and cooling, can initiate creep, as can the expansion and contraction of clay minerals with changes in moisture levels (see Chapter 10). In a related phenomenon, the freeze-thaw cycle can push rocks upward through the soil profile as revealed by farm fields in New England and other northern climates where the fields seem to grow boulders. Farmers clear the fields of rocks, but years later, the same fields are filled with numerous boulders at the surface. In these cases, the freezing forms ice crystals below the boulders that push them upward; during the thaw cycle, the ice around the edges of the boulder melt first, and mud and soil seep down into the crack and find their way beneath the boulder. This process, repeated over years, is able to lift boulders to the surface, keeping the northern farmer busy.

The operation of the freeze-thaw cycle makes *rates of creep* faster on steep slopes than on gentle slopes, and faster with more water and greater numbers of freeze-thaw cycles. Rates of creep of up to half an inch per year are common.

Solifluction

Solifluction is the slow, viscous, downslope movement of waterlogged soil and debris. Solifluction is most common in polar latitudes where the top layer of permafrost melts, resulting in a water-saturated mixture resting on a frozen base. It is also common in very wet climates, as found in the tropics. Rates of movement are typically an inch or two per year, which is slightly faster than downslope flow by creep. Solifluction results in distinctive surface features such as lobes and sheets carrying the overlying vegetation; sometimes the lobes override each other, forming complex structures (Figure 5.4). Solifluction lobes are relatively common sights on mountainous slopes in wet climates, especially in areas with permafrost. The frozen layer beneath the soil prevents drainage of water deep into the soil or into the bedrock, so the uppermost layers in permafrost terrains tend to be saturated with water, aiding solifluction.

Debris Flows

Debris flows involve the downslope movement of unconsolidated regolith, most of which is coarser than sand. Some debris flows begin as slumps, but then continue to flow downhill as debris flows. They typically fan out and come to rest when they emerge out of steeply sloping mountain valleys onto lower-sloping plains. Rates of movement in debris flows vary from several feet per year to several hundred miles per hour. Debris flows are commonly shaped like a tongue with numerous ridges and depressions. Many form after heavy rainfalls in mountainous areas, and the number of debris flows is increasing with increasing deforestation of

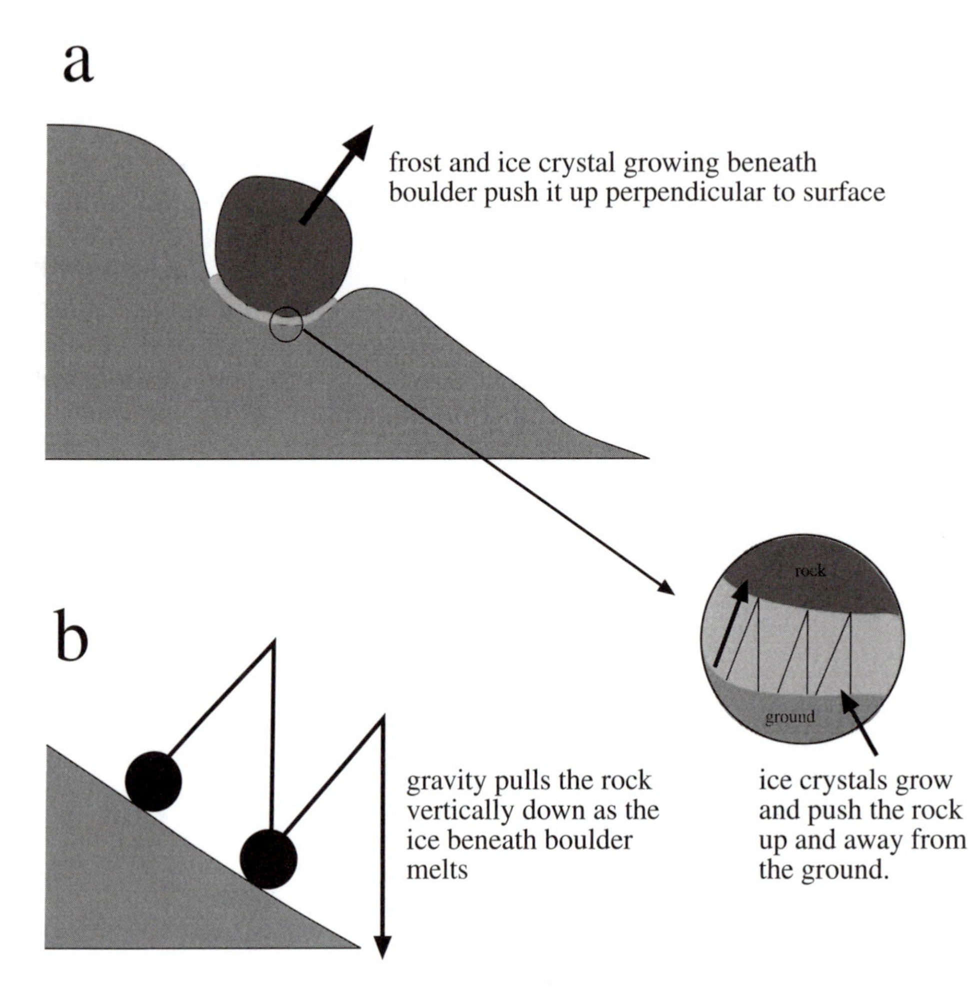

Figure 5.3. Creep by frost heaving. In areas subjected to freeze-thaw cycles, boulders may creep down hill very effectively by being repeatedly pushed up perpendicular to the surface by the force of ice crystals that grow beneath the boulder, then sinking vertically back down under the force of gravity, being displaced slightly in the process. A similar phenomenon also explains why many boulders rise to the surface in cleared farmers' fields in areas with active freeze-thaw cycles.

mountainous and hilly areas. This is particularly obvious in Madagascar, where deforestation in places has taken place at an alarming rate, removing most of the island's trees. What was once a tropical rain forest is now a barren (but geologically spectacular) landscape, carved by numerous landslides and debris flows that bring the terra rosa soil to rivers, making them run red to the sea.

Most debris flows that begin as rock falls or avalanches move outward in relatively flat terrain less than twice the distance they fell. Internal friction between particles in

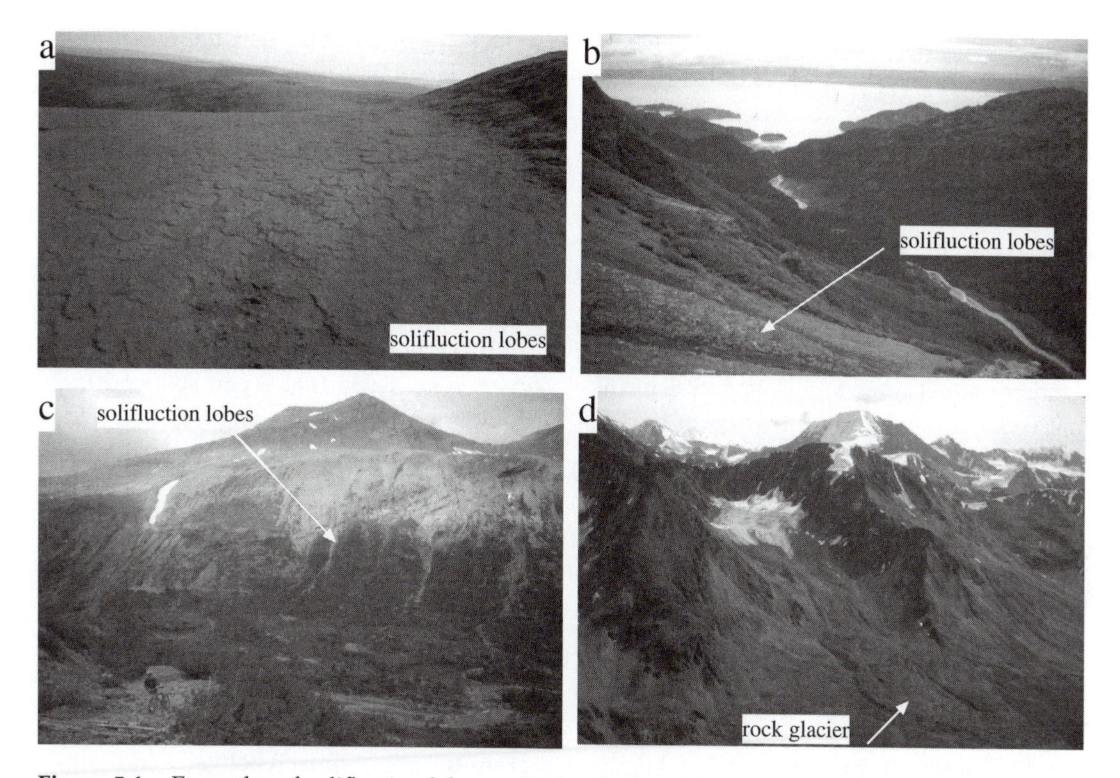

Figure 5.4. Examples of solifluction lobes and related phenomena in Alaska. (a) Solifluction lobes on a tundra-covered hillslope in northern Alaska form tongue-shaped lobes, slowly moving downhill. (b) Oblique view of solifluction lobe on mountain slope in southern Alaska. (c) View of scars in mountainside partly initialized by solifluction, and amplified by other mass wasting processes. (d) A rock glacier on the Kenai Peninsula of Alaska, formed by the rocks and ice flowing out of cirque (bowl-shaped hollow) in the background. Photos by T. Kusky

the flow, and external friction especially along the base of the flow, slow them. However, some of the largest debris flows that originated as avalanches or debris falls travel exceptionally large distances at high velocities. These are described below under the discussion of debris avalanches.

Mudflows

Mudflows resemble debris flows except that they have higher concentrations of water (up to 30 percent), which makes them more fluid, with a consistency ranging from soup to wet concrete. Mudflows often start as a muddy stream in a dry mountain canyon, which, as it moves, picks up more and more mud and sand until eventually the front of the stream is a wall of moving mud and rock. When this comes out of the canyon, the wall commonly breaks open, spilling the water behind it in a gushing flood, which moves the mud around on the valley floor. These types of deposits form many of the gentle slopes at the bases of mountains in the southwest United States.

Mudflows have also become a hazard in highly urbanized areas such as Los Angeles, where most of the dry river beds have been paved over and development has moved into the mountains surrounding the basin. The rare rainfall events in these areas then have no place to infiltrate and rush rapidly into the city, picking up all kinds of street mud and debris and forming walls of moving mud that cover streets and low-lying homes in debris. Unfortunately, after the storm rains and water recede, the mud remains and hardens in place. Mudflows are also common with the first heavy rains after prolonged droughts or fires, as residents of many California towns know. After the drought and fires of 1989 in Santa Barbara, heavy rains brought mudflows down out of the mountains, filling the riverbeds and inundating homes with many feet of mud. Similar mudflows followed the heavy rains in Malibu in 1994, which remobilized barren soil exposed by the fires of 1993. Three to four feet of mud filled many homes and covered parts of the Pacific Coast Highway. Mudflows are part of the natural geologic cycle in mountainous areas, and they serve to maintain equilibrium between the rate of uplift of the mountains and their erosion. Mudflows are only catastrophic when people have built homes, highways, and businesses in places that mudflows must go.

Volcanoes too can produce mudflows: Layers of ash and volcanic debris, sometimes mixed with snow and ice, are easily remobilized by rain or by an eruption and may travel many tens of kilometers. Volcanic mudflows are known as lahars, which were covered in Chapter 3. Mudflows have killed tens of thousands of people in single events and have been some of the most destructive of mass movements.

Slurry Flows

A *slurry flow* is a moving mass of sediment saturated in water that is transported with the flowing mass. The mixture, however, is so dense that it can suspend large boulders or roll them along the base. When slurry flows stop moving, the resulting deposit therefore consists of a nonsorted mass of mud, boulders, and finer sediment.

Granular Flows and Earthflows

Granular flows are unlike slurry flows because in granular flows, the full weight of the flowing sediment is supported by grain-to-grain contact between individual grains. *Earthflows* are relatively fast granular flows with velocities ranging from one m/day to 360 m/hour.

Rockfalls and Debris Falls

Rockfalls occur when detached bodies of bedrock free-fall from a cliff or steep slope. They are common in areas of very steep slopes where rockfall deposits may form huge deposits of boulders at the base of the cliff. Rockfalls can involve a single boulder or the entire face of a cliff. *Debris falls* are similar to rockfalls, but these consist of a mixture of rock and weathered debris and regolith.

Rockfalls have been responsible for the destruction of parts of many villages in the Alps and other steep mountain ranges, and rockfall deposits have dammed many a river valley, creating lakes behind the newly fallen mass. Some of these natural dams have been extended and heightened by engineers to make reservoirs, with examples including Lake Bonneville on the Columbia River and the Cheakamus Dam in British Columbia. Smaller examples abound in many mountainous terrains. The next time you come across a long lake in the bottom of a mountain valley, look to see if one side of the lake is dammed by a large lumpy mass of regolith. If it is, see if you can locate the source of the debris on one of the adjacent mountains.

A spectacular rockfall in 1965 covered the Canadian town of Hope with 52,000,000 cubic yards of debris (Matthews and McTaggert, 1978). This rockfall was caused by the construction of a highway through steep mountains made of rocks with a strong layering (schistosity) oriented roughly parallel to the slope of the mountains. After construction of the highway, minor ground shaking caused by snow avalanches apparently initiated the huge landslide that covered a two-mile-long section of the road with more than 250 feet of debris, killing four people.

A large rockfall recently rocked Yosemite Valley, California, killing one visitor (Figure 5.5). At 6:52 P.M. on July 10, 1996, 162,000 tons of granite suddenly fell off Glacier Point, first sliding down a steep slope for more than 500 feet, then falling off an overhang and flying through the air at 270 miles per hour another 1,640 feet before hitting the valley floor. Upon impact, the granite was pulverized and formed a huge dust cloud that rose more than half a mile high and shot across the valley, knocking down 1,000 trees in its path and covering fifty acres of the scenic valley with an inch-thick layer of granitic dust.

Rockslides and Debris Slides

Rockslide is the term given to the sudden downslope movement of newly detached masses of bedrock; these are *debris slides* if the rocks are mixed with other material or regolith common in glaciated mountains with steep slopes and also in places where planes of weakness exist, such as bedding planes or fracture planes that dip in the direction of the slope. Like rockfalls, rockslides may form fields of huge boulders coming off mountain slopes. The movement to this talus slope is by falling, rolling, and sliding.

A spectacular rockslide buried the coal-mining town of Frank in Alberta, Canada, on April 29, 1903. The town was built in the Oldham River Valley along the base of a beautiful mountain known as Turtle Mountain. Turtle Mountain is made of steeply sloping limestone, underlain by weak shale, sandstone, and coal. At 4:10 A.M., 90 million tons of Turtle Mountain suddenly slid 3,000 feet down the mountain into the valley, becoming instantly pulverized, then moving two miles as a debris avalanche across the valley and coming to rest 400 feet up the opposite mountain slope. The slide took about one and a half minutes to fall from the mountain and move up the next slope. Seventy people lost their lives in the burial of Frank, and the town virtually vanished beneath the debris.

Figure 5.5. Photo of 1996 avalanche in Yosemite Valley, California. On July 11, 1996, at 7:00 P.M. Pacific Daylight Time, a huge rock weighing 200 tons broke away from Granite Point, near Happy Isles, a popular trailhead and concession stand. The rock disintegrated when it landed, creating an air blast that was so powerful that it flattened as many as 2,000 trees in the area. The rockfall deposition killed one person at a concession stand, and seriously injured 14 others. The dust kicked up from the pulverized granite blocked out sunlight and coated tents and recreational vehicles, similar to an ashfall from a volcano. Massive rockfalls continue to occur. In 1999, a rockfall in the same area killed one climber and injured several others. Photo Credit: Edwin Harp, U.S. Geological Survey

Debris Avalanches

Debris avalanches are granular flows moving at very high velocity and covering large distances. These are rare but incredibly destructive and spectacular events. Some have ruined entire towns, killing tens of thousands of people in them without warning. Some have been known to move as fast as 250 miles per hour. These avalanches thus can move so fast that they move down one slope, then thunder right up and over the next slope and into the next valley. One theory of why these avalanches move so fast is that when the rocks first fall, they trap a cushion of air and then travel on top of it like a hovercraft.

Two of the worst debris avalanches in recent history originated from the same mountain, Nevados Huascaran, the highest peak in the Peruvian Andes. Nevados Huascaran is made of granite and is cut by many vertical joint surfaces, forming many planes of weakness. At 6:13 P.M. on January 10, 1962, a huge mass of rock and ice fell off Nevados Huascaran. It formed a debris flow that moved into the valley below at 105 miles per hour, coming to rest as a several-square-mile pile of rock and debris that was thirty to fifty feet thick and covered parts of the town of Ranrahirca. Four thousand people died in this event, and the survivors looked toward the sacred mountain with a feeling of foreboding. They were correct in their fear.

On May 31, 1970, a large earthquake off the coast of Peru initiated another larger debris flow from Nevados Huascaran. The entire face of the mountain between elevations of 18,000 and 21,000 feet collapsed, releasing more than 300 million cubic feet of rock and ice that moved down the valley at 175 to 210 miles per hour. This avalanche and debris flow covered a larger area, including many villages and the city of Yungay, killing more than 18,000 people in a few short minutes.

BASE OF SLOPE DEPOSITS

Weathered rock debris accumulates at the bases of mountain slopes and is deposited there by rock falls, slides, and other downslope movements. The entire body of rock waste sloping away from the mountains is known as talus, and the sediment comprising it is known as *sliderock*. Sliderock tends to accumulate at the angle of repose, which depends on the way the particles are packed together, their shapes, and the size of the particles.

SUBAQUEOUS MASS WASTING

Mass wasting is not confined to the land. Submarine mass movements are common and widespread on the continental shelves, slopes, and rise, and also in lakes. Mass movements under water, however, typically form *turbidity currents,* which leave large deposits of graded sand and shale. Under water, these slope failures can begin with very gentle slopes, even those of $< 1°$. Other submarine slope failures are similar to slope failures on land.

Slides and slumps are also common in the submarine realm, as are debris flows. Submarine deltas, deep-sea trenches, and continental slopes are common sites of submarine slumps, slides, and debris flows. Some of these are huge, covering hundreds of square miles. Many of the mass wasting events that produced these deposits must have produced large tsunami (see Chapter 4). The continental slopes are cut by many canyons, produced by submarine mass wasting events that carried material eroded from the continents into the deep ocean basins.

INITIATION OF MASS WASTING

Earthquakes or long periods of rain typically trigger landslides and other mass wasting events. When highways are built through mountains, slopes may be modified so that the new slope exceeds the angle of repose, and this is just asking for trouble. Retaining walls can delay catastrophic failure but rarely prevent it (Figure 5.6).

Undercutting

Streams and ocean waves often undercut waterside cliffs, making them very unstable, especially during storms when waves bash against the cliffs. The undercutting increases erosion from the top of the cliff as the slope tries to regain its angle of repose. In other cases, the slope may suddenly and catastrophically fail in a slump or slide. It is advisable to stay away from cliffs after storms.

Volcanic Eruptions

Shaking and slope oversteepening associated with volcanic eruptions typically trigger large numbers of mass wasting events. If ice and glaciers lie on the volcano, they can melt and trigger more mass wasting.

Submarine Mass Wasting

As discussed, submarine mass wasting events are triggered by many phenomena, some similar to those in the abovewater realm and some different. Shaking by earthquakes and displacement by faulting can initiate submarine mass wasting events, as can rapid release of water during sediment compaction. High sedimentation rates in deltas, continental shelves, continental slopes, and other depositional environments may create unstable slopes that can fail spontaneously or through triggers including agitation by storm waves. In some instances, sudden release of methane and other gases in the submarine realm may trigger mass wasting events.

HAZARDS TO HUMANS

From the descriptions of mass wasting processes and specific events above, it should be apparent that mass wasting presents a significant hazard to humans. The greatest hazards are from building on mountain slopes, which when oversteepened may fail catastrophically. Gradual creep moves cultural and natural features downhill, which accounts for the greatest cumulative amount of material moved through mass wasting events. Human-built structures are not designed to move downhill or to be covered in debris, so mass wasting needs to be appreciated and accounted for when designing communities, homes, roads, pipelines, and other cultural features. The best planning involves not building in areas that pose a significant hazard, but if this can not be done, the haz-

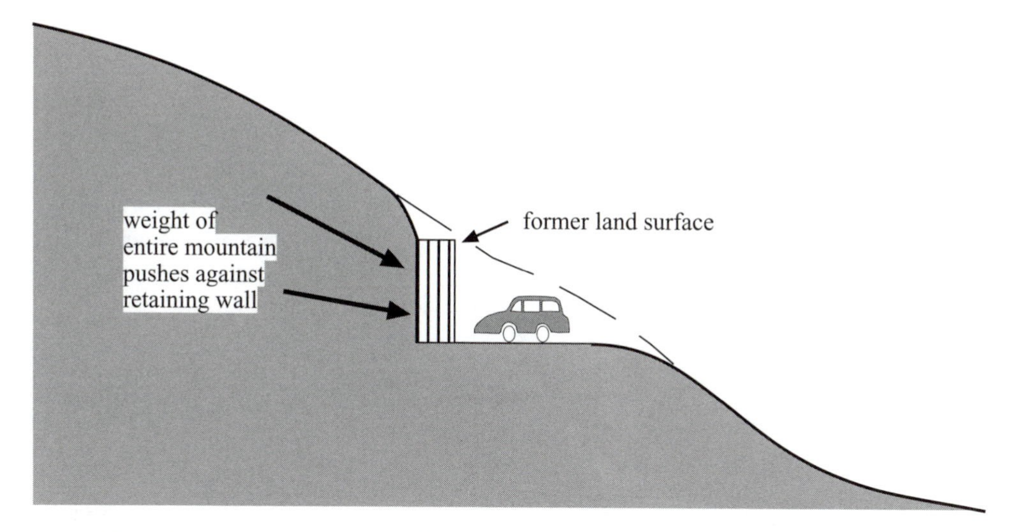

Figure 5.6. Forces on a retaining wall—the weight of the entire uphill portion of a hillslope or mountain may push against a retaining wall, potentially causing it to collapse if not constructed well.

ards should be minimized through slope engineering, as described below.

MASS WASTING STATISTICS

Mass wasting is one of the most costly natural hazards, with the slow downslope creep of material causing billions of dollars of damage to properties every year in the United States. Earth movements do not kill many people in most years, but occasionally massive landslides take thousands or even hundreds of thousands of lives (Table 5.1). Mass wasting is becoming more of a hazard in the United States as population increases and people move in great numbers from the plains and into mountainous areas. This trend is expected to continue in the future, and more mass wasting events like those described in this chapter may be expected every year. Good engineering practices and understanding of the driving forces of mass wasting will hopefully prevent many mass wasting events, but it will be virtually impossible to stop the costly gradual downslope creep of material, especially in areas with freeze-thaw cycles.

WHAT CAN BE DONE TO REDUCE DOWNSLOPE FLOW HAZARDS?

To reduce the hazards from mass wasting, it is necessary to first recognize which areas may be most susceptible to mass wasting and then to recognize the early warning signs that a catastrophic mass wasting event may be imminent. As with many geological hazards, a past record of downslope flows is a good indicator that a given area is prone to additional landslide hazards. Geological surveys and hazard assessments should be completed in mountainous and hilly terrain before construction of homes, roads, railways, power lines, and other features.

Prediction, Prevention, and Mitigation of Mass Movement Damage

What can be done to reduce the damage and human suffering inflicted by mass movements? Greater understanding of the dangers and specific triggers of mass movements can help reduce casualties from individual catastrophes, but long-term planning is needed to reduce the costs from damage to structures and our nation's infrastructure that are inflicted by even slow downslope movements. One approach to reducing the hazards is to produce maps that show areas that have, or are likely to suffer from, mass movements. These maps should clearly show hazard zones and areas of greatest risk from mass movements, and what types of events may be expected in any given area. These maps should be made publicly available and used for planning communities, roads, pipelines, and other constructions.

Several factors need to be considered when making risk maps for areas prone to mass movements. First, slopes play a large role in mass movements, so anywhere there is a slope, there is a potential for mass movements. In general, the steeper the slope, the greater the potential for mass movements. In addition, any undercutting or oversteepening of slopes (such as from coastal erosion or construction) increases the chances of downslope movements, and anything that loads the top of a slope (like a heavy building) also increases the chances of initiating a downslope flow. Slopes that are in areas prone to seismic shaking are particularly susceptible to mass flows, and the hazards are increased along these slopes. Slopes that are wet and have a buildup of water in the slope materials are well lubricated, exert extra pressure on the slope material, and are thus more susceptible to failure.

A slope's underlying geology is also a strong factor that influences whether or not it may fail. The presence of joints, bedding planes, or other weaknesses increases the chances of slope failure. Additionally, rocks that are water-soluble may have large open spaces and are also more susceptible to slope failure.

These features need to be considered when preparing landslide potential maps, and once a significant landslide potential is determined for an area, it should be avoided for building. If this is not possible, several engineering projects can be undertaken to reduce the risk. The slope could be engineered to remove excess water, decreasing the potential for failure. This can be accomplished through the installation of drains at the top of the slope and/or the installation of perforated pipes into the slope that help drain the excess water from the slope material, decreasing the chances of slope failure.

Slopes can be reduced by removing material, thus reducing the potential for landslides. If this is not possible, the slope can be terraced, which decreases runoff and stops material from falling all the way to the base of the slope. Slopes can also be covered with stone, concrete, or other materials that can reduce infiltration of water and reduce erosion of the slope material.

Retaining walls can be built to hold loose material in place, and large masses of rocks can be placed along the base of the slope (called base loading), which serves to reduce the potential of the base of the slope slumping out by increasing the resistance to the movement. Unvegetated slopes can be planted, as plants and roots greatly reduce erosion and may help soak up some of the excess water in the soil.

If a slope can not be modified and people must use the area, there are several other steps that may be taken to help reduce the risk to people in the area. Cable nets and wire fences may be constructed around rocky slopes that are prone to rock falls, and these wire meshes will serve to catch falling rocks before they hit passing cars or pedestrians. Large berms and ditches may be built to catch falling debris or to redirect mudflows and other earthflows. Rock sheds and tunnels may be built for shelter in areas prone to avalanches, and people can seek shelter in these structures during snow and/or rock avalanches.

What are the signs that may warn people of an imminent mass wasting event? Areas that have previously suffered mass wasting events may be most prone to repeated events, so geomorphological evidence for ancient slumps and landslides should be viewed as a warning. It is recognized that seismic activity and periods of heavy rainfall destabilize slopes and are times of increased hazards. The activity of springs can be monitored to detect when the slopes may be saturated and unstable, and features such as wet areas or puddles oriented parallel to an escarpment should be viewed as potential warnings that the slope is saturated and perhaps ready to slide. In some cases, slopes or whole mountains have experienced accelerated rates of creep soon before large mass wasting events, such as the Vaiont Dam disaster described above.

WHAT IF YOU ARE IN A DOWNSLOPE FLOW?

If you are in an area prone to landslides or other mass wasting movements, then it is important to keep an eye on your surroundings and watch for falling rocks, debris, and sliding slopes. If authorities warn of an imminent slope failure, leave immediately. If you see a rapid downslope flow moving toward you, you may have only a few seconds or minutes to respond and get to higher ground. In many places prone to avalanches, civic authorities have constructed avalanche shelters designed to withstand the force of the avalanche. If you are able to get into one of these structures during an avalanche, do so, as they are much safer than hiding in your car or behind boulders.

RESOURCES AND ORGANIZATIONS

Print Resources

Armstrong, B. R., and Williams, K. *The Avalanche Book.* Armstrong, Colo.: Fulcrum Publishing, 1992.

Brabb, E. E., and Harrod, B. L. *Landslides: Extent and Economic Significance.* Proceedings of the 28th International Geological Congress: Symposium on Landslides, Washington, D.C., July 17 1989. Rotterdam, The Netherlands: A. A. Balkema, 1989.

Coates, D. R., ed. "Landslides: Geological Society of America Reviews in Engineering." *Geology* 3 (1977), 278.

Hsu, K. J. "Catastrophic Debris Streams (Sturzstorms) Generated by Rockfalls." *Geological Society of America Bulletin* 86 (1989), 129–40.

Kiersch, G. A. "Vaiont Reservoir Disaster." *Civil Engineering* (1964), 32–39.

Matthews, W. H., and McTaggert, K. C. "Hope Rockslides, British Columbia," in B. Voight, ed., *Rockslides and Avalanches.* Amsterdam: Elsevier, 1978, 259–75.

Nilsen, T. H. and Brabb, E. E. "Landslides," in R. D. Borcherdt, ed., *Studies for Seismic Zonation of the San Francisco Bay Region.* Washington, D.C.: U.S. Geological Survey Professional Paper, 1975, 941A.

Norris, R. M. "Sea Cliff Erosion." *Geotimes* 35 (1990), 16–17.

Pinter, N., and Brandon, M. "How Erosion Builds Mountains," in J. Rennie, ed., *Earth from the Inside Out.* Scientific American (2000), 24–29.

Plafker, G., and Ericksen, G. E. "Nevados Huascaran Avalanches, Peru." In B. Voight, ed., *Rockslides and Avalanches.* Amsterdam: Elsevier, 1978, Chapter 8.

Schultz, A. P., and Southworth, C. S., eds. "Landslides in Eastern North America." Washington, D.C.: U.S. Geological Survey Circular 1008, 1987, 43 pp.

Schuster, R. L., and Fleming, R. W. "Economic Losses and Fatalities Due to Landslides." *Bulletin of the Association of Engineering Geologists* 23 (1986), 11–28.

Shaefer, S. J., and Williams, S. N. "Landslide Hazards." *Geotimes* 36 (1991), 20–22.

Varnes, D. J. "Slope Movement Types and Processes" in R. L. Schuster and R. J. Krizek, eds., *Landslides, Analysis and Control.* Washington, D.C.: National Academy of Sciences, 1978, Chapter 2.

Non-Print Resources

Videos

Debris Flow Dynamics, 1984, U.S. Geological Survey, 23 mins.

Landslide: Gravity Kills, 1999, Discovery Channel, 52 mins.

Raging Planet: Avalanche, 1997, Discovery Channel, 50 mins.

Web Sites

U.S. Geological Survey Web site:
http://www.usgs.gov/themes/hazard.html
Has pages on landslide hazards (Fact Sheet FS-0071–00) and downslope flow hazards.

Organizations

Federal Emergency Management Agency (FEMA)
500 C Street, SW
Washington, DC 20472
202–646–4600
http://www.fema.gov

U.S. Army Corps of Engineers Headquarters
441 G. Street, NW
Washington, DC 20314
202–761–0008
http://www.usace.army.mil/

U.S. Geological Survey
Federal Center
Box 25046, MS 967
Denver, CO 80225–0046
303–273–8500

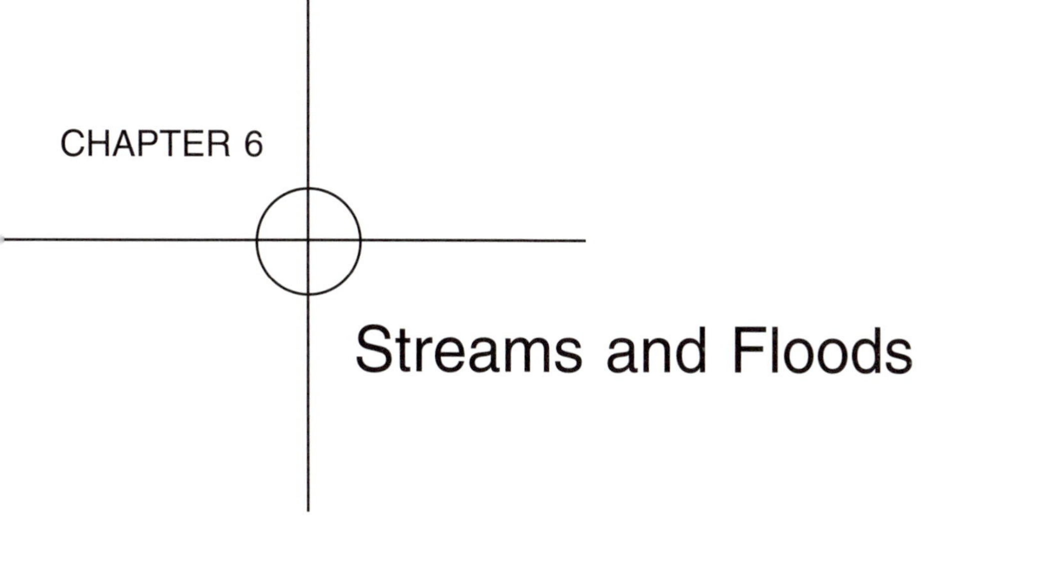

CHAPTER 6

Streams and Floods

INTRODUCTION

Stream and river valleys have been preferred sites for human habitation for millions of years, going back as far as our earliest known ancestors in Turkana Gorge in the rift valley of Kenya. Stream and river valleys provide routes of easy access through rugged mountainous terrain, and they also provide water for drinking, watering animals, and irrigation. The soils in river valleys are also some of the most fertile that can be found, as they are replenished by yearly or less frequent floods. The ancient Egyptians, whose entire culture developed in the Nile River Valley, appreciated this, and their agriculture revolved around the flooding cycles of the river. Rivers now provide easy and relatively cheap transportation on barges, and river valleys are preferred routes for roads and railways as they are relatively flat and easier to build in than over mountains. Many streams and rivers have also become polluted as industry has dumped billions of gallons of chemical waste into our nation's waterways.

However, streams and rivers are dynamic environments. Their banks are prone to erosion, and rivers periodically flood over their banks. During floods, rivers typically cover their floodplains with several or more feet of water and drop layers of silt and mud. This is part of a river's normal cycle and the ancient Egyptians relied upon this for replenishing and fertilizing their fields. Now, since many flood plains are industrialized or populated by residential neighborhoods, the floods are no longer welcome and natural floods are regarded as disasters. On average, floods kill a couple of hundred people each year in the United States. Dikes and levees have been built around many rivers in attempts to keep the floodwaters out of towns. However, this tends to make the flooding problem worse because it confines the river to a narrow channel, and the waters rise more quickly and can not seep in the ground of the floodplain.

Streams are important geologic agents that are critical for other earth systems. They carry most of the water from the land to the sea, they transport billions of tons of sediment to the beaches and oceans, and they erode and reshape the land's surface, forming deep valleys, floodplains, and passes through mountains.

This chapter examines the processes that control the development and flow in stream and river systems, and discusses how and when these processes become hazardous. Examples of flooding and other river disasters are presented throughout the text and at the end of the chapter.

EROSION BY RUNNING WATER

Water is an extremely effective erosional agent, including when it falls as rain and runs across the surface in finger-sized tracks called rivulets and when it runs in organized streams and rivers. Water begins to erode as soon as the raindrops hit the surface—the raindrop impact moves particles of rock, breaking them free from the surface and setting them in motion.

During heavy rains, the runoff is divided into overland flow and stream flow. *Overland flow* is the movement of runoff in broad sheets. Overland flow usually occurs through short distances before it concentrates into discrete channels as streamflow. Erosion performed by overland flow is known as sheet erosion. *Streamflow* is the flow of surface water in a well-defined channel. Vegetative cover strongly influences the erosive power of overland flow by water. Plants that offer thicker ground cover and have extensive root systems prevent erosion much more than thin plants and those crops that leave exposed barren soil between rows of crops. Ground cover between that found in a true desert and savanna grasslands tends to erode the fastest, whereas tropical rainforests offer the best land cover to protect from erosion: The leaves and branches break the force of the falling raindrops, and the roots form an interlocking network that holds soil in place.

Under normal flow regimes, streams attain a kind of equilibrium, eroding material from one bank and depositing it on another. Small floods may add material to overbank and floodplain areas, typically depositing layers of silt and mud over wide areas. However, during high-volume floods, streams may become highly erosive, even removing entire floodplains that may have taken centuries to accumulate. The most severely erosive floods are found in confined channels with high flow, such as where mountain canyons have formed downstream of many small tributaries that have experienced a large rainfall event. Other severely erosive floods have resulted from dam failures and, in the geological past, from the release of large volumes of water from ice-dammed lakes about 12,000 years ago. The erosive power of these floodwaters dramatically increases when they reach a velocity known as supercritical flow, at which time they are able to cut through alluvium like butter and even erode bedrock channels. Luckily, supercritical flow can not be sustained for long periods of time, as the effect of increasing the channel size causes the flow to self-regulate and become subcritical.

Cavitation in streams can also cause severe erosion. Cavitation occurs when the stream's velocity is so high that the vapor pressure of water is exceeded and bubbles begin to form on rigid surfaces. These bubbles alternately form and then collapse with tremendous pressure, and they form an extremely effective erosive agent. Cavitation is visible on some dam spillways, where bubbles form during floods and high discharge events, but it is different from the more common and significantly less erosive phenomena of air entrapment by turbulence, which accounts for most air bubbles observed in whitewater streams.

GEOMETRY OF STREAMS

Streams are dynamic, ever-changing systems. Streams are defined primarily by their channels, which are the elongate depressions where the water flows. There are several different types of stream banks that separate the stream channels from the adjacent flat floodplains. Streams create their broad flat floodplains by erosion and redeposition during floods, and these plains serve as the stream bottom during large floods. Even though floodplains may not have any water over them for many tens of years, they are part of the stream system and the stream will return. Many communities in the United States and elsewhere have built extensively on the floodplains, and these communities will eventually be flooded.

Stream channels are self-adjusting features that modify their shapes and sizes to best accommodate the amount of water flowing in the stream. A stream's *discharge* is the amount of water passing a given point per unit time. During floods, the discharge may be two, three, ten, or more times normal levels. The stream channel may then overflow, causing the water to spread across the adjacent floodplain and inundating any towns or farms built on the floodplain.

The cross-sectional shape of streams changes with time and amount of water flow, and also changes upstream and downstream as the slope and volume of water changes. Small narrow streams are typically as deep as they are wide, whereas large streams and rivers are much wider than they are deep.

The *gradient* or slope of a stream is a measure of the vertical drop over a given horizontal distance, and the average gradient decreases downstream (Figure 6.1). Going downstream, several changes also occur. First, the discharge increases, which causes the width and depth to increase as well. Less intuitively, as you move downstream and the gradient decreases, the velocity increases. Although one might expect the velocity of a stream to decrease with a decrease in slope (gradient), anyone who has seen the Mississippi at New Orleans or the Nile at Cairo can testify to their great velocity as compared to their upstream sources. There are two reasons for this increase in velocity. First, the upstream portions of these mighty rivers have courses with many obstacles and more friction per stream volume, reducing velocity. Second, there is more water flowing in the downstream portions of the streams, and this has to move quickly to allow the added discharge from the various tributaries that merge with the main stream.

The *base level* of a stream is that stream's limiting level, below which a stream can not erode the land. The ultimate base level is sea level, but in many cases streams entering a lake or dammed region (Figure 6.1) form a local base level.

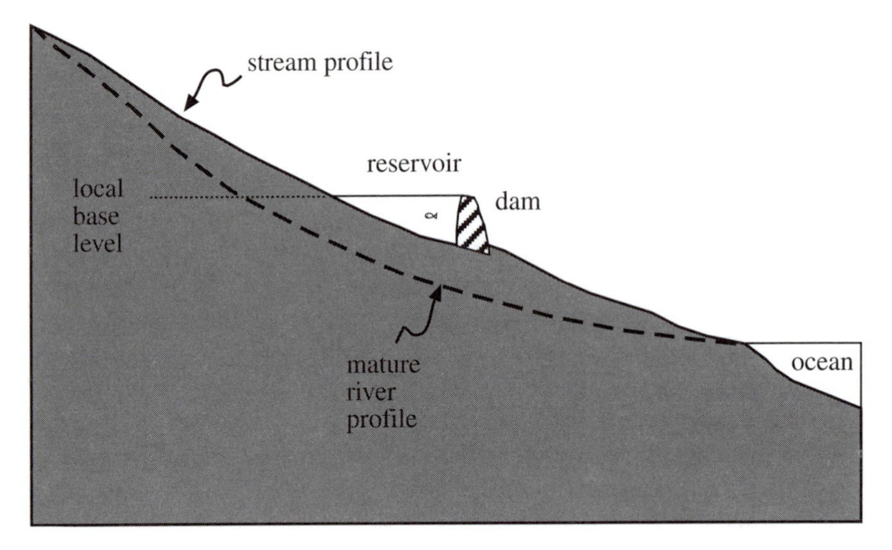

Figure 6.1. The stream gradient or profile is a measure of the change of slope with distance. It gradually decreases toward the sea, where it reaches base level, and may be interrupted by local base levels in lakes and dams.

Channel Patterns

Stream channels are rarely straight, and the velocity of flow changes in different places. Friction makes the flow slower on the bottom and sides of the channel, and bends in the river make the zone of fastest flow swing from side to side.

Straight channels are very rare, and those that do occur have many of the properties of curving streams (Figure 6.2). The *thalweg* is a line connecting the deepest parts of the channel. In straight segments, the thalweg typically meanders from side to side of the stream. In places where the thalweg is on one side of the channel, a bar may form on the other side. A *bar* (for example, a sand bar) is a deposit of alluvium in a stream.

Meandering Streams

Most streams move through a series of bends known as meanders (Figure 6.2). Meanders are always migrating across the floodplain by the process of the deposition of the point bar deposits, which are wedge-shaped deposits of sand and gravel that form on the inside curves of meandering streams. Deposition of the point bars happens along with the erosion of the bank on the opposite side of the stream with the fastest flow, providing material for the point bar deposit, and causing the stream channel to migrate laterally. The erosion typically occurs through slumping of the stream bank (Figure 6.3). Meanders typically migrate back and forth and also downvalley at a slow rate. If the downstream portion of a meander encounters a slowly erodable rock, the upstream part

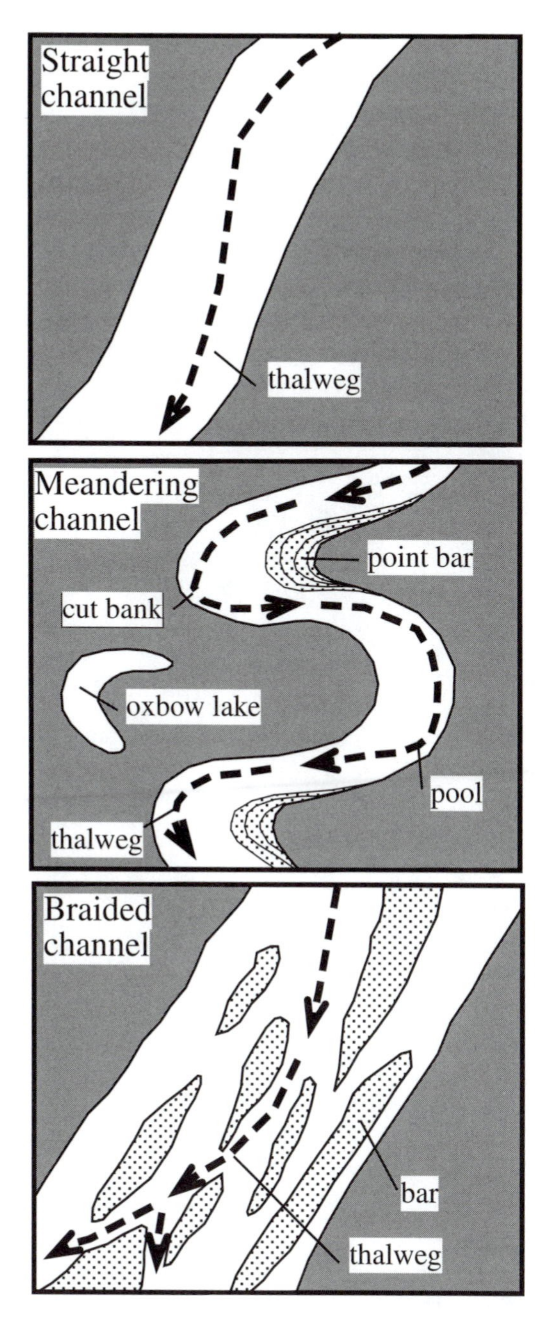

Figure 6.2. Patterns of braided, straight and meandering streams. Heavy dashed line shows the path of the thalweg, the line connecting the deepest and fastest moving parts of the stream.

may catch up and cut off the meander. This forms an *oxbow lake*, which is an elongate and curved lake formed from the former stream channel (Figure 6.2).

Braided Stream Channels

Braided stream channels consist of two or more adjacent but interconnected channels separated by bars or islands (Figure 6.2). Braided streams have constantly shifting channels, which move as the bars are eroded and redeposited during large fluctuations in discharge. Most braided streams have highly variable discharge in different seasons, and they carry more load than meandering streams.

DYNAMICS OF STREAMFLOW

Streams are very dynamic systems, and they constantly change their patterns and the amount of water (discharge) and sediment being transported in the system. Streams may transport orders of magnitude more water and sediment in times of spring floods, as compared to low-flow times of winter or drought. Since streams are dynamic systems, as the amount of water flowing through the channel changes, the channel responds by changing its size and shape to accommodate the extra flow. Five factors control how a stream behaves:

1. Width and depth of channel, measured in feet (meters)
2. Gradient, measured in feet per mile (m/km)
3. Average velocity, measured in feet per second (m/sec)
4. Discharge, measured in cubic feet per second (m^3/s)

Figure 6.3. One major problem to people living downstream of Mount St. Helens is the high sedimentation rates resulting from stream erosion of the volcanic deposits. Streams are continuously downcutting channels, eroding their banks, and eating away at the avalanche and lahar deposits. This material is eventually transported downstream and deposited on the streambeds, decreasing the carrying capacity of the channels and increasing the chances of floods. Photo Credit: USGS/Cascades Volcano Observatory

5. Load, measured as tons per cubic yard (metric tons/m^3)

All these factors are continually interplaying to determine how a stream system behaves. As one factor such as discharge changes, so do the others, expressed as $Q = w \times d \times v$. All factors vary across stream, so they are expressed as averages. If one term changes, then all or one of the others must change, too. For example, with increased discharge, the stream erodes, widens, and deepens its channel. With increased discharge, the stream may also respond by increasing its sinuosity through the development of meanders, effectively creating more space for the water to flow in and occupy by adding length to the stream. The meanders may develop quickly during floods because the increased stream velocity adds more energy to the stream system, and this can rapidly erode the cut banks, enhancing the meanders.

The amount of sediment load available to the stream is also independent of the stream's discharge, so different types of stream channels develop in response to different amounts of sediment load availability. If the sediment load is low, streams tend to have simple channels, whereas braided stream channels develop in which the sediment load is greater than the stream's capacity to carry that load. If a large amount of sediment is dumped into a stream, the stream will respond by straightening, thus increasing the gradient and stream velocity and increasing the stream's ability to remove the added sediment. When streams enter lakes or reservoirs along their path to the sea, the velocity of the stream will suddenly decrease. This causes the sediment load of the stream or river to be dropped as a delta on the lake bottom, and the stream attempts in this way to fill the entire lake with sediment. The stream is effectively attempting to regain its gradient by filling the lake, then eroding the dam or ridge that created the lake in the first place. When the water of the stream flows over the dam, it does so without its sediment load and therefore has greater erosive power and can erode the dam more effectively.

FLOODS

Seasonal variations in rainfall cause stream discharge to rise and sometimes overflow the stream's banks. Both stream discharge and velocity increase during floods, so during floods the streams carry larger particles. Floods of specific magnitude have a probable interval of recurrence. Small floods occur quite often, typically every year. Larger floods occur less frequently, and the largest floods occur with the longest time interval between them. The time interval between floods of a specific discharge is known as the recurrence interval, and some floods are commonly known as the fifty-year flood, the 100-year flood, etc.

Curves of the discharge versus flood recurrence interval can be drawn for every stream and river to determine its characteristic flooding frequency. This is important information for everyone living near a stream or on a river floodplain: to know how likely it is that a flood of a certain height will occur again within a certain time frame. For instance, if a flood of 150 cubic feet/second covered a small town with ten feet of water thirty years ago, is it safe to build a new housing development on the floodplain on the outskirts of town? Using the flood frequency curve for that river, planners could determine that floods of 150 cubic feet/second are expected on average every forty years, and floods of two times that magnitude are expected every 100 years. Planners and insurers might (in the best of situations) conclude from that information that it is unwise to build extensively on the floodplain and that the new community should be located on higher ground.

Understanding flood frequency and the chances of floods of specific magnitude occurring along a river is also essential for planning many other human activities. It is necessary to know how much water bridges and drainage pipes must be built to handle floods, and to know how to plan for land use across the floodplain. In many cases, bigger, more expensive bridges should be built, even if it seems unlikely that a small stream will ever rise high enough to justify such a high bridge. In other cases, structures are built with a short lifetime of use expected, and planners must calculate whether the likelihood of flood warrants the extra cost of building a flood-resistant structure.

STREAM LOAD

Streams carry a variety of materials as they make their way to the sea. These materials range from minute dissolved particles and pollutants to giant boulders moved only during the most massive floods. A stream's *bed load* consists of the coarse particles that move along or close to the bottom of the stream bed. Particles move more slowly than the stream, by rolling or sliding. Saltation is the movement of a particle by short intermittent jumps caused by the current lifting the particles. Bed load typically constitutes between 5–50 percent of the total load carried by the stream, with a greater proportion carried during high-discharge floods. A stream's *suspended load* consists of the fine particles suspended in the stream. It makes many streams muddy, and it consists of silt and clay that moves at the same velocity as the stream. The suspended load generally accounts for 50–90 percent of the total load carried by the stream. The *dissolved load* of a stream consists of dissolved chemicals, such as bicarbonate, calcium, sulfate, chloride, sodium, magnesium, and potassium. The dissolved load tends to be high in streams fed by groundwater. Pollutants such as industrial chemicals and fertilizers and pesticides from agriculture also tend to be carried as dissolved load in streams.

A wide range in sizes and amounts of material can be transported by a stream. The *competence* of a stream is the size of particles a stream can transport under a given set of hydraulic conditions, measured in diameter of largest bed load. A stream's *capacity* is the potential load it can carry, measured in the amount (volume) of sediment passing a given point in a set amount of time. The amount of material carried by streams depends on a number of factors. *Climate* studies show erosion rates are greatest in climates between a true desert and grasslands. *Topography* affects stream load: Rugged topography contributes more detritus, and some rocks are more erodable than others. *Human activity* such as farming, deforestation, and urbanization all strongly affect erosion rates and stream transport. Deforestation and farming greatly increase erosion rates and supply more sediment to streams, increasing their loads. Urbanization has complex effects, including deceased infiltration and decreased times between rainfall events and floods, as discussed in detail below.

DEPOSITIONAL FEATURES

During great floods, streams flow way out of their banks and fill adjacent floodplains. During these times, when the water flows out of the channel, its velocity suddenly decreases and it drops its load, forming levees and overbank silt deposits. *Terraces* are abandoned floodplains formed when a stream flowed above its present channel and floodplain level. These form when a stream erodes downward through its deposits to a new lower level. *Paired terraces* are terrace remnants that lie at the same elevation on either side of the present floodplain. *Nonpaired terraces* form at different levels on either side of the current floodplain and imply several episodes of erosion (Figure 6.4).

Deltas

When a stream enters the relatively still water of a lake or ocean, its velocity and its capacity to hold sediment drop suddenly. Thus, the stream dumps its sediment load there, and the resulting deposit is known as a delta. Where a coarse sediment load of an alluvial fan dumps its load in a delta, the deposit is known as a *fan-delta. Braid-deltas* are formed when braided streams meet local base levels and deposit their coarse-grained load. When a stream deposits its load in a delta, it first drops the coarsest material, then progressively finer material further out, forming a distinctive sedimentary deposit. The resulting *foreset layer* is thus graded from coarse material nearshore to fine material offshore. The *bottomset layer* consists of the finest material, deposited far out. As this material continues to build outward, the stream must extend its length and forms new deposits, known as *topset layers,* on top of all this. Most of the world's large rivers have built huge deltas at their mouths, such as the Mississippi, the Nile, and the Ganges, yet all of these are different in detail.

Drainage Systems

A *drainage basin* is the total area that contributes water to a stream, and the line that divides different drainage basins is known as a divide (such as the United States's Continental Divide). Streams are arranged in an orderly fashion, known as the *stream order,* in drainage basins. Smallest segments lack tributaries and are known as first order streams, second order streams form where two first order streams converge, third order streams form where two second order streams converge, and so on (Figure 6.5).

Stream capture occurs when headland erosion diverts one stream and its drainage into another drainage basin.

Several categories of streams reflect different geologic histories. A *consequent stream* is one whose course is determined by the direction of the slope of the land. A *subsequent stream* is one whose course has become adjusted so that it occupies a belt of weak rock or another geologic structure. An *antecedent stream* is one that has maintained its course across topography that is being uplifted by tectonic forces; these streams cross high ridges. *Superposed streams* are those whose course was laid down in overlying strata onto unlike strata below.

HAZARDS OF STREAMS AND FLOODS

Floods may occur in most parts of the world, from coastal plains to tropical, arctic, desert, and even mountainous environments. Floods are a stream or river system's natural response to greater than normal amounts of rainfall, snowmelt, or storms. Small streams respond by overflowing their banks quickly to high amounts of local rainfall, whereas large rivers rise and recede more slowly, generally in response to above average rainfalls in a large drainage basin for an extended period of time. Such long-term regional floods have caused some of the highest

Figure 6.4. Photographs of stream terraces. *Top:* Shows a 100,000-year-old stream terrace, consisting of tens of meters of cemented river gravel, from the United Arab Emirates. The present-day active wadi channel has cut through this older terrace, and now resides at a lower topographic level (people are in active stream channel). *Center:* Shows a similar terrace, forming a flat area for a modern village in Oman, which replaced the old village of Ghul, built high on the hillside, visible in the foreground. *Bottom:* Shows an uncemented gravel terrace from the northern Oman Mountains, with the active wadi downcut through the old terrace deposits. Photos by T. Kusky

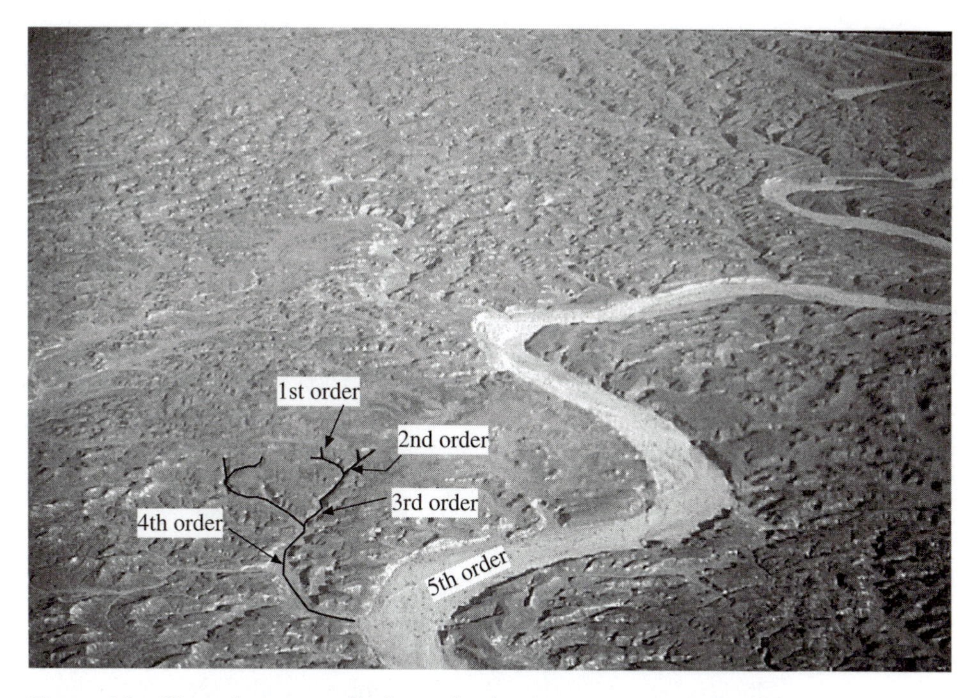

Figure 6.5. Photo from a small plane of a dry stream system in the Dhofar region of southern Oman and Yemen, showing how two first order streams merge to form a second order stream, two second order streams merge to form a third order stream, and so on. Photo by T. Kusky

death tolls resulting from any natural hazard. Floods have also caused rivers to change course, wreaking havoc on transportation routes and on urban and agricultural land use, and causing drastic changes to natural ecosystems.

TYPES OF FLOODS

Floods are of several kinds, including those associated with hurricanes and tidal surges in coastal areas, those caused by rare large thunderstorms in mountains and canyon territory, and those caused by prolonged rains over large drainage basins.

Flash Floods

Flash floods result from short periods of heavy rainfall and are common near warm oceans, along steep mountain fronts that are in the path of moist winds, and in areas prone to thunderstorms. They are well known from the mountains and canyonlands of the United States's desert southwest and many other parts of the world. Some of the heaviest rainfalls in the United States have occurred along the Balcones escarpment in Texas. Atmospheric instability in this area often forms along the boundary between dry desert air masses to the northwest and warm moist air masses rising up the escarpment from the Gulf of Mexico to the south and east. Up to twenty inches of rain have fallen along the Balcones escarpment in as little as three hours from this weather situation. The Balcones escarpment also seems to trap tropical hurricane rains, such as those from Hurricane Alice, which dumped more than forty inches of rain on the escarpment in 1954. The resulting floodwaters were sixty-five feet deep, one of the largest floods ever recorded in Texas. Approximately 25 percent of the catastrophic flash flooding events in the United States have occurred along the Balcones escarpment. On a slightly longer time scale, tropical hurricanes, cyclones, and monsoonal rains may dump several feet of rain over periods of a few days to a few weeks, resulting in fast but not quite flash flooding.

The national record for the highest, single-day rainfall is held by the south Texas region, when Hurricane Claudette dumped forty-three inches of rain on the Houston area in 1979. The region was hit again by devastating floods during June 8–10, 2001, when an early-season tropical storm suddenly grew off the coast of Galveston and dumped 28–35 inches of rain on Houston and surrounding regions. The floods were among the worst in Houston's history, leaving 17,000 people homeless and twenty-two dead. More than 30,000 laboratory animals died in local hospital and research labs, and the many university and hospital research labs in the flood area experienced hundreds of millions of dollars in damage. Fifty million dollars were set aside to buy out the properties of homeowners who had built on particularly hazardous floodplains. Total damages have exceeded $5 billion and are likely to rise. The standing water left behind by the floods became breeding grounds for disease-bearing mosquitoes, and the humidity led to a dramatic increase in the release of mold spores, which cause allergies in some people and are sometimes toxic.

The Cherrapunji region in northern India, at the base of the Himalayan Mountains, has received the world's highest rainfalls. Moist air masses from the Bay of Bengal move toward Cherrapunji, where they begin to rise over the high Himalayan Mountains. This produces a strong *orographic effect*, in which the air mass can not hold as much moisture as it rises and cools, so heavy rains result. Cherrapunji has received as much as thirty feet of rain in a single month (July 1861) and more than seventy-five feet of rain for all of 1861.

Flash floods typically occur in localized areas where mountains cause atmospheric upwelling leading to the development of huge convective thunderstorms, which can pour several inches of rain per hour onto a mountainous terrain, which focuses the water into steep-walled canyons. The result can be frightening, with floods raging down canyons as steep, thundering walls of water that crash into and wash away all in their paths. Flash floods can severely erode the landscape in arid and sparsely vegetated regions, but do much less to change the landscape in heavily vegetated humid regions.

Many canyons in mountainous regions have fairly large upriver parts of their drainage basins. Sometimes, the storm that produces a flash flood with a wall of water may be located so far away that people in the canyon do not even know that it is raining somewhere or that they are in immediate and grave danger. Such was the situation in some of the examples described below.

The severity of a flash flood is determined by a number of factors other than the amount of rainfall. The shape of the drainage basin is important because it determines how quickly rainfall from different parts of the basin converge at specific points. The soil moisture and previous rain history are important, as are the amounts of vegetation, urbanization, and slope.

A final type of flood occurs in areas where rivers freeze over. The annual spring breakup can cause severe floods, initiated when blocks of ice get jammed behind islands, bridges, or along bends in rivers. These ice dams can create severe floods, causing the high spring waters to rise quickly, bringing the ice cold waters into low-lying villages. When ice dams break up, the force of the rapidly moving ice is sometimes enough to cause severe damage, knocking out bridges, roads, and homes. Ice-dam floods are fairly common in parts of New England, including New Hampshire, Vermont, and Maine.

Big Thompson Canyon, Colorado, 1976 Big Thompson Canyon is a popular recreation area located about fifty miles northwest of Denver in the Front Ranges of the Rocky Mountains. On July 31, 1976, a large thunderhead cloud had grown over the front ranges, and it suddenly produced a huge cloudburst (rainfall) instead of doing the usual thing of blowing eastward over the plains. Approximately seven and a half inches of rain fell in a four-hour period, an amount approximately equal to the average yearly rainfall in the area. The steep topography focused the water into Big Thompson Canyon, where a flash flood with a raging twenty-foot-high wall of water rushed through the canyon narrows at fifteen miles per hour, killing 145 people who were driving into and out of the canyon. As the wall of water roared through the canyon, many people abandoned their cars and scrambled up the canyon walls to safety, only to watch their cars get washed away by the floods. Those people who climbed the canyon walls to escape the flash flood survived, but those who did not or could not perished in the flood. In addition to the deaths, this flash flood destroyed 418 homes, fifty-two businesses, and washed away 400 cars. Damage totals are estimated at $36 million.

Flash Floods in the Northern Oman Mountains The Northern Oman (Hajar) Mountains are a steep rugged mountain range on the northeastern Arabian Peninsula, with deep and long canyons that empty into the Gulf of Oman and Arabian Sea. These normally dry canyons are known as wadis, and the local villagers dig wells in the wadi bottoms to reach the groundwater table for use in homes and agriculture. The region is normally very dry, but occasional thunderstorms grow and explode over parts of the mountains. Occasionally, a typhoon works its way from the Indian Ocean across the Arabian Peninsula and may also dump unusual amounts of rain on the mountains. In either situation, the canyons become extremely unsafe places to be, and local villagers have tales of flash floods with hundred-foot-tall walls of water wiping away entire settlements, leaving only coarse gravel in their place. The inhabitants of this region have learned to place their villages on high escarpments above the wadis, out of the reach of the rare but devastating flash flood. Older destroyed villages are visible in some wadi floors, but the wisdom acquired from experiencing a devastating flash flood has encouraged these people to move to higher ground. The inconvenience of being located a hundred feet or more above their water source is avoided by building long aqueduct-like structures (known as falaj) from water sources located at similar elevations far upstream, and letting gravity bring the water to the elevated village site.

Flash Floods in the Southern Alps and Algeria, 2000 and 2001 In November 2001, parts of Algeria in North Africa received heavy rainfalls over a period of two days that led to the worst flooding and mudslides in the capital city of Algiers in more than forty years. It is estimated that close to 1,000 people died in Algiers, being buried by fast-moving mudflows that swept out of the Atlas Mountains to the south and moved through the city, hitting some of the poorest neighborhoods with the worst flooding. The Bab El Oued District, one of the poorest in Algiers, was hit the worst when 600 people there were buried under several foot thick mudflows.

These floods followed similar heavy rains, floods, and mudslides that devastated parts of southern Europe in October 2000. Northern Italy and Switzerland were among the worst-hit areas; there, water levels reached their highest in thirty years, killing about fifty people. In Switzerland, the southern mountain village of Gondo was devastated when a 120-foot-wide mudflow ripped through the town center, removing ten homes (one-third of the village) and killing thirteen people. Numerous roads, bridges, and railroads have been washed away throughout the region, stretching from southern France through Switzerland and Italy and to the Adriatic Sea. Crops have been destroyed on a massive scale. Tens of thousands of people had to be evacuated from throughout the region, and total damage estimates are in the range of many billions of dollars.

Regional Floods from Prolonged Rainfall and Snowmelt

Mississippi River Basin and the Midwest of the United States The Mississippi River is the largest river basin in the United States and the third-largest river basin in the world. It is the site of frequent, sometimes devastating floods. All of the eleven major tributaries of the Mississippi River have also experienced major floods, including events that have at least quadrupled the normal river discharge in 1993, 1973, 1927, 1909, 1903, 1892, and 1883. Three of the major rivers (Mississippi, Missouri, and Illinois) meet in St. Louis, which has seen some of the worst flooding along the entire system.

Floods along the Mississippi River in the eighteenth and nineteenth centuries prompted the formation of the Mississippi River Commission, which oversaw the construction of high levees along much of the length of the river from New Orleans to Iowa. By the year 1926, over 1,800 miles of levees had been constructed, many of them over twenty feet tall. The levees gave people a false sense of security against the floodwaters of the mighty Mississippi, and the levees restricted the channel, causing floods to rise more quickly and forcing the water to flow faster.

Many weeks of rain in the late fall of 1926, followed by high winter snowmelts in the upper Mississippi River Basin, caused the river to rise to alarming heights by the spring of 1927. Residents all along the Mississippi were worried and were strengthening and heightening the levees and dikes along the river, hoping to avert disaster. The crest of water was moving through the upper Midwest and had reached central Mississippi, and the rains continued. In April, levees began collapsing along the river, sending torrents of water over thousands of acres of farmland, destroying homes, livestock, and leaving 50,000 people homeless. One of the worst-hit areas was Washington County, Mississippi, where an intense late April storm dumped an incredible fifteen inches of rain in eighteen hours, causing additional levees along the river to collapse. One of the most notable was the collapse of the Mounds Landing Levee, whose collapse caused a ten-foot-deep lobe of water to cover the Washington County town of Greenville on April 22. The river reached fifty miles in width and had flooded approximately 1 million acres, washing away an estimated 2,200 buildings in Washington County alone. Many people perished trying to keep the levees from collapsing and were washed away in the deluge. The floodwaters remained high for more than two months, and people were forced to leave the area (if they could afford to) or to live in refugee camps on the levees, which were crowded and unsanitary. An estimated 1,000 people perished in the floods of 1927, some from the initial flood and more from famine and disease in the months following the initial inundation by the floodwaters.

1972 was another wet year along the Mississippi, with most tributaries and reservoirs being filled by the end of the summer. The rains continued through the winter of 1972–1973, and the snowpack thickened over the northern part of the Mississippi basin. The combined snowmelts and continued rains caused the river to reach flood levels at St. Louis in early March, before the snow had even finished melting. Heavy rain continued throughout the Mississippi basin and the river continued to rise through April and May, spilling into fields and low-lying areas. The Mississippi was so high that it rose to more than fifty feet above its average levels for much of the lower river basin, and these river heights caused many of the smaller tributaries to back up until they too were at this height. The floodwaters rose to levels not seen for 200 years. At Baton Rouge, Louisiana, the river nearly broke through its banks and established a new course to the Gulf of Mexico, which would have left New Orleans without a river.

The floodwaters began peaking in late April, causing 30,000 people in St. Louis, and close to 70,000 people in the region, to be evacuated by April 28th. The river remained at record heights throughout the lower drainage basin through late June. Damage estimates exceeded $750 million (1973 dollars).

In the late summer of 1993, the Mississippi River and its tributaries in the upper basin rose to levels not seen in more than

130 years. The discharge at St. Louis was measured at more than 1 million cubic feet per second. The weather situation that led to these floods was remarkably similar to that of the floods of 1927 and 1973, only worse. High winter snowmelts were followed by heavy summer rainfalls caused by a low-pressure trough that stalled over the Midwest, because it was blocked by a stationary high-pressure ridge that formed over the east coast of the United States. The low-pressure system drew moist air from the Gulf of Mexico that met the cold air from the eastern high-pressure ridge, initiating heavy rains for much of the summer. The rivers continued to rise until August, when they reached unprecedented flood heights. The discharge of the Mississippi was the highest recorded, and the height of the water was even greater because all the levees that had been built restricted the water from spreading laterally and caused the water to rise more rapidly than it would have without the levees in place. More than two-thirds of all the levees in the Upper Mississippi River basin were breached, overtopped, or damaged by the floods of 1993. Forty-eight people died in the 1993 floods, and 50,000 homes were damaged or destroyed. Total damage costs are estimated at more than $20 billion.

The examples of the floods of 1927 and 1993 on the Mississippi reveal the dangers of building extensive levee systems along rivers. First, levees adversely affect the natural processes of the river and may actually make floods worse. The first effect they have is to confine the river to a narrow channel, causing the water to rise faster than if it were able to spread across its floodplain. Additionally, since the water can no longer flow across the floodplain, it can not seep into the ground as effectively, and a large amount of water that would normally be absorbed by the ground now must flow through the confined river channel. The floods are therefore larger because of the levees. A third hazard of levees is associated with their failure. When a levee breaks, it does so with the force of hundreds or thousands of acres of elevated river water pushing it from behind. The force of the water that broke through the Mounds Landing Levee in the 1927 flood is estimated to be equivalent to the force of water flowing over Niagara Falls. If the levees were not in place, the water would have risen gradually and would have been much less catastrophic when it eventually came into the farmlands and towns along the Mississippi River basin.

The U.S. Army Corps of Engineers is mitigating another hazard and potential disaster where the Atchafalaya River branches off the Mississippi. The Mississippi River has, over geological time, altered its course so that its mouth has migrated east and west by hundreds of miles (Figure 6.6). Each course of the river has produced its own delta, which subsides below sea level after the river migrates to another location (Figure 6.7). Subsidence of the delta deposit occurs primarily because the river no longer replenishes the top of the delta, and the buried muds gradually compact as the water is expelled from the pore spaces by the weight of the overlying sediments. As the delta subsides to sea level, waves add to the erosion, keeping the delta surface below sea level. At the present time, the lower Mississippi River follows a long and circuitous course from where the Atchafalya River branches off from it, past New Orleans to its mouth near Venice, Louisiana. The Mississippi River is ready to switch its course back to its earlier position of following the Atchafalaya, which would offer it a shorter course to the sea and would take less energy to transport sediment to the Gulf of Mexico. If this were to occur, it would be devastating to the lower delta, which would quickly subside below sea level. The city of New Orleans is currently below sea level and is only protected from the river, storms, and the Gulf of Mexico by high levees built around the city. To prevent this disaster from occurring, the Army Corps has been constructing an extensive system of diversions, levees, and dams at the Mississippi/Atchafalaya junction, with the aim of keeping the Mississippi in its channel.

EFFECTS OF LEVEES ON FLOODING

The examples of Mississippi River Basin flooding have taught us some valuable lessons on how to manage flood control on river basins. Levees are commonly built along riverbanks to protect towns and farmlands from river floods. These levees are usually successful at the job they were intended to do, but they also cause some other collateral effects. First, the levees do not allow waters to spill onto the floodplains, so the floodplains do not receive the annual fertilization by thin layers of silt, and they may begin to deflate and slowly degrade as a result of this loss of nourishment by the river. The ancient Egyptians relied on such yearly floods to maintain their fields' productivity, which has declined since the Nile has been dammed and altered in recent times. Another effect of levees is that they constrict the river to a narrow channel, so that floodwaters that once spread slowly over a large region are now focused into a narrow space. This causes floods to rise faster, reach greater heights, have a greater velocity, and reach downstream area faster than rivers without levees. The extra speed of the river is in many cases enough to erode the levees and return the river to its natural state.

One of the less appreciated effects of building levees on the sides of rivers is that they sometimes cause the river to slowly rise above the height of the floodplain. Many rivers naturally aggrade or accumulate sediment along their bottoms. In a natural system without levees, this aggradation is accompanied by lateral or sideways migration of the channel so that the river stays at the same height with time. However, if a levee is constructed and maintained, the river is forced to stay in the same location as it builds up its bottom. As the bottom rises, the river naturally adds to the height of the levee, and people will build up the height of the levee as well as the river rises, to prevent further flooding. The net result is that the river may gradually rise above the floodplain until some catastrophic flood causes the levee to break and the river establishes a new course.

The process of breaking through a levee happens naturally as well, and it is known as avulsion. Avulsion has occurred seven times in the last 6,000 years along the lower Mississippi River. Each time, the river has broken through a levee a few hundred

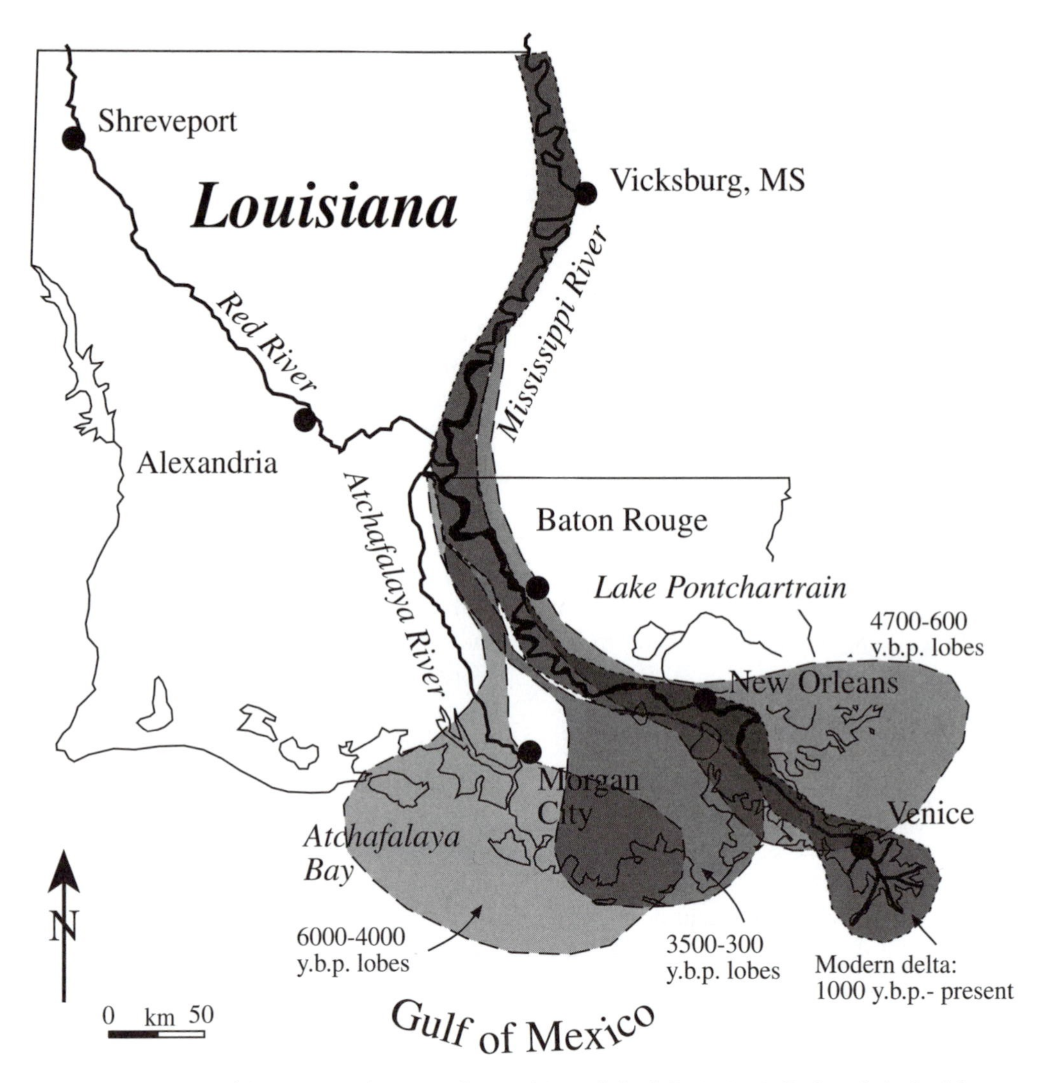

Figure 6.6. Map of Louisiana, showing the position of the Mississippi, Red and Atchafalaya Rivers, and the present and past positions of the Mississippi River delta lobes. New Orleans and the active lobe are currently subsiding, and the Mississippi seems poised to attempt to find a shorter route to the sea, such as along the Atchafalaya River. The U.S. Army Corps of Engineers is actively preventing this through the construction of a series of river control features. Figure 6.7 shows a detailed view of the end of the modern delta lobe.

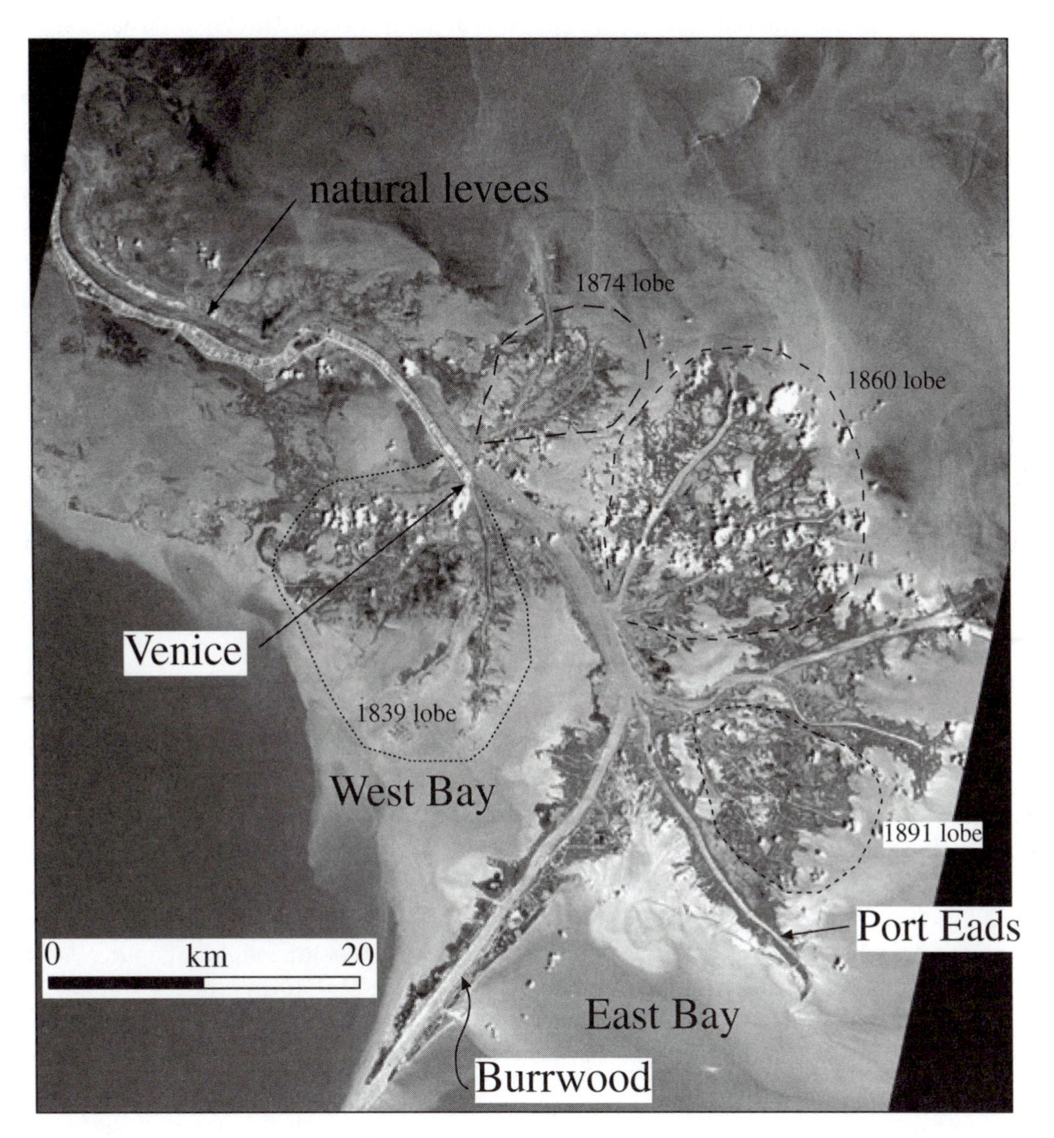

Figure 6.7. Landsat view and map annotations of the active lobe of the Mississippi River delta, showing the fragility of the system. Natural levees are easily breached during storms, and many older lobes of the delta are visible. Background image from NASA (http://earthobservatory.nasa.gov/Newsroom/NewImages/images.php3?img_id=9304)

miles from the mouth of the river and has found a new, shorter route to the Gulf of Mexico (Figure 6.7). The old river channel and delta is then abandoned, and the delta subsides below sea level as the river no longer replenishes it. A new channel is established and this gradually builds up a new delta until it too is abandoned in favor of a younger, shorter channel to the Gulf.

The Yellow (Huang) River, China

More people have been killed from floods along the Yellow River in China than from any other natural feature in the world, whether river, volcano, fault, or coastline. It is estimated that millions of people have died as a result of floods and famine generated by the Yellow River, which has earned it the Chinese nickname "River of Sorrow."

The Yellow River flows out of the Kunlun Mountains across much of China into the wide lowland basin between Beijing and Shanghai. The river has switched courses in its lower reaches at least ten times in the last 2,500 years. It currently flows into Chihli (Bohai) Bay and then into the Yellow Sea.

The Chinese have attempted to control and modify the course of the Yellow River since dredging operations in 2356 B.C. and the construction of levees in 602 B.C. One of the worst modern floods along the Yellow River was in 1887 when the river rose over the top of the seventy-five-foot-high (twenty-two meters) levees and covered the lowlands with water. Over 1 million people died from the floods and subsequent famine. Crops and livestock were destroyed, and sorrow returned to the river.

The Yellow River was also the site of an unnatural disaster in 1938. As part of the war effort, in 1938 Japan attacked and bombed the levees along the Yellow River. The river escaped and took another million lives.

The Yellow River is continuing its natural process of building up its bottom, and the people along the river continue to raise the level of the levees in an attempt to keep the river's floods out of their fields. Today, the river bottom rests an astounding sixty-five feet (twenty meters) above the surrounding floodplain (Figure 6.8), a testament to the attempts of the river to find a new lower channel and to abandon its current channel in the process of avulsion. What will happen if heavy rains cause another serious flood along the River of Sorrow? Will another million people perish?

URBANIZATION AND FLASH FLOODING

When heavy rains fall in an unaltered natural environment, the land surface responds to accommodate the additional water. Desert regions may experience severe erosion in response to the force of falling raindrops that dislodge soil and also by overland flow during heavy rains. This causes upland channel areas to enlarge and become able to accommodate larger floods. Areas that frequently receive heavy rains may develop lush vegetative cover, which helps to break the force of the raindrops and reduce soil erosion; also, the extensive root system of such vegetation holds the soil in place against erosion by overland flow. Stream channels may be large so that they can accommodate large-volume floods.

Figure 6.8. The Yellow River in China has built natural levees much higher than surrounding floodplains, which has caused catastrophic floods in the past. The top two photos show views of the Yellow River and surrounding floodplains, and the bottom photo shows details of the reinforcements that local villagers have built to keep the levee from collapsing. Photos by T. Kusky.

When the natural system is altered in urban areas, the result can be dangerous. Many municipalities have paved over large parts of drainage basins and covered much of the recharge area with roads, buildings, parking lots, and other structures. The result is that much of the water that used to seep into the ground and infiltrate the groundwater system now flows overland into stream channels, which may themselves be modified or even paved over. The net effect of these alterations is that flash floods may occur much more frequently than in a natural system since more water flows into the stream system than before the alterations. The floods may occur with significantly lower amounts of rainfall as well, and since the water flows overland without slowly seeping into the ground, the flash floods may reach urban areas more quickly than the floods did before the alterations to the stream system. Overall the effect of urbanization is faster, stronger, bigger floods, which have greater erosive power and do more damage. It is almost as if the natural environment responds to urban growth by increasing its ability to return the environment to its natural state.

Many examples of the effects of urbanization on flood intensity have been documented from California and the desert southwest. Urban areas such as Los Angeles, San Diego, Tucson, Phoenix, and other cities have documented the speed and severity of floods from similar rainfall amounts along the same drainage basin. What these studies have documented is that the floodwaters rise much more quickly after urbanization, and they rise up to four times the height of pre-urbanization, depending on the amount of paving over of the surface. The increased speed at which the floodwaters rise, and the increased height to which they rise, are directly correlated with the amount of land surface that is now covered over by roads, houses, and parking lots that block infiltration.

In natural systems, floods gradually wane after the highest peak passes, and the slow fall of the floodwaters is related to the stream system being recharged by groundwater that seeped into the shallow surface area during the heavy rainfall event. However, in urbanized areas the floodwaters not only rise quickly, but also recede faster than in the natural environment. This is attributed to the lack of groundwater continuing to recharge the stream after the flood peak in urbanized areas.

Many other modifications to stream channels have been made in urbanized areas with limited success in changing nature's course to suit human needs. Many stream channels have been straightened, which only causes the water to flow faster and have more erosive power. Straightening the stream course also shortens the stream length and thereby steepens the gradient. The stream may respond to this by aggrading and filling the channel with sediment in an attempt to regain the natural gradient.

RESOURCES AND ORGANIZATIONS

Print Resources

Arnold, J. G., Boison, P. J., and Patton, P. C. "Sawmill Brook—An Example of Rapid Geomorphic Change Related to Urbanization." *Journal of Geology* 90 (1982), 115–66.

Baker, V. R. "Stream-Channel Responses to Floods, with Examples from Central Texas." *Geological Society of America Bulletin* 88 (1977), 1057–71.

Bamford, D. "Algeria Army Helps Flood Victims." *BBC News*, November 18, 2001.

Belt, C. B., Jr. "The 1973 Flood and Man's Constriction of the Mississippi River." *Science* 189 (1975), 681–84.

Berger, E., and Freemantle, T. "Tropical Storm Alison Threatens Texas and Louisiana Coasts." *Houston Chronicle*, June 9, 2001.

Booth, D. B. "Stream Channel Incision Following Drainage Basin Urbanization." *Water Resources Bulletin* 26 (1990), 407–17.

CNN. "Rain Eases As Italy, Switzerland Battle Floods." October 20, 2000.

Collier, M. P., Webb, R. H., and Andrews, E. D. "Experimental Flooding in the Grand Canyon." *Scientific American* (January 1997), 82–89.

Coomarasamy, J. "Floods Leave Trail of Destruction." *BBC News*, November 15, 2001.

Gordon, N. D., McMahon, T. A., and Finlayson, B. L. *Stream Hydrology: An Introduction for Ecologists*. New York: John Wiley and Sons, 1992, 526 pp.

Hesse, L. W., and Sheets, W. "The Missouri River Hydrosystem." *Fisheries* 18 (1993), 5–14.

Jacobson, R. B., Femmer, S. R., and McKennery, R. A. "Land Use Changes and the Physical Habitat of Streams: A Review with Emphasis on Studies within the U.S. Geological Survey Federal-State Cooperative Program." U.S. Geological Survey Circular 1175, 2001, 63 pp.

Junk, W. J., Bayley, P. B., and Sparks, R. E. "The Flood Pulse Concept in River-Floodplain Systems." *Canadian Special Publication Fisheries and Aquatic Sciences* 106 (1989), 110–27.

Leopold, L. B. *A View of the River*. Cambridge, Mass.: Harvard University Press, 1994, 29 pp.

Leopold, L. B., and Wolman, M. G. *River Channel Patterns—Braided, Meandering, and Straight*. U.S. Geological Survey Professional Paper 282-B, 1957, 39 pp.

Lewin, J., Bradley, S., and Macklin, G. "Historical Valley Alluviation in Mid-Wales." *Geological Journal* 18 (1983), 331–50.

Maddock, T., Jr. "A Primer on Floodplain Dynamics." *Journal of Soil and Water Conservation* 31 (1976), 44–47.

Masterman, S. "Deadly Waters, Rains Lead to Floods, Mudslides in Europe." *ABC News*, October 16, 2000.

National Oceanographic and Atmospheric Administration (NOAA). *The Great Flood of 1993*. Silver Spring, Md.: NOAA, 1994.

Noble, C. C. "The Mississippi River Flood of 1973," in D. R. Coates, ed., *Geomorphology and Engineering*. London: Allen and Unwin, 1980, 79–98.

Parsons, A. J., and Abrahams, A. D. *Overland Flow-Hydraulics and Erosion Mechanics*. London: UCL Press Ltd., 1992, 391 pp.

Rosgen, D. *Applied River Morphology*. Pasoga Springs, Colo.: Wildland Hydrology, 1996, 352 pp.

Schumm, S. A. *The Fluvial System*. New York: Wiley—Interscience, 1977, 338 pp.

U.S. Geological Survey. *Storm and Flood of July 31–August 1, 1976, in the Big Thompson and Cache la Poudre River Basins, Larimer and Weld Counties, Colorado*. U.S. Geological Survey Professional Paper, 1115, 1979.

Non-Print Resources

Videos

The Earth Revealed—Rivers, 1992, Annenbergh/CPH, 30 mins.

The Earth Revealed—Running Water, Annenberg/CPH, 1992, 30 mins.

Fatal Flood, WGBH/PBS Home Video, 2001, 60 mins.

Flood!, 1996, NOVA/WGBH, 60 mins.

Web Sites

Federal Emergency Management Agency
http://www.fema.gov/
This agency responds to major flooding and other disaster events. This site contains information on risks, current hazards, and weather.

FEMA (Federal Emergency Management Agency) and ESRI (Environmental Systems Research Institute)
http://www.esri.com/hazards/
FEMA and ESRI have formed a national part-

nership in part aimed at providing multi-hazard maps and information to U.S. residents, business owners, schools, community groups, and local governments via the Internet. The information provided here is intended to assist in building disaster resistant communities across the country by sharing geographic knowledge about local hazards.

The National Weather Service, FEMA, and the Red Cross Flood Information
http://www.nws.noaa.gov/om/brochures/ffbro.htm
These organizations maintain a Web site dedicated to providing information on how to prepare for floods, describing floods of various types, and providing in-depth descriptions of flood warnings and the types of home emergency kits that families should keep in their homes.

NOVA Online
http://www.pbs.org/wgbh/nova/flood/
NOVA has produced an interactive Web site that features text, graphics, and sounds of floods. It discusses floods from some of the world's major rivers including the Yellow, Nile, and Mississippi, and talks about the climate and weather systems that bring fertile soils and fatal floods. A list of resources is available on the site.

U.S. Geological Survey
http://water.usgs.gov/
The survey monitors weather and stream flow conditions nationwide, as well as groundwater levels. Their Web site also contains information on water quality and water use and maps and charts of water use related issues. The Web site has connections to other related Web sites. The site also has a number of fact sheets about specific floods and regional issues (http://water.usgs.gov/wid/index-hazards.html). Floods in California, on the Colorado and Mississippi Rivers are highlighted.

Organizations

National Weather Service
National Operational Hydrologic Remote Sensing Center
1735 Lake Drive W.
Chanhassen, MN 55317
952–361–6610
http://www.nohrsc.nws.gov

Red Cross Headquarters
P.O. Box 37243
Washington, D.C. 20013
877–272–7337
http://www.redcross.org/
The American Red Cross responds and provides assistance to victims of disasters ranging from apartment fires to floods, earthquakes, hurricanes and tornadoes.

U.S. Army Corps of Engineers
US Army Engineer Research and Development Center
U.S. Army Corps of Engineers
3909 Halls Ferry Road
ATTN: CEERD-PA-Z
Vicksburg, MS 39180-6199
http://www.wes.army.mil/EL/flood/fl93home.htmlERDC research provides support in: mapping and terrain analysis; infrastructure design, construction, operations, and maintenance; structural engineering; cold regions and ice engineering; coastal and hydraulics engineering; environmental quality; geotechnical engineering; and high performance computing and information technology. The Environmental Laboratory, http://www.wes.army.mil/EL/flood/fl93home.html, monitors floods and other environmental disasters.

CHAPTER 7

Coastal Hazards

INTRODUCTION: HAZARDS ASSOCIATED WITH THE INTERACTION OF OCEAN, LAND, AND ATMOSPHERE

Coastlines offer some of the best scenery and recreation areas of our country, yet they also rank among the most hazardous areas to live. Coastal regions are prone to strikes by hurricanes, are subject to beach erosion, and are constantly changing dynamic environments. Despite this, coastal areas have become increasingly densely populated, even areas along fragile coastal islands. More than half of the population of the United States now lives within a one-day drive of the coast. Many of these areas are disasters waiting to happen, as one strike by a moderate-sized hurricane could totally devastate these regions, removing homes and potentially killing many people.

Coastal regions are prone to changes at several different geologic time scales. Sea level is presently gradually rising. Sea level rises and falls by hundreds of feet over periods of millions of years, forcing the position of the coastline to move inland or seaward by many tens of miles over long time periods. Most people do not think that changes over these time frames will affect their lives, but a sea level rise of even a foot or two, which is possible over decades, can cause extensive flooding, increased severity of storms, and landward retreat of the shoreline.

Most changes to coastal areas are noticeable on shorter time scales and may occur in single disastrous events. A large winter storm or summer hurricane can remove tens of feet from coastal cliffs, can cause new channels to open up in barrier beach complexes, can cause extensive flooding and removal of buildings with strong tidal and storm surges, and can remove the vegetation from coastal areas. Damage, loss of property, and loss of life have risen dramatically along U.S. coastlines during each of the past few decades as more and more of the population moves to coastal areas that are prone to hurricanes and storm damage. Many long-range climate forecasts predict that hurricanes will become more severe and more frequent over the next few decades, so these hazards will only grow more severe, and more planning is needed to mitigate the effects of storm damage (Figure 7.1).

Other hazards along coastlines are associated with changes on intermediate time scales—the slow but steady passage of coastal currents along a shoreline moves tons of sand daily along the coast, removing headlands and depositing the sand as bars and in both natural and manmade harbors. Cliffs gradually collapse, and the shoreline gradually retreats. Waves constantly pound the coast, eroding cliffs and seawalls and carrying the sand and rock particles along the coast or out to sea.

CHARACTERISTIC LANDFORMS OF THE COASTAL ENVIRONMENT

The coastal zone is defined as the area between the highest point at which tides influence the land to the point at which the first breakers form offshore. The area of interaction of the land, sea, and atmosphere is actually much larger and includes much of the continental shelf, where long-wavelength storm waves feel the sea bottom, and extends inland to areas where onshore winds blow sand. In some cases, the coastal zone may be considered to extend far into the continental interior when coastal storms and hurricanes bring ocean energy far from the shoreline. Many subenvironments are characteristic of the coastal zone, including beaches, bays, estuaries, tidal wetlands, lagoons, and barrier islands (Figure 7.2).

Beaches are accumulations of sediment exposed to wave action along a coastline. The beach extends from the limit of the low tide line to the point inland where the vegetation and landforms change to that which is typical of the surrounding region. Many beaches merge imperceptibly with grasslands and forests, whereas others end abruptly at cliffs or other permanent features. Beaches may occupy bays between headlands, they may form elongate strips attached (or detached, in the cases of barrier islands) to the mainland, or they may form spits that project out into the water. Beaches are very dynamic environments and are always changing, being eroded and redeposited constantly from day to day and from season to season. Beaches are typically eroded to thin strips by strong winter storms, and built up considerably during summer when storms tend to be less intense. The processes controlling this seasonal change are related to the relative amounts of energy in summer and winter storms: Summer storms (except for hurricanes) tend to have less energy than winter storms, so they have waves with relatively short wavelengths and heights. These waves gradually push the offshore sands up to the beachface, building the beach throughout the summer (Figure 7.2). In contrast, winter storms have more energy with longer wavelength and higher amplitude waves. These large waves break on the beach, erode the beach face, and carry the sand seaward, depositing it offshore (Figure 7.2). Some beaches are bordered by steep cliffs, many of which are experiencing active erosion (Figure 7.1). The erosion is a function of waves undercutting the base of the cliffs and oversteepening the slopes, which attempt to recover to the angle of repose by rainwater erosion or slumping from the top of the cliff. This ero-

Figure 7.1. Erosion of coastal cliffs at Plymouth, Massachusetts. (a) Shows steep cliffs made of sand and gravel, being eroded by erosion of the cliff base during storms. This oversteepens the slope beyond the angle of repose, leading to erosion at the top of the cliff. (b) Shows landowners' attempts at reducing erosion by building rock walls, constructing snow fences, and planting vegetation. (c) Shows how ineffective these remedial measures can be if only applied to small parts of the regional cliff—if one land owner builds a rock wall and the neighbor does not, erosion will work its way around the rock wall and cause the cliff and wall to continue to collapse. (d) Illustrates slumps at top of cliff undermining structures and making them uninhabitable. Photos by T. Kusky

sion can be dramatic, with many tens of feet removed during single storms. The material that is eroded from the cliffs replenishes the beaches, and without the erosion the beaches would not exist. Coarser material is left behind as it can not be transported by the waves or tidal currents, and these typically form a rocky beach with a relatively flat platform known as a wave-cut terrace.

Estuaries are transitional environments between rivers and the sea, where fresh water mixes with seawater and is influenced by tides. Most were formed when sea levels were lower, and rivers carved out deep valleys that are now flooded by water. They are typically bordered by tidal wetlands and are sensitive ecological zones that are prone to disturbances by pollution, storms, and overuse by people.

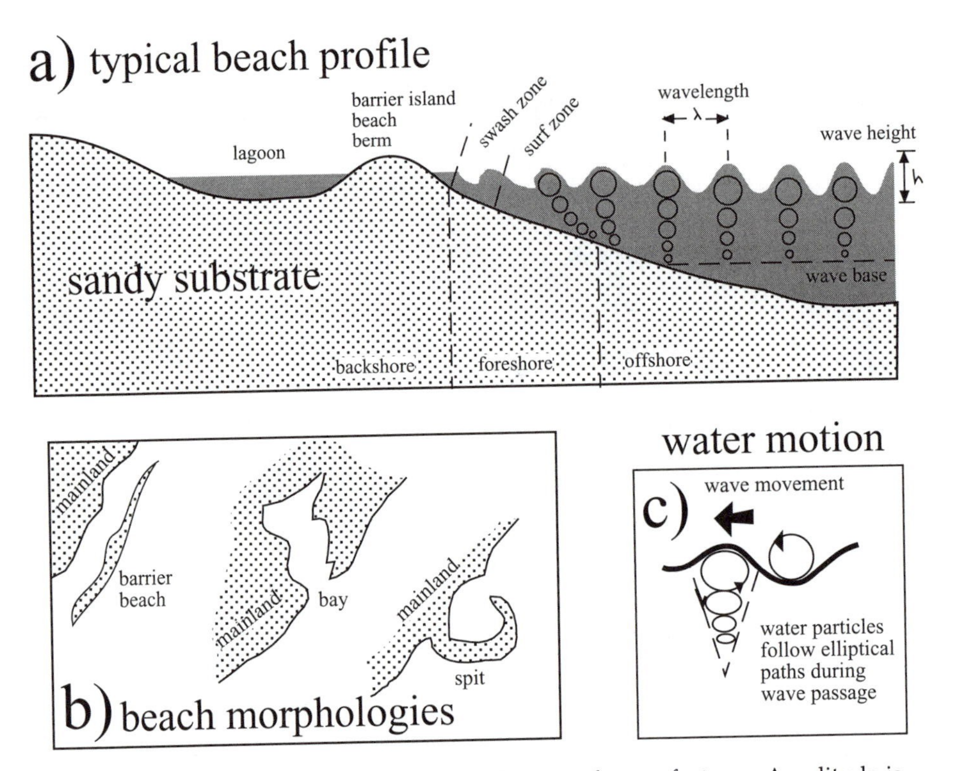

Figure 7.2. (a) Typical beach profile and definition of wave features. Amplitude is the distance between two successive crests, wave height is the elevation difference between the tops and bottoms of crests or troughs, and wave base is the distance at which the elliptical orbits of water particles dies out is half the wavelength. Diagram also shows main elements of the beach and (b) shows general forms of barrier beaches, bays, and spits. (c) Illustrates particle motion paths during passage of a wave.

Lagoons, bays, and sounds separate the mainland from barrier islands that are long, narrow offshore beaches. Barrier islands are common along the east coast of the United States (e.g., the south shore of Long Island, New York; Atlantic City, New Jersey; Outer Banks, North Carolina; and Galveston, Texas).

ORIGIN OF COASTAL HAZARDS: COASTAL PROCESSES

To appreciate and mitigate the hazards posed by living along coastlines, the processes that affect the interaction of the land, water, and atmosphere in this critical, dynamic, and ever-changing environment must be understood. There are several major factors that influence the development and potential of hazards along the coastline, including waves, tides, storms, changing sea levels, and human-induced changes to the shoreline.

Waves

Waves are the most important contributor of energy to the shoreline and are thus the most important process for understanding coastal changes. Waves are generated by winds that blow across the water surface, and the frictional drag of the wind on the surface transfers energy from the air to the sea, which is expressed as waves. The waves may travel great distances across entire ocean basins, and they may be thought of as energy in motion. This energy is released or transferred to the shoreline when the waves crash on the beach. It is this energy that is able to move entire beaches and erode cliffs grain-by-grain, slowly changing the appearance of the beach environment.

When waves are generated by winds over deep water, often from distant storms, they develop a characteristic spacing and height known as the wavelength and wave height (Figure 7.2). The wave crest is the highest part of the wave, and the wave trough is the low point between waves. Wavelength is the distance between successive crests or troughs, the wave height is the vertical distance between troughs and crests, and a wave's amplitude is one-half of its total height. Wave fronts are (imaginary) lines drawn parallel to the wave crests, and the wave moves perpendicular to the wave fronts. The time (in seconds) that it takes successive wave crests to pass a point is known as the wave period.

The height, wavelength, and period of waves is determined by how strong the wind is that generates the waves, how long it blows, and the distance over which it blows (known as the fetch). As any sailor can tell you, the longer and stronger and the greater the distance the wind blows over, the longer the wavelength, the greater the wave height, and the longer the period.

It is important to remember that waves are energy in motion, and the water in the waves does not travel along with the waves. The motion of individual water particles as a wave passes is roughly a circular orbit that decreases in radius with depth below the wave. You have probably experienced this motion when sitting in waves at the ocean's shore, feeling yourself moving roughly up and down or in a circular path as the waves pass you.

The motion of water particles in waves changes as the water depth decreases and the waves approach shore. At a depth equal to roughly one-half of the wavelength, the circular motion induced by the wave begins to feel the sea bottom, which exerts a frictional drag on the wave. This changes the circular particle paths to elliptical paths and causes the upper part of the wave to move ahead of the deeper parts. Eventually, the wave becomes oversteepened and begins to break as the wave crashes into shore in the surf zone (Figure 7.2). In this surf zone, the water is actually moving forward, causing the common erosion, transportation, and deposition of sand along beaches.

Most coastlines are irregular and have many headlands, bays, bends, and changes in water depth from place to place. These variables cause waves that are similar in deep water to approach the shoreline differently in different places. You may have noticed, on beaches, how waves may come ashore gently in one place yet form nice breakers down the beach. These changes can be attributed to changes in water depth and the shape of the beach. Wave refraction occurs when a straight wave front approaches a shoreline obliquely. The part of the wave front that first feels shallow water (with a depth of less than one-half of the wavelength, known as the wave base) begins to slow down while the rest of the wave continues at its previous velocity. This causes the wave front to bend, or be refracted (Figure 7.3). This effect tends to cause waves to bend toward headlands and concentrate energy in those places while concentrating less energy in bays. Material is eroded from the headlands and is transported to and deposited in the bays.

When waves approach a beach obliquely, a similar phenomenon occurs. Even though the waves are refracted, they may still crash onto the beach obliquely, moving sand particles sideways up the beach with each wave. As the wave returns to the sea in the backwash, the wave energy has been transferred to the shoreline (and has moved the sand grains); gravity is the driving force moving the water and sand, which then moves directly downhill. The net result is that sand particles move obliquely up and straight down the beach slope, moving slightly sideways along the beach with each passing wave. It is common for individual sand grains to move almost a mile per day along beaches through a process known as longshore drift. If the supply or transportation route of the sand particles is altered by human activity, such as the construction of sea walls or groins, the beach will respond by dramatically changing in some way, as discussed below.

Tides

In some places, currents induced by tides are the most significant factor controlling development of the beach and shoreline environment. These places include tidal inlets, passages between islands and the mainland, and areas with exceptionally large tidal ranges (e.g., Bay of Fundy, Gulf of Alaska; and Cook Inlet, North Sea). Tides are responsible for depositing deltas on the lagoonal and oceanward sides of tidal inlets, and for moving large amounts of sediment in regions with high tidal ranges.

Storms

Storms can cause some of the most dramatic and rapid changes to the coastal zones, and represent one of the largest, most unpredictable hazards to people living along coastlines. Storms include hurricanes, which form in the late summer and fall, and extratropical lows, which form in the late fall through spring. Hurricanes originate in the tropics and, for North America, migrate westward and northwestward before turning back to the northeast to return to the cold North Atlantic, weakening the storm. North Atlantic hurricanes are driven to the west by the trade winds and bend to the right because the Coriolis force makes moving objects curve to the right in the Northern hemisphere. Hurricane paths are further modified by

Figure 7.3. Wave Refraction, location unknown. ©Joel W. Rogers/CORBIS

other weather conditions, such as the location of high and low pressure systems and their interaction with weather fronts. Extratropical lows (also known as coastal storms and nor'Easters) move eastward across North America and typically intensify when they hit the Atlantic and move up the coast. Both types of storms rotate counterclockwise, and the low pressure at the centers of the storms raises the water several to several tens of feet. This extra water moves ahead of the storms as a storm surge that represents an additional height of water above the normal tidal range. The wind from the storms adds further height to the storm surge, with the total height of the storm surge being determined by the length, duration, and direction of wind, plus how low the pressure gets in the center of the storm. The most destructive storm surges are those that strike low-lying communities at high tide, as the effects of the storm surge and the regular astronomical tides are cumulative. Add high winds and large waves on top of the storm surge, and coastal storms and hurricanes are seen as very powerful agents of destruction. They are capable of removing entire beaches and rows of homes, causing great amounts of cliff erosion and causing significant redistribution of sands in dunes and the back beach environment. Very precise prediction of the height and timing of the approach of the storm surge is necessary to warn coastal residents of when they need to evacuate and when they may not need to leave their homes.

Like many natural catastrophic events, the heights of storm surges to strike a coastline are statistically predictable. If the height of storm surges is plotted on a semilogarithmic plot, with the height plotted in a linear interval and the frequency (in years) plotted on a logarithmic scale, then a linear slope results. What this means is that communities can plan for storm surges of certain heights to occur once every fifty, 100, 300, or 500 years, although there is no way to predict when the actual storm surges will occur. It must be remembered that this is long-term statistical average, and that one, two, three, or more 500-year events may occur over a relatively short period, but averaged over a long time, the events average out to once every 500 years.

Storms are known to open new tidal inlets where none existed previously (without regard to whether or not any homes were present in the path of the new tidal inlet) and to close inlets previously in existence. Storms also tend to remove large amounts of sand from the beach face and redeposit it in the deeper water offshore (below wave base), but this sand tends to gradually move back on to the beach in the intervals between storms when the waves are smaller. In short, storms are extremely effective modifiers of the beach environment, although they are unpredictable and dangerous.

Changing Sea Levels

Sea level has risen and fallen by hundreds of feet many times in Earth's history, and it is presently slowly rising at about one foot per century, but may be accelerating due to the effects of global warming. The causes of sea level rise and fall are complex (see Chapter 9). These include growth and melting of glaciers, changes in the volume of the mid-ocean ridges, and other complex interactions of the distribution of the continental landmass in mountains and plains during periods of orogenic and anorogenic activity.

Rising sea levels cause the shoreline to move landward, whereas a sea level fall causes the shoreline to move oceanward. With the present sea level rise, coastal cliffs are eroding, barrier islands are migrating (or being submerged if they were heavily protected from erosion), beaches are moving landward, and estuaries are being flooded by the sea. At some point in the not-too-distant future, low-lying coastal cities will be flooded under several feet of water, and eventually the water could be hundreds of feet deep. Cities including New York, Washington, Houston, London, Shanghai, Tokyo, and Cairo will be inundated, and the world's nations need to begin to plan how to handle this inevitable geologic hazard. This topic is discussed at length in Chapter 9.

About 70 percent of the world's sandy beaches are being eroded. The reasons for this include rising sea levels, increased storminess, decrease in sediment transport to beaches from the damming of rivers, and perhaps shifts in global climate belts. Construction of seawalls to reduce erosion of coastal cliffs also causes a decreased supply of sand to replenish the beach, so it also increases beach retreat. Pumping of groundwater from coastal aquifers also results in coastal erosion, because pumping causes the surface to subside, leading to a relative sea level rise.

When sea level rises, beaches try to maintain their equilibrium profile and move each beach element landward. A sea level rise of one inch is generally equated with a landward shift of beach elements of more than four feet. Most sandy beaches worldwide are retreating landward at rates of twenty inches to three feet per year, consistent with sea level rises of an inch every ten years.

Human-induced Changes to the Shoreline

People are modifying the shoreline environment on a massive scale with the construction of new homes, resorts, and structures that attempt to reduce or prevent erosion along the beach. These modifications have been changing the dynamics of the beach in drastic ways, and, most often, result in degradation of the beach. In many cases, obstacles are constructed that disrupt the transportation of sand along the beach in longshore drift. This causes sand to build up at some locations and to be removed from other locations further along the beach. Some of the worst culprits are groins, or walls of rock, concrete, or wood built at right angles to the shoreline that are designed to trap sand from longshore drift and replenish a beach. The problem with groins is that they stop the longshore drift, causing the sand to accumulate on the updrift side and to be removed from the downdrift side. Groins also set up conditions favorable for the formation of riptides, which tend to take sand (and unsuspecting swimmers) offshore out of the longshore drift system. The result of groin construction is typically a few triangular areas of sand next to rocky groins along what was once a continuous beach. Little or no sand will remain in the areas

Figure 7.4. Groins built for erosion protection jut out from the shores of the sandy beach on the Baltic Sea, Osteebad Ahrenshoop, Mecklenburg, Germany. ©Wolfgang Kaehler/CORBIS

on the downdrift sides of the groins (Figure 7.4). Therefore, when groins are constructed, it usually becomes necessary to begin an expensive program of artificial replenishment of beach sands to fill in the areas that were eroded by the new pattern of longshore drift set up by the groins.

Construction or stabilization of inlets though barrier islands or beaches often includes the construction of jetties on either side of the channel to prevent sand from entering and closing the channel. Like groins, these jetties prevent sand transportation by longshore drift, causing beaches to grow on the updrift side of the jetty. Sand that used to replenish the beach on the downdrift side gets blocked or washed around the jetty into the tidal channel, where it moves into the lagoon to form tidal deltas. The result is that the beaches on the downdrift side of the jetties become sand-starved and thin, eventually disappearing.

Seawalls are also hazardous to maintaining a healthy beach. Seawalls have been constructed along many U.S. beaches in attempts to limit damage to buildings from high storm surges and waves. If a seawall is built along the base of a cliff at the backside of a beach, it may protect the cliff from erosion for some time, but it also accelerates erosion of the beach because it removes the source of sediment that previously replenished the beach. Eventually, the beach becomes narrower and lower, and the waves attack the seawall by undermining it until it collapses. If seawalls are only built along small sections of cliffs, such as along single plots owned by individual homeowners, the problems may be worse, such as in the case of Plymouth, Massachusetts, which is discussed later in this chapter. Seawalls have also been constructed along large sections of beaches by municipalities that suffered severe storm damage and want to try to reduce the effects of the next storm. These extensive seawalls are a sentence of death for the beach, because they not only remove the source of sediment replenishment for the beach, but they also accelerate erosion by causing the wave energy to be reflected off the beach and down into the sand, causing it to be removed. Eventually, the erosion undermines the seawall, causing it to collapse.

Sand dunes in the back beach area are naturally vegetated and covered with sea grasses and other plants. These dunes form a natural and effective barrier to high storm surges and waves, protecting the back beach environment from the worst effects of these storms. Recreational overuse of many beaches has resulted in people trampling many of these grasses or driving vehicles through them, creating many unvegetated paths and patches. When wind blows over these patches and paths, the wind removes sand from these areas in a process called *deflation,* resulting in a lowering of the dune height. When this happens, the dunes are no longer effective barriers for storm surges and waves, and the waves begin to wash over the dunes during storms. The storm waters will seek out the lowest point in the dune field and surge through this gap, deeply eroding a storm surge channel through the dunes. The result then is that the dunes are no longer able to protect the back beach and lagoon from the storm's effects, and a storm channel and delta is formed from the sand that used to form the dunes.

Another human-induced change to the coastal environment results from the construction of dams along rivers that feed into the ocean. These rivers once fed sediment to the ocean, replenishing the beach. Since the dams have been built, there is more fresh water available for new coastal residents to drink, but the beaches are receiving less replenishment and are rapidly shrinking. The reservoirs behind the dams are filling up with the sediment that used to feed the beaches, so eventually the reservoirs will not be effective and will have to be dredged (thus replenishing the beaches) or abandoned.

HOW ARE COASTAL HAZARDS STUDIED, AND WHO STUDIES THEM?

Coastal hazards are studied by a large number of organizations, including those from the government, university, private, and industry sectors. The Federal Emergency Management Agency (FEMA) studies coastal hazards and makes contingency plans on how to deal with specific anticipated hazards and catastrophes. The U.S. Army Corps of Engineers performs extensive studies of coastal hazards, and designs and builds structures such as seawalls, groins, and dams to try to prevent or reduce the hazards at specific sites. Many coastal cities and towns commission private studies of coastal hazards and contract with construction firms to help reduce erosion. The university research sector is very active in coastal studies, with many geologists and other scientists who have research specialties in various aspects of the coastal zone.

EXAMPLES OF COASTAL HAZARDS—EROSION OF COASTAL CLIFFS AT PLYMOUTH, MASSACHUSETTS

The town of Plymouth, Massachusetts, is located between Cape Cod and Boston and has many high coastal cliffs composed largely of old glacial deposits (Figure 7.1). These cliffs are being actively eroded at a rate of a few inches to a few tens of feet per year, and the residents and communities have attempted a wide variety of techniques to try to eliminate or reduce the erosion. So far all have failed, and houses perched on top of the coastal cliffs continue to lose acreage and eventually fall down the cliffs. Some of the efforts of local residents to prevent erosion on their property have actually compounded the problems for the area as a whole. Some residents have built seawalls on their property, others have constructed jetties, others have built fences or planted vegetation, and other have built pipes and drainages to divert water to try to cut the erosion from runoff (Figure 7.1).

When seawalls are built in one small, restricted area but not in adjacent areas, the rate of erosion on the coastal cliffs may actually be increased because the energy that hits the wall is reflected downward and removes sand from the beach in front of the wall. The waves also hit the edges of the wall and severely erode the unprotected adjacent cliff section. The erosion induced by the seawall eventually causes the beach in front of it to become steeper and narrower, which allows bigger waves to run up the now narrower beach and crash directly into the wall. This will eventually destroy the wall with repeated pounding by waves, sand, and rocks. Eventually the cliffs erode behind the seawall, causing it to collapse (Figure 7.1). Other residents have constructed fences to try to reduce the velocity of the waves, but these are typically washed away by the first nor'Easter.

All of these attempts have locally and temporarily slowed the erosion of the cliffs, but over the course of many years they have been insignificant or have even accelerated erosion. These problems and attempts at solutions are not unique to Plymouth: Many places in New England have similar problems, including Cape Cod, Situate, and many others, yet none have been able to completely halt the erosion.

TIDES

Tidal currents can be hazardous, particularly to those who may not be familiar with an area and not know the tidal range and the speed at which the incoming tides can fill an area. High tidal ranges with a 20–40 foot difference between the low tide and high tide level are known from a number of locations around the world, including the Bay of Fundy (Canada), the North Sea, and parts of the Gulf of Alaska. These high tidal ranges produce a number of unusual "mushroom" shaped sea stacks in the Bay of Fundy (Figures 7.5a and 7.5b), giving visitors a sense of the tidal range and danger. Other areas with high coastal ranges are flat, with little expression of the tidal variation, and other places exhibit high rates of coastal cliff erosion (Figure 7.5c).

A good example of the hazards associated with tidal currents is provided by parts of Cook Inlet, Alaska. Just outside of Anchorage, a part of Cook Inlet known as Turnagain Arm has tidal ranges in the 30–35 foot range. At low tides, Turnagain Arm consists of mudflats with tidal channels, and when the tide comes in it typically does so with a tidal bore, a wave front that moves quite rapidly up the drowned fiord—faster than a person can run. The mud that forms the tidal flats is composed largely of glacial flour, a mixture of mud and fine silt that has some unusual properties. The mud is *thixotropic,* meaning that it appears to be fairly rigid when it is held or is still, but when it is shaken or disturbed, it rapidly turns into a fluid. Every year, tourists or other unsuspecting residents walk out onto the mud, which can suddenly start to grip your boots. Most people's reaction is to wiggle to get loose, but this only makes the mud more fluid, and causes the person to sink further into the mud. The mud surrounds the person and has an incredibly strong suction force such that makes it virtually impossible to pull the person out of the mud without tearing off their limbs. One possible way to escape is seek emergency assistance to pump water next to the person, and remove them with a helicopter. But Turnagain Arm is fairly remote, even though it is close to a major city, and it takes time for emergency crews to arrive. The tides are fast and they come in at 20–40 miles per hour and quickly rise many feet, drowning the person stuck in the mud.

LONGSHORE SEDIMENT TRANSPORT AND MANAGEMENT

Longshore drift of sand is strongly affected by changes to the beach, including the construction of groins and jetties. There are many examples of where groins were constructed and beaches downdrift from the groins disappeared. Likewise, there are many examples of where jetties were constructed at the ends of inlets, disrupting the flow of sand to the downdrift beaches, which then gradually disappear. The sand that used to replenish these beaches either moves into the channel and forms deltas in the lagoon or is carried to deep water by rip currents set up by the jetties.

One interesting example of the consequences of building jetties and breakwaters is provided by the boat harbor at Santa Barbara, California. In the downtown of scenic Santa Barbara, a beautiful harbor with many well-cared-for pleasure boats sits behind rock walls built as breakwaters, which are designed to keep large waves out of the harbor. The breakwaters, however, caused sand to build up and form a spit on the breakwaters. The spit continued to grow by longshore drift and the sand curved around the end of the spit, filling the harbor with sand. Now the harbor needs nearly constant dredging to keep it navigable, and a dredging barge sweeps back and forth in the harbor, taking the sand from one side of the harbor entrance

Figure 7.5. Erosion of coastal cliffs is a severe problem in many parts of the world, and also produces some spectacular landforms. The top two photos show mushroom-shaped seastacks produced by cliff erosion by 40–70 foot high tides in the Bay of Fundy, New Brunswick Canada. The bottom photo shows rapid erosion of steeply tilted beds in coastal cliffs at Old Wives Point, Nova Scotia. Photos by T. Kusky.

to the other, artificially keeping the process of longshore drift operating.

EFFECTS OF STORMS AND HURRICANES

Some of the most rapid and severe damage to coastal regions is inflicted by hurricanes (Figures 7.6 and 7.7), and these storms are responsible for the largest numbers of deaths in coastal disasters. The number of deaths from hurricanes has been reduced dramatically in recent years with our increased ability to forecast the strength and landfall of hurricanes and our ability to monitor their progress with satellites. However, the costs of hurricanes in terms of property damage have greatly increased as more and more people build expensive homes along the coast. The greatest number of deaths from hurricanes has been from effects of the storm surge. Storm surges typically come ashore as a wall of water that rushes onto land at the forward velocity of the hurricane, as the storm waves on top of the surge are pounding the coastal area with additional energy. For instance, when Hurricane Camille hit Mississippi in 1969 with 200 m.p.h. winds, a twenty-four-foot-high storm surge moved into coastal areas, killing most of the 256 people that perished in this storm.

Winds and tornadoes account for more deaths, and heavy rains from hurricanes also cause considerable damage. Flooding and severe erosion is often accompanied by massive mudflows and debris avalanches, such as those caused by Hurricane Mitch in Central America in 1998. In a period of several days, Mitch dropped twenty-five to seventy-five inches of rain on Nicaragua and Honduras, initiating many mudslides that were the main cause of the more than 11,000 deaths from this single storm. One of the worst events was the filling and collapse of a caldera on Casitas Volcano: When the caldera could hold no more water, it gave way, sending mudflows (lahars) cascading down on several villages and killing 2,000 people.

Galveston Island, Texas, 1900

The deadliest natural disaster to affect the United States was when a category 4 hurricane hit Galveston Island, Texas, on September 8, 1900. Galveston is a low-lying barrier island located south of Houston, and in 1900 it served as a wealthy port city. Residents were warned of an approaching hurricane, and many evacuated the island to move to relative safety inland. However, many remained on the island. In the late afternoon, the hurricane moved in to Galveston and the storm surge hit at high tide, covering the entire island with water. Even the highest point on the island was covered with one foot of water. Winds of 120 miles per hour destroyed wooden buildings as well as many of the stronger brick buildings. Debris from destroyed buildings crashed into other structures, demolishing them and creating a moving mangled mess for residents trapped on the island. The storm continued through the night, battering the island and city with thirty-foot-high waves. In the morning, residents who found shelter emerged to see half of the city totally destroyed and the other half severely damaged. Worst of all, thousands of bodies were strewn everywhere, 6,000 in

Figure 7.6. Remains of Atlantic House Restaurant at Folly Beach, North Carolina, after Hurricane Hugo, 1989. Photo Credit: NOAA

all. There was no way off the island as all boats and bridges were destroyed, so survivors were in danger of disease from the decaying bodies. When help arrived from the mainland, the survivors needed to dispose of the bodies before cholera set in, so they put the decaying corpses on barges and dumped them at sea. However, the tides and waves soon brought the bodies back, and they eventually had to be burned in giant funeral pyres built from wood from the destroyed city. Galveston was rebuilt with a seawall that was supposed to protect the city—however, in 1915, another hurricane struck Galveston, claiming 275 additional lives.

The Galveston seawall has since been reconstructed and is higher and stronger, although some forecasters believe that even this seawall will not be able to protect the city from a category 5 hurricane. The possibility of a surprise storm hitting Galveston again is not so remote, as shown by the surprise tropical storm of early June 2001. Weather forecasters were not successful in predicting the rapid strengthening and movement of this storm (see Chapter 6), which dumped 23–48 inches of rain on the Galveston-Houston area and attacked the seawall and coastal structures with huge waves and thirty mile per hour winds. Twenty-two people died in the area from the surprise storm of 2001.

Netherlands Storm Surge, 1953

On February 1, 1953, a huge storm surge inundated parts of the Netherlands, drowning 1,800 people. The Netherlands has a low-lying coastal plain in the north,

Table 7.1
Saffir-Simpson Scale of Hurricane Damage Potential

Category	Wind speeds (m.p.h.)	Storm surge height (feet)	Typical damage
1	74–95	4–5	Trees damaged, unanchored mobile homes moved
2	96–110	6–8	Trees uprooted, roofs torn off some buildings
3	111–130	9–12	Foliage removed from trees, large trees uprooted, structural damage to some buildings
4	131–155	13–18	All signs blown down, mobile homes completely destroyed, extensive damage to roofs, doors, windows, extensive inland flooding, damage to lower floors of coastal structures
5	>155	>18	Severe damage to windows, doors, roofs, of large sturdy buildings, small buildings blown away, major damage or total destruction of all buildings less than 15 feet above sea level, within 2,000 feet of shore

and the flood covered this. This unexpectedly high surge rose nine feet above anything that was anticipated, causing the collapse of dikes that were constructed to keep high tides and storm surges out of the coastal environment.

Bangladesh

Bangladesh is a densely populated low-lying country sitting mostly at or near sea level between India and Myanmar (Burma). It is a delta environment, built where the Ganges and Brahmaputra Rivers drop their sediment eroded from the Himalayan Mountains. Bangladesh is frequently flooded from high river levels, with up to 20 percent of the low-lying country being underwater in any year. It also sits directly in the path of many Bay of Bengal tropical cyclones (hurricanes) and has been hit by seven of the nine most deadly hurricane disasters in the history of the world. On November 12 and 13, 1970, a category 5 typhoon hit Bangladesh with 155 mile per hour winds and a twenty-three-foot-high storm surge that struck at the astronomically high tides of a full moon. The result was devastating, with 400,000 human deaths and half a million farm animals perishing. Again in 1990, another cyclone hit the same area, this time with a twenty-foot storm surge and 145 mile per hour winds, killing another 140,000 people and another half-million farm animals.

Why do so many people continue to move to an area that is prone to repeated strikes by tropical cyclones? Bangladesh is an overpopulated country, with a population density fifty times as great as that of farm lands typical of the Midwestern United States. Its per capita income is only $200 per year. Despite these drawbacks, the delta region of Bangladesh is the most fertile in the country and farmers can expect to yield three crops of rice per year, making it an attractive place to live despite the risk of perishing in a storm surge. With the continued population explosion in Bangladesh and the paucity of fertile soils in higher grounds, the delta region will continue to be farmed by millions and it will continue to be hit by tropical cyclones like the 1970 and 1990 events.

Hurricane Andrew, Florida, 1992

Hurricane Andrew was the most destructive hurricane in the United States's history, causing more than $30 billion in damage in August 1992. Andrew began to form over North Africa and grew in strength as it was driven across the Atlantic by the trade winds. On August 22, Andrew had grown to hurricane strength and moved across the Bahamas with 150 mile per hour winds, killing four people. On August 24, Andrew smashed into southern Florida with a nearly seventeen-foot-high storm surge, steady winds of 145 miles per hour and gusts to 200 miles per hour. Andrew's path took it across a part of south Florida that had hundreds of thousands of poorly constructed homes and trailer parks, and the hurricane's wind caused the most destruction (Figure 7.7). Andrew destroyed 80,000 buildings, severely damaged another 55,000, and demolished thousands of cars, signs, and trees. Thirty-three people died in south Florida. By August 26, Andrew had traveled across Florida and was losing much of its strength, but then moved back into the warm waters of the Gulf of Mexico and regained that strength. On August 26, Andrew hit land again, this time in Louisiana with 120 mile per hour winds, where it killed another fifteen people. Andrew's winds stirred up the fish-rich marshes of southern Louisiana, where the muddied waters were agitated so much that the decaying organic material overwhelmed the oxygen-rich surface layers, suffocating millions of fish. Andrew then continued to lose strength, but dumped flooding rains over much of Mississippi.

New Orleans and the Atchafalaya River

The city of New Orleans in southern Louisiana rests several feet below sea level and is surrounded by dikes to keep the Gulf of Mexico and Mississippi River water out of the city streets and homes. There is limited access to the city, including a long narrow bridge over Lake Pontchartrain. If a fast moving and powerful hurricane ever makes a direct hit on New Orleans, it will likely initiate one of the most serious natural disasters that this nation has ever seen. Authorities estimate that it will take seventy-two hours to evacuate the city—and even with today's satellites and ability to forecast weather, it is unlikely that an accurate hurricane trajectory could give residents of "The Big Easy" enough time to evacuate. The city will be flooded, dikes will be breached and destroyed, and thousands will perish. But even worse may be in store.

If the hurricane is powerful enough to add flooding rains to areas upriver of New Orleans, the course of the Mississippi River could be changed. The Mississippi has switched courses many times over the past several thousand years, and when that

Figure 7.7. Hurricane Andrew devastated much of southern Florida in 1992. This photo shows Dadeland Mobile Home Park, Florida, after passage of Andrew. Photo Credit: NOAA

happens the active part of the delta stops receiving sediment replenishment and subsides below sea level. Right now, the Mississippi follows a sinuous path past New Orleans and drains into the Gulf of Mexico. Years ago, the Mississippi started to build a new shorter course to the Gulf of Mexico along the Atchafalaya River (Figure 6.6). The river would now be flowing along this shorter, lower energy route if the Army Corps of Engineers had not built a series of dikes and dams to keep the flow in the Mississippi. However, if a hurricane causes a large flood in New Orleans, the engineering works could be breached and the very course of the river could be changed. This would cause New Orleans and the surrounding part of the delta to subside well below sea level forever.

EFFECTS OF RISING SEA LEVELS

Sea level is rising presently at a rate of one foot per century, although this rate seems to be accelerating. This rising sea level will obviously change the coastline dramatically—a one-foot rise in sea level along a gentle coastal plain can be equated with a 1,000-foot landward migration of the shoreline. What will the world look like when sea levels rise significantly? Think of our low-lying cities, like New York, London, and most other cities in the world. Then think of Venice. The world's rich farmlands on coastal plains, like the east coast of the United States, northern Europe, Bangladesh, and much of China will be covered by shallow seas. If sea levels rise more significantly, like they have in the past, then vast parts of the interior plains of North America will be covered by inland seas and much of the world's climate and vegetation zones will be shifted to different latitudes. These subjects are covered in Chapter 9.

It is clear that governments must begin to plan for how to deal with rising sea levels, yet very little has been done so far. It is time that groups of scientists and government planners begin to meet to first understand the magnitude of the problem, then to study and recommend which tactics they must initiate to mitigate the effects of rising sea levels. Can massive dikes and seawalls be built? Will our cities look like Venice, with abandoned lower levels, submerged subways, and boats in the street? These scenes will probably not become reality for a long time, but these events are inevitable on geological time scales.

WHAT CAN BE DONE TO REDUCE COASTAL HAZARDS?

Great progress has been made in predicting when and where certain coastal hazards, especially hurricanes, may become disasters. These advances have been made primarily because of our ability to monitor the progress of these superstorms with satellites and reconnaissance aircraft. These changes have caused a significant decrease in the number of deaths from hurricanes during the last century. However, the population of the United States is continuing to migrate to the coastal regions in record numbers, and the dollar value of property losses has been steadily increasing. People who build on the coast must come to realize that their structures are temporary, and society in general should not be made to bear the costs of rebuilding the structures that individuals knowingly constructed in hazardous zones. Insurance companies, states, and the federal government should stop insuring and subsidizing the rebuilding of communities devastated by storms, coastal erosion, and other coastal hazards, and begin to let these areas return to their natural, preconstruction state. Many communities and states have begun to initiate such policies, such as the Cape Cod National Seashore and the Outer Banks of North Carolina.

To better understand which areas are prone to the most severe coastal hazards, we as a nation should begin a serious program of mapping hazard zones and areas of high risk. This could be carried out effectively by the U.S. Geological Survey and university groups, perhaps working together to complete the work on a national scale. When the country's coasts are categorized as to where it is relatively safe to build and where buildings should never be built, then people will be able to enjoy the beauty of the coast in a safer way that will not cost billions of dollars when disaster strikes.

WHAT IF YOU ARE IN A COASTAL DISASTER?

It is very likely that some day you will be in a hurricane and will hear warnings on the radio and television of where the approaching storm may hit, how powerful it may be, and what you should do. If you are in a coastal area that is difficult to evacuate, it is best to leave early. Experiences

Table 7.2
The Ten Worst Coastal Disasters (1970–1999) in Terms of Deaths and Destruction

Type, Location	Date	Deaths
Typhoon, Bangladesh	1970	300,000
Hurricane Gorky, Bangladesh	1991	140,000
Storm flooding, Venezuela	1999	50,000
Typhoon, Orissa, India	1999	15,000
Typhoon, Bangladesh	1985	10,000
Typhoon, India	1977	10,000
Hurricane Mitch, Honduras	1998	10,000
Typhoons, Thema and Uring, Philippines	1991	6,304
Typhoon Linda, Vietnam	1977	3,840
Hurricane, Réunion	1978	3,200

from recent hurricanes have shown that monstrous traffic jams can block highways leading away from the coast for days, and a car is not a very safe place to try to ride out a storm in. Take the advice of the authorities and leave early, give up your beach vacation or home, and get to higher ground so that you can live to visit the coast again.

Coastal cliffs should be avoided during and after storms and wet periods, as they are most prone to collapse and slumping when the cliff material is saturated in water. Buildings should not be placed near cliff edges or at their bases.

Other coastal hazards do not usually strike so rapidly. Longshore drift is a slow process, and greater understanding of how the beaches are a dynamic environment will help engineers understand that it is not effective to build groins, seawalls, and other artificial barriers in attempts to slow erosion or divert waves. It is best to leave the beach alone, realizing it is a dynamic and complex environment. Changing one thing on a beach will change many others, and it is impossible to isolate and change one characteristic without changing the very nature of the beach.

COASTAL HAZARD STATISTICS

What Are the Costs of Coastal Hazards?

Coastal hazards cost billions of dollars each year in the United States, and when a coastal disaster strikes, a single event such as a hurricane can cost tens of billions of dollars (Table 7.3), or roughly the same amount that the Gulf War cost the United States and its allies combined. Despite this, recent federal budgets have allocated no funds to be set aside for coastal (or any other) natural disasters and have used this to help build estimates of the federal budget surplus.

Coastal disasters also take their toll in lives, with thousands of lives lost in a typ-

Table 7.3
The Ten Worst Coastal Disasters (1970–1999) in Terms of Costs to Insurance Companies

Type, Location	Date	Cost in 1999 $USD (millions of dollars)
Hurricane Andrew, United States	1992	30,068
Typhoon Mireille, Japan	1991	14,122
Winter Storm Daria, Europe	1990	5,882
Hurricane Hugo, United States	1989	5,664
Winter Storm Lothar, Europe	1999	4,500
October 15 Storm, Europe	1987	4,415
Winter Storm Vivian, Europe	1990	4,088
Hurricane Georges, Caribbean and United States	1998	3,622
Typhoon Bart, Japan	1999	2,980
Hurricane Floyd, United States and Bahamas	1999	2,360

ical year of hurricanes and coastal disasters. In some years the toll is astronomical, such as when severe typhoons hit Bangladesh, claiming hundreds of thousands of lives (Table 7.2). Loss of life in hurricanes has been gradually decreasing with better forecasting, whereas at the same time the cost of physical damages to the coastal infrastructure has increased because of the increased population and use of the coastal region.

What Agencies Deal with Coastal Hazards?

Several U.S. government agencies help forecast, manage, and mitigate coastal disasters. Included are NOAA, the National Oceanographic and Atmospheric Administration, which is responsible for long-range forecasts and short-term prediction of storm paths. NOAA's mission is to conserve and wisely manage the nation's coastal and marine resources. The Federal Emergency Management Agency (FEMA) deals with emergency management and preparation, and issues warnings and evacuation orders when coastal storms and disasters appear imminent. The U.S. Army Corps of Engineers is charged with keeping the nation's waterways navigable and typically constructs barriers that are designed to reduce beach erosion, cliff erosion, and river and estuarine flooding. The U.S. Geological Survey performs scientific surveys of the coastal environment, prepares maps and reports about coastal hazards, and works with other government agencies to help civic authorities know how to deal with coastal hazards and disasters.

RESOURCES AND ORGANIZATIONS

Print Resources

Dean, C. *Against the Tide: The Battle for America's Beaches*. New York: Columbia University Press, 1999, 279 pp.

Dolan, R., Godfrey, P. J., and Odum, W. E. "Man's Impact on the Barrier Islands of North Carolina." *American Scientist* 61 (1973), 152–62.

Federal Emergency Management Agency. *Coastal Construction Manual.* Washington, D.C.: FEMA, 1986, 104 pp.

Horton, T. "Hanging in the Balance—Chesapeake Bay." *National Geographic* 183, no. 6 (1993), 2–35.

Kaufman, W., and Pilkey, O. H., Jr. *The Beaches Are Moving.* Durham, N.C.: Duke University Press, 1983.

King, C.A.M. *Beaches and Coasts.* London: Edward Arnold Publishers, 1961, 403 pp.

Komar, P. D., ed. *CRC Handbook of Coastal Processes and Erosion.* Boca Raton, Fla.: CRC Press, 1983, 123–50.

Miele, P. T. "Air Force Plans for Fourth Cliff Ride Wave of Worries over Erosion," July 29, 2001, sec. 1.

Williams, S. J., Dodd, K., and Gohn, K. K. "Coasts in Crisis." U.S. Geological Survey Circular 1075, 1990, 32 pp.

Non-Print Resources

Videos

The Beach: A River of Sand, 1996, Encyclopedia Britannica, 21 mins.

The Beaches Are Moving: The Drowning of America's Shoreline, 1990, Environmental Media, 55 mins.

Organizations

Federal Emergency Management Agency (FEMA)
500 C Street, SW
Washington, D.C. 20472
202–646–4600
http://www.fema.gov
FEMA is the nation's premier agency that deals with emergency management and preparation, and it issues warnings and evacuation orders when coastal storms and disasters appear imminent. FEMA maintains a Web site that is updated at least daily and includes information on hurricanes, floods, fires, national flood insurance, and disaster prevention, preparation, emergency management. Divided into national and regional sites. Also contains information on maps, costs of disasters, and how to do business with FEMA.

National Oceanographic and Atmospheric Association (NOAA)
Department of Commerce
14th Street & Constitution Avenue, NW, Room 6013
Washington, DC 20230
202–482–6090
http://www.noaa.gov/
NOAA conducts research and gathers data about the global oceans and atmosphere, and about space and the sun, and applies this knowledge to science and service that touches the lives of all Americans. NOAA's mission is to describe and predict changes in the earth's environment, and to conserve and wisely manage the nation's coastal and marine resources. NOAA's strategy consists of seven interrelated strategic goals for environmental assessment, prediction, and stewardship. These include 1) providing advance short-term warnings and forecast services, 2) implementing seasonal to interannual climate forecasts, 3) assessing and predicting decadal to centennial change, 4) promoting safe navigation, 5) building sustainable fisheries, 6) recovering protected species, and 7) sustaining healthy coastal ecosystems. NOAA runs a Web site that includes links to current satellite images of weather hazards, issues warnings of current coastal hazards and disasters, and has an extensive historical and educational service.

U.S. Army Corps of Engineers
441 G. Street, NW
Washington, DC 20314
202–761–0008
202–761–1683
http://www.erdc.usace.army.mil/
http://www.usace.army.mil/public.html#REstate
The U.S. Army Corps of Engineers has an emer-

gency response unit, set for responding to environmental, coastal, and other disasters. The Headquarters Office (http://www.usace.army.mil/where.html#Headquarters) is a good place to start a search for any specific problem.

U.S. Geological Survey
U.S. Department of the Interior
Reston, VA 20192
703–648–4000
http://www.usgs.gov/
The USGS is responsible for making maps of the coastal zone and assessing hazard risks.

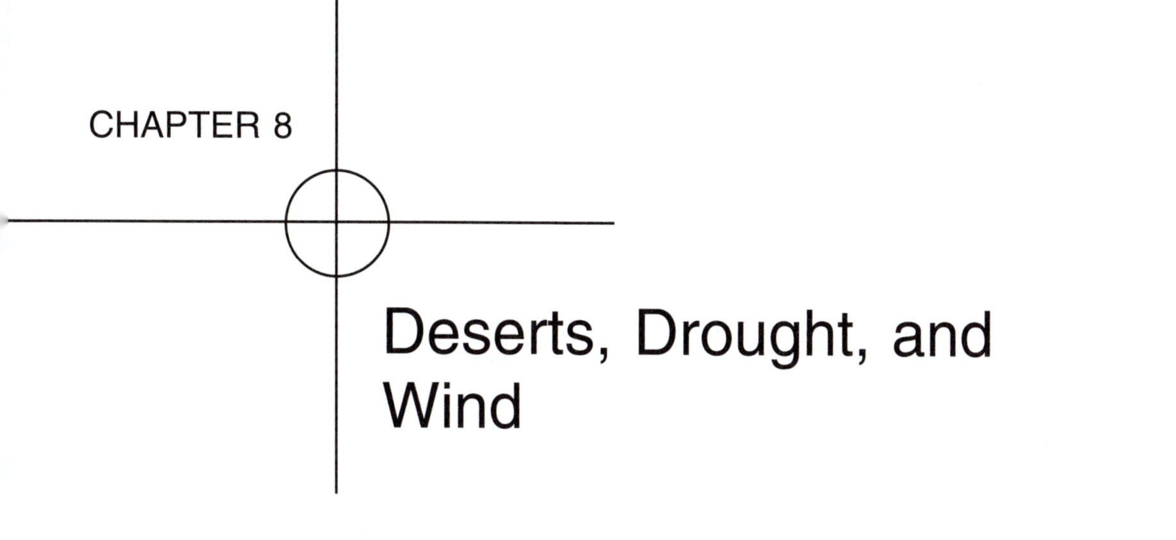

CHAPTER 8

Deserts, Drought, and Wind

Deserts are the driest places on Earth, by definition receiving less than one inch (250 mm) of rain per year. Most deserts are so dry that more moisture is able to evaporate than falls as precipitation. At present, about 30 percent of the global landmass is desert, and the United States has about 10 percent desert areas. With changing global climate patterns and shifting climate zones, much more of the planet is in danger of becoming desert.

Most deserts are also hot, with the highest temperature on record being 136° F in the Libyan Desert. With high temperatures, the evaporation rate is high; many deserts are capable of evaporating twenty times the amount of rain that falls, and some places, like much of the northern Sahara, are capable of evaporating 200–300 times the amount of rain that falls in rare storms. Deserts are also famous for large variations in the daily temperature, sometimes changing as much as 50–70° between day and night (called a diurnal cycle). These large temperature variations are enough to shatter boulders in some cases. Deserts are also windy places and are prone to sand and dust storms. The winds arise primarily because the heat of the day causes warm air to rise and expand, and other air must rush in to take its place. Airflow directions also shift frequently between day and night in response to the large temperature difference between these times, and between any nearby water bodies, which tend to stay at the same temperature between day and night.

Many different types of deserts are located in all different parts of the world. Some deserts are associated with patterns of global air circulation, and others form because they are in continental interiors far from any sources of moisture. Deserts may form on the "back" or leeward side of mountain ranges, where downwelling air is typically dry, or they may form along coasts where cold upwelling ocean currents lower the air temperature and lower its ability to hold moisture. Deserts may also form in polar regions where extremely dry and cold air has the ability to evaporate (or sublimate) much more moisture than falls as snow in any given year. Parts of Antarctica have not had any significant ice or snow cover for thousands of years!

Deserts have a distinctive set of landforms and hazards associated with these landforms. The most famous desert landform is a sand dune, which is a mobile accumulation of sand that shifts in response to wind. Deserts tend to be very windy places, and some of the hazards in deserts are associated with sand and dust carried by the wind. Dust eroded from deserts can be carried around the globe and is a significant factor in global climate and sedimentation. Some sandstorms can be so fierce that they can remove the paint from cars or the skin from an unprotected person. Other hazards in deserts are associated with flash floods, debris flows, avalanches, extreme heat, and extreme fluctuations in temperature.

Droughts are different from deserts. A drought is a prolonged lack of rainfall in a region that typically gets more rainfall. If a desert normally gets a small amount of rainfall and it still is getting little rainfall, then it is not experiencing a drought. In contrast, a different area that receives more rainfall than the desert may be experiencing a drought if it normally receives significantly more rainfall than it is at present. A drought-plagued area may become a desert if the drought is prolonged. Droughts are the most severe natural hazard in terms of their severity, area affected, social impact, and other long-term effects. Droughts can cause widespread famine, loss of vegetation, loss of life and livelihood, and eventual death or mass migrations of entire populations.

Droughts may lead to conversion of previously productive lands to desert, in a process called desertification. Desertification may occur if the land is stressed prior to or during the drought, typically from poor agricultural practices, overuse of groundwater and surface water resources, and overpopulation. Global climate goes through several different variations that can cause belts of aridity to shift back and forth with time. The Sahel region of Africa has experienced some of the more severe droughts in recent times—for instance, in November 2002 the Prime Minister of Ethiopia warned that a catastrophic drought and famine is threatening more than 15 million people with starvation in his country. This was followed in December 2002 by a warning from the Director of the U.N.'s World Food Program that at least 38 million people in the region are at risk of starvation. The Middle East and parts of the desert southwest of the United States are overpopulated and their environments are stressed. If major droughts occur in these regions, major famines could result and the land may be permanently desertified.

HOW AND WHERE DO DESERTS FORM?

More than 35 percent of the land area on the planet is arid or semiarid, and these deserts form an interesting pattern on the globe that reveals clues about how they form. There are six main categories of desert based on their geographic location with respect to continental margins, oceans, and mountains (Figure 8.1).

Trade Wind or Hadley Cell Deserts

Many of the world's largest and most famous deserts are located in two belts be-

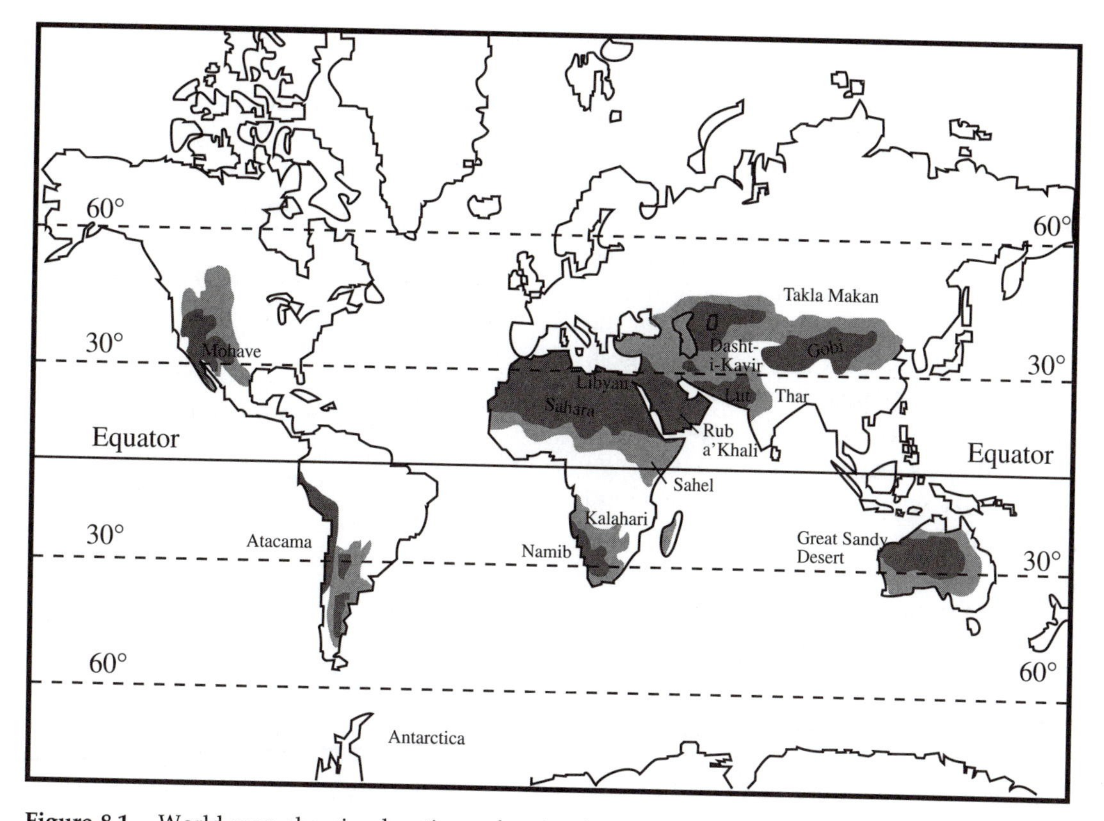

Figure 8.1. World map showing locations of major deserts. Note that most deserts are located between 20° and 40° latitude, being controlled by the global atmospheric circulation system.

tween 15° and 30° north and south latitude (Figure 8.1). Included in this group of deserts are the Sahara, the world's largest desert, and the Libyan Desert of North Africa. Other members of this group include the Syrian Desert; the Rub a'Khali (Empty Quarter) and Great Sand Desert of Arabia; the Dasht-i-Kavir, Lut, and Sind Deserts of southwest Asia; the Thar Desert of Pakistan; and the Mojave and Sonoran Deserts of the United States. In the southern hemisphere, deserts that fall into this group include Kalahari Desert of Africa and the Great Sandy Desert of Australia, and this effect contributes to the formation of the Atacama Desert in South America, the world's driest place.

The location of these deserts is controlled by a large-scale atmospheric circulation pattern driven by energy from the sun. The sun heats equatorial regions more than high-latitude areas, which causes large-scale atmospheric upwelling near the equator. This air rises, and as it rises it becomes less dense and can hold less moisture, which helps form large thunderstorms in equatorial areas. This drier air then moves away from the equator at high altitudes, cooling and drying more as it does until it eventually forms two circum-global downwelling belts between 15–30° N and S latitude (Figure 1.2). This cold downwelling air is dry and has the ability to hold much more water than it has brought with it on its circuit from the equator. These belts of circulating air are known as Hadley cells and are responsible for the formation of many of the world's largest, driest deserts. As this air completes its circuit back to the equator, it forms dry winds that heat up as they move toward the equator. The dry winds dissipate existing cloud cover and allow more sunlight to reach the surface, which then warms even more.

Deserts formed by global circulation patterns are particularly sensitive to changes in global climate, and seemingly small changes in the global circulation may lead to catastrophic expansion or contraction of some of the world's largest deserts. For instance, the sub-Saharan Sahel has experienced several episodes of expansion and contraction of the Sahara, displacing or killing millions of people in this vicious cycle. When deserts expand, croplands are dried up and livestock and people can not find enough water to survive. Desert expansion is the underlying cause of some of the world's most severe famines.

Continental Interior/Midlatitude Deserts

Some places on Earth are so far from ocean moisture sources that by the time weather systems reach them, most of the moisture they carry has already fallen. This effect is worsened if the weather systems have to rise over mountains or plateaus to reach these areas, because cloud systems typically lose moisture as they rise over mountains. These remote areas therefore have little chance of receiving significant rainfalls. The most significant desert in this category is the Taklimakan-Gobi region of China, resting south of the Mongolian steppe on the Alashan plateau, and the Karakum of western Asia (Figure 8.1). The Gobi is the world's northernmost desert, and it is characterized by 1,000-foot-high sand dunes made of coarser-than-normal sand and gravel, built up layer by layer by material moved and deposited by the wind. It is a desolate region, conquered successively by Genghis Khan, warriors of the Ming Dynasty, and then the People's Army of China. The sands are still littered with remains of many of these battles, such as the abandoned city of Khara Khoto. In 1372, Ming Dynasty warriors conquered this walled city by cutting off its water supply from the Black River, waiting, then massacring any remaining people in the city (see Webster, 2002).

Rainshadow Deserts

A third type of desert is found on the leeward (back) side of some large mountain ranges, such as the sub-Andean Patagonian Gran Chaco and Pampas of Argentina, Paraguay, and Bolivia (Figure 8.1). A similar effect is partly responsible for the formation of the Mojave and Sonoran deserts of the United States. These deserts form because as moist air masses move toward the mountain ranges, they must rise to move over the ranges. As the air rises it cools, and cold air can not hold as much moisture as warm air. The clouds thus drop much of their moisture on the windward (front) side of the mountains, explaining why places like the western Cascades and western Sierras of the United States are extremely wet, as are the western Andes in Peru. However the eastern lee sides of these mountains are extremely dry. The reason for this is that as the air rose over the fronts or windward sides of the mountains, it dropped its moisture as it rose. As the same air descends on the lee side of the mountains, it gets warmer and is able to hold more moisture than it has left in the clouds. The result is that the air is dry and it rarely rains. This explains why places like the eastern sub-Andean region of South America and the Sonoran and Mojave deserts of the western United States are extremely dry.

Rainshadow deserts tend to be mountainous because of the way they form, and they are associated with a number of mass wasting hazards such as landslides, debris flows, and avalanches (see Chapter 5). Occasional rainstorms that make it over the blocking mountain ranges can drop moisture in the highlands, leading to flash floods coming out of mountain canyons and into the plains or intermountain basins on the lee side of the mountains.

Coastal Deserts

There are some deserts that are located along coastlines, where intuition would seem to indicate that moisture should be plentiful. However, the driest place on Earth is the Atacama Desert located along the coast of Peru and Chile (Figure 8.1). The Namib Desert of southern Africa is another coastal desert, which is known legendarily as the Skeleton Coast, because it is so dry that many of the large animals that roam out of the more humid interior climate zones perish there, leaving their bones sticking out of the blowing sands.

How do these coastal deserts form adjacent to such large bodies of water? The answer lies in the ocean currents, for in these places cold water is upwelling from the deep ocean, which cools the atmosphere. The effect is similar to rainshadow deserts, where cold air can hold less moisture, and the result is no rain.

Monsoon Deserts

In some places on the planet, seasonal variations in wind systems bring alternating dry and wet seasons. The Indian Ocean is famous for its monsoonal rains in the summer as the southeast trade winds bring moist air on shore. However, as the moisture moves across India, it loses moisture and must rise to cross the Aravalli Mountain Range. The Thar Desert of Pakistan and the Rajasthan Desert of India are located on the back or lee side of these mountains and do not generally receive this seasonal moisture (Figure 8.1).

Polar Deserts

A final class of deserts is the polar desert, found principally in the Dry Valleys and other parts of Antarctica; in parts of Greenland; and in northern Canada, including the Native American territory of Nunavut. Approximately 3 million square miles on Earth consists of polar desert environments. In these places, cold downwelling air lacks moisture, and the air is so dry that the evaporation potential is much greater than the precipitation. Temperatures do not exceed 10° C in the warmest months, and precipitation is less than one inch per year. As mentioned earlier, there are places in the Dry Valleys of Antarctica that have not been covered in ice for thousands of years.

Polar deserts are generally not covered in sand dunes but are marked by gravel plains or barren bedrock. Hazards to travelers in polar deserts include the effects of extreme cold such as hypothermia, frostbite, and dehydration. Polar deserts may also have landforms shaped by frost wedging, where alternating freeze-thaw cycles allow small amounts of water to seep into cracks and other openings in rocks. When the water freezes it expands, pushing large blocks of rock away from the main mountain mass. In polar deserts and other regions affected by frost wedging, large talus slopes may form adjacent to mountain fronts, and these are prone to frequent rock falls from frost wedging.

DESERT LANDFORMS

Desert landforms are some of the most beautiful on Earth, often presenting bizarre sculpted mountains, steep walled canyons, and regional gravel plains. They can also be some of the most hazardous landscapes on the planet. The regolith in deserts is thin, discontinuous, and much coarser-grained than in moist regions, and is produced predominantly by mechanical weathering (chemical weathering is only of minor importance because of the rare moisture in deserts). Also, the coarse size of particles produced by mechanical weathering produces steep slopes, eroded from steep cliffs and escarpments.

Much of the regolith that sits in deserts is coated with a dark coating of manganese and iron oxides, known as *desert varnish,* which is produced by a combination of microorganism activity and chemical reactions with fine manganese dust that settles from the wind.

Desert Drainage Systems

Most streams in deserts evaporate before they reach the sea. Most are dry for long periods of time and are subject to *flash floods* (see Chapter 6) during brief but intense rains. A joke about many small desert crossroads is that they get one inch of rain a year—and you should be there the day it happens! These flash floods transport most of the sediment in deserts and form fan-shaped deposits of sand, gravel, and boulders, which are found at the bases of many mountains in desert regions. These flash floods also erode deep, steep-walled canyons through the upstream mountain regions (Figure 8.2), which is the source of the boulders and cobbles found on the mountain fronts. Intermountain areas in deserts typically have finer-grained material, which is deposited by slower moving currents that represent the waning stages of floods as they expand into open areas between mountains after they escape out of mountain canyons.

Flash floods can be particularly hazardous in desert environments, especially when the floods are the result of distant rains. More people die in deserts from drowning in flash floods than from thirst or dehydration. In many cases, rain in faraway mountains may occur without people in downstream areas even aware that it is raining upstream. Rain in deserts is typically a brief but intense thunderstorm, and these can drop a couple of inches of rain in a short time. The water may then quickly move downstream as a wall of water in mountain canyons, sweeping away all loose material in its path. Any people or vehicles caught in such a flood are likely to be lost and swept away by the swiftly moving torrent.

Dry lake beds in low-lying flat areas, which may only have water in them once every few years, characterize many deserts. These are known as playas or hardpans, and they typically have deposits of white salts that formed when water from storms evaporated, leaving the lakes dry. There are more than 100 playas in the American southwest, including Lake Bonneville, which formed during the last ice age and now covers parts of Utah, Nevada, and Idaho. When there is water in these basins, they are known as playa lakes. Playas are very flat surfaces that make excellent racetracks and runways. The United States's space shuttles commonly land on Rogers Lake playa at Edwards Air Force Base in California.

The coarse-grained deposits of alluvium that accumulate at the fronts of mountain canyons are known as alluvial fans. Alluvial fans are very common in deserts, where they are composed of both alluvium and debris flow deposits. Alluvial fans are quite important for people in deserts because they are porous and permeable and they contain large deposits of groundwater. In many places, alluvial fans so dominate the land surface that they form a *bajada* (or slope) along the base of the mountain range, formed by fans that have coalesced to form a continuous broad alluvial apron.

Pediments represent different kinds of desert surfaces. They are surfaces sloping away from the base of a highland and covered by a thin or discontinuous layer of alluvium and rock fragments. These are erosional features formed by running water and typically cut by shallow channels. Pediments grow as mountains are eroded.

Inselbergs are steep-sided mountains or ridges that rise abruptly out of adjacent monotonously flat plains in deserts. Ayres Rock (also called Uluru) in central Australia is perhaps the world's best-known inselberg. These are produced by differential erosion that leaves behind, as a mountain, rocks that for some reason are more resistant to erosion.

WIND IN DESERTS

Wind plays a significant role in the evolution of desert landscapes. Wind erodes in two basic ways. *Deflation* is a process whereby wind picks up and removes material from an area, resulting in a reduction in the land surface. The process is akin to deflating a balloon below the surface, hence its name. *Abrasion* is a different process that occurs when particles of sand and of other sizes impact each other. Exposed surfaces in deserts are subjected to frequent abrasion, which is akin to sandblasting.

Yardangs are elongate, streamlined wind-eroded ridges, which resemble an overturned ship's hull sticking out of the water. These unusual features are formed

Figure 8.2. Photographs of desert streams and wadis. (a) Upper Wadi Dayqah, northern Oman, shows water ponded behind a small dam. Flash floods occur frequently here, generated by rains in the mountains in the background. (b) Shows a steep-walled canyon type of wadi in the Sinai (Wadi Zaghra). (c) Lower Wadi Dayqah, northern Oman, showing ephemeral flow, and steep-walled canyon, dangerous during flash floods. (d) Shows lower reaches of a wadi in northern United Arab Emirates, with the Persian (Arabian) Gulf in the background. (e) Wadi Watir, which drains into the Gulf of Aqaba at Nuweiba (Sinai), showing downcutting of alluvial fan by recent flash flood. (f) Exit of Wadi from Jabal Akhdar, Oman, where ancient village of Tanuf was constructed for its closeness to water sources. Photos by T. Kusky

by abrasion, which affects the ridge as long-term sandblasting along specific corridors. The sandblasting leaves erosionally resistant ridges but removes the softer material, which itself will contribute to sandblasting in the downwind direction and eventually contribute to the formation of sand, silt, and dust deposits.

Deflation is important on a large scale in places where there is no vegetation, and in some places the wind has excavated large basins known as deflation basins. Deflation basins are common in the United States from Texas to Canada as elongate (several kilometer long) depressions that are typically only three to ten feet deep. However, in some places like in the Sahara, deflation basins may be as much as several hundred feet deep.

Deflation by wind can only move small particles away from the source, since the size of the particle that can be lifted is limited by the strength of the wind, which rarely exceeds a few tens of miles per hour. Deflation therefore leaves boulders, cobbles, and other large particles behind. These get concentrated on the surface of deflation basins and other surfaces in deserts, leaving a surface known as desert pavement (Figure 8.3).

Desert pavements represent a long-term stable desert surface, and they are not particularly hazardous. However, when the desert pavement is broken by, for instance, being driven across, the coarse cobbles and pebbles get pushed beneath the surface and the underlying sands get exposed to wind action again. Driving across a desert pavement can raise a considerable amount of sand and dust, and if many vehicles drive across the surface, then it can be destroyed and the whole surface becomes active. This may lead to the generation of dust storms, and contribute to the formation of dunes and sand and dust covered surfaces (Figure 8.4).

A very striking large-scale example of this process was provided by events in the Gulf War of 1991. After Iraq invaded Kuwait in 1990, United States and Allied Forces massed hundreds of thousands of troops on the Saudi Arabian side of the border with Iraq and Kuwait and eventually mounted a multipronged counterattack on Kuwait City that led to its liberation. Several of the prongs circled far to the north, then turned around and came back south to Kuwait City. These prongs took many thousands of heavy tanks, artillery, and vehicles across a region of stable desert pavement, and the weight of these military vehicles destroyed the pavement to free Kuwait. Since the liberation, the steady winds from the northwest have continued, and this area that was once stable desert pavement and stable dune surfaces (covered with desert pavement and minor vegetation) has been remobilized, causing large sand dunes to form from the sand previously trapped under the pavement. Other dunes that were stable have been reactivated. Now, Kuwait City residents are bracing for what they call the second invasion of Kuwait, but this time the invading force is sand and dust, not a foreign army.

Several things have been considered to try to stabilize the newly migrating dunes. One consideration is to try to reestablish the desert pavement by spreading cobbles across the surface, but this is unrealistic because of the large area involved. Another

Figure 8.3. (a) Desert pavement, formed by the winnowing away of fine-grained sand by the wind, leaving a resistant veneer of heavier cobbles, boulders, and wind-scoured rocks known as ventifacts. (b) Driving across the fragile desert pavement disrupts the surface, exposing underlying sand to the wind, and causing sand mobility. Photos by T. Kusky

Figure 8.4. Examples of some of the hazardous effects of blowing sand and dust. (a) and (b) show two photos, separated by about 10 minutes, of a dust storm moving south down the Gulf of Suez. The sand covers everything and works its way into all open spaces. At times, the sand blasts through with enough power to remove paint from buildings and vehicles. (c) Shows camel in a wind and sand-swept terrain in the United Arab Emirates, with giant sand dunes visible in the background. (d) Shows wind blown sand and loess covering vast expanses of flat-lying sedimentary rocks of the Egma Plateau in southern Sinai. Photos by T. Kusky

proposition, being tested, is to spray petroleum on the migrating dunes to effectively create a blacktop or tarred surface that would be stable in the wind. This is feasible in oil-rich Kuwait, but not particularly environmentally friendly.

In China's Gobi and Taklimakan Deserts (Figure 8.1), a different technique to stabilize dunes has proven rather successful. Bales of hay are initially placed in a grid pattern near the base of the windward side of dunes, which decreases the velocity of the air flowing over the dune and reduces the transportation of sand grains over the slip surface. Drought-resistant vegetation is planted between the several-foot-wide grid of hay bales, and then when the dune is more stabilized, vegetation is planted along the dune crest. China is applying this technique across much of the Gobi and Taklimakan Deserts, protecting railways and roads. In northeastern China, this technique is being applied in an attempt to reclaim some lands that became desert through human activity, and they are constructing a 5,700-mile-long line of hay bales and drought-resistant vegetation. China is thus said to be building a new "Green Wall," which will be longer than the famous Great Wall of China and will hopefully prove more effective at keeping out invading forces (in this case, sand) from Mongolia.

Wind-blown Sand

Most people think of deserts as areas with lots of big sand dunes and continual swirling winds of dust storms. Really, dunes and dust storms (Figure 8.4) are not as common as depicted in popular movies, and rocky deserts are more common than sandy deserts. For instance, only about 20 percent of the Sahara Desert is covered by sand, and the rest is covered by rocky, pebbly, or gravel surfaces. However, sand dunes are locally very important in deserts (Figures 8.5 and 8.6), and wind is one of the most important processes in shaping deserts worldwide. Shifting sands are one of the most severe geologic hazards of deserts. In many deserts and desert border areas, the sands are moving into inhabited areas, covering farmlands, villages, and other useful land with thick accumulations of sand (Figure 8.7). This is a global problem, as deserts are currently expanding worldwide. The Desert Research Institute in China has recently estimated that in China alone, 950 square miles are encroached on by migrating sand dunes from the Gobi desert each year, costing the country $6.7 billion per year and affecting the lives of 400 million people (Webster, 2002).

Wind moves sand by *saltation*, which means that it moves in arced paths in a series of bounces or jumps. You can see this often by looking close to the surface in dunes on beaches or deserts (Figure 8.5). Wind typically sorts different sizes of sedimentary particles, forming elongate small ridges known as sand ripples that are very similar to ripples found in streams. Sand dunes are larger than ripples, up to 1,500 feet high, and are made of mounds or ridges of sand deposited by wind. These may form where an obstacle distorts or obstructs the flow of air, or they may move freely across much of a desert surface. Dunes have many different forms, but all are asymmetrical. They have a gentle slope that faces into the wind and a steep face that faces away from the wind. Sand particles move by saltation up the windward side and fall out near the top where the pocket of low-velocity air can not hold the sand anymore. The sand avalanches or slips down the leeward slope, known as the slip face. This keeps the slope at the angle of repose, 30–34°. The asymmetry of old dunes is used to tell the directions ancient winds blew.

The steady movement of sand from one side of the dune to the other causes the whole dune to migrate slowly downwind, typically about 80–100 feet per year, burying houses, farmlands, temples, and towns. Rates of dune migration of up to 350 feet per year have been measured in the Western Desert of Egypt and the Ningxia Province of China.

A combination of many different factors lead to the formation of very different

Figure 8.5. Photographs of some of the variation in dune morphology and size. (a) Shows a star dune from Ismalia, on the outskirts of Cairo, Egypt. (b) Shows giant, hundred-meter-tall dune from the Northern Sinai Peninsula. (c) and (d) show small and giant barchan dunes, from the United Arab Emirates. Photos by T. Kusky

types of dunes, each with a distinctive shape, potential for movement, and hazards. The main variables that determine a dune's shape are the amount of sand that is available for transportation, the strength (and directional uniformity) of the wind, and the amount of vegetation that covers the surface (Hack, 1941). If there is a lot of vegetation and little wind, no dunes will form. In contrast, if there is very little vegetation, a lot of sand, and moderate wind strength (conditions that you might find on a beach), then a group of dunes known as transverse dunes form, with the dune crests aligned at right angles to the dominant wind.

Barchan dunes (Figure 8.5) have crescent shapes and have horns pointing downwind; they form on flat deserts with steady winds and a limited sand supply. *Parabolic dunes* have a U-shape with the U facing upwind. These form where there is significant vegetation that pins the tails of migrating transverse dunes, with the dune being warped into a wide U-shape. These dunes look broadly similar to barchans, except that the tails point in the opposite direction. They can be distinguished because in

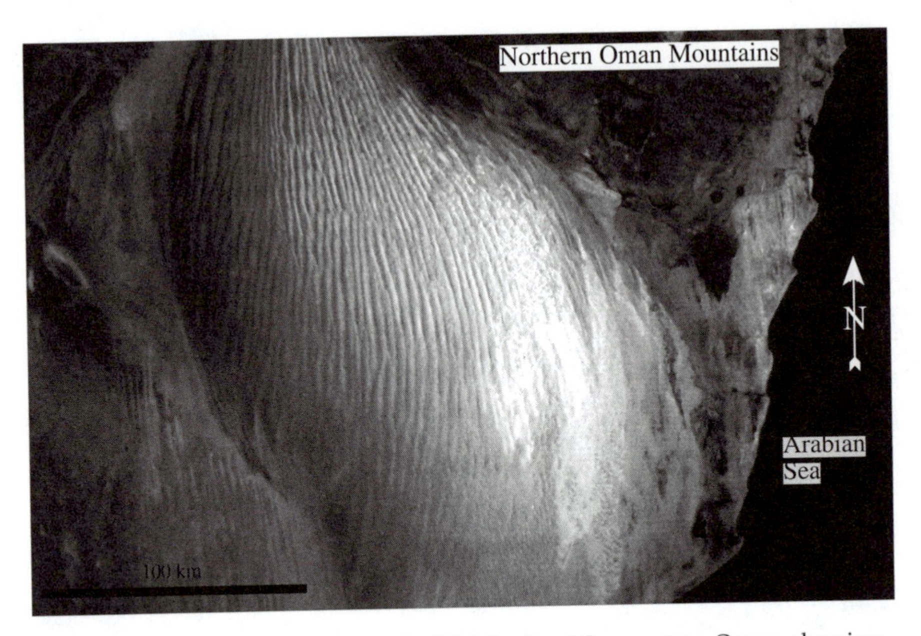

Figure 8.6. Landsat TM image of the Wahiba Sand Sea, eastern Oman, showing giant linear dunes elongate parallel to the dominant wind direction. The wind brings sand from south to north, where a wadi picks up the sand and returns it to the Arabian Sea. Landsat image processed at Boston University Center for Remote Sensing.

both cases, the steep side of the dune points away from the dominant winds direction. *Linear dunes* (Figure 8.6) are long, straight, ridge-shaped dunes elongate parallel to the wind direction. These occur in deserts with little sand supply and strong, slightly variable winds, and they are elongate parallel to the wind direction. *Star dunes* (Figure 8.5a) form isolated or irregular hills where the wind directions are irregular.

Wind-blown Dust

Strong winds that blow across desert regions sometimes pick up dust made of silt and clay particles and transport them thousands of kilometers from their source. For instance, dust from China is found in Hawaii, and the Sahara Desert commonly drops dust in Europe. This dust is a nuisance, has a significant influence on global climate, and has, at times such as in the Dust Bowl of the 1930s, been known to nearly block out the sun.

Loess is a name for silt and clay deposited by wind. It forms a uniform blanket that covers hills and valleys at many altitudes, which distinguishes it from deposits of streams. In Shaanxi Province, China, an earthquake that killed 830,000 people in 1556 had such a high death toll mostly because the people in the region built their homes out of loess. The loess formed an easily excavated material that hundreds of thousands of villagers cut homes into, essentially living in caves. When the earthquake struck, the fine-grained loess proved to be a poor building material and led to the large-scale collapse of homes.

Recently, it has been recognized that wind-blown dust contributes significantly to global climate. Dust storms that come out of the Sahara can be carried around the world and can partially block out some of the sun's radiation. The dust particles may also act as small nucleii for raindrops to form around, perhaps acting as a natural cloud seeding phenomenon. One interesting point to ponder is that as global warming increases global temperatures, the amount and intensity of storms increase and some of the world's deserts expand. Dust storms may serve to reduce global temperatures and increase precipitation. Might the formation of dust storms represent some kind of self-regulating mechanism, whereby the Earth moderates its own climate?

SHIFTING DESERTS AND DESERTIFICATION

Deserts expand and contract, reflecting global environmental changes. Many cultures and civilizations on the planet are thought to have met their demise because of desertification of the lands they inhabited and their inability to move with the shifting climate zones. *Desertification* is defined as the degradation of formerly productive land, and it is a complex process involving many causes. Climates may change, and land use on desert fringes may make fragile ecosystems more susceptible to becoming desert. Among civilizations thought to have been lost to the sands of encroaching deserts are several Indian cultures of the American southwest such as the Anasazi, and many peoples of the Sahel, where up to 250,000 people are thought to have perished in droughts in the late 1960s. Expanding deserts are associated with shifts in other global climate belts, and these shifts too are thought to have brought down several societies. Included are the Mycenaean civilization of Greece and Crete, the Mill Creek Indians of North America, and perhaps even the Viking colony in Greenland (e.g., Bryson and Murray, 1977). Many deserts are presently expanding, creating enormous drought and famine conditions like those in Ethiopia and Sudan.

Desertification is the invasion of a desert into nondesert areas (Figure 8.7), and is an increasing problem in the southwestern United States, in part due to human activities. This decreases water supply, vegetation, and land productivity; about 10 percent of the land in this country has been converted to desert in the last 100 years, and nearly 40 percent is well on the way. Desertification is also a major global problem, costing hundreds of billions of dollars per year. China, as discussed earlier, estimates that the Gobi Desert alone is expanding at a rate of 950 square miles per year, an alarming increase since the 1950s when the desert was expanding at less than 400 square miles per year.

Desertification is a global problem, which could drastically alter the distribution of agriculture and wealth on the globe. If deserts continue to expand, the wheat belt of the central United States will be displaced to Canada, the sub-Saharan Sahel

Figure 8.7. Fence Erected to Avert Dune Encroachment. ©Julia Waterlow; Eye Ubiquitous/CORBIS

will become part of the Sahara, and the Gobi Desert may expand out of the Alashan Plateau and Mongolian Plateau.

Desertification is a multistage process beginning with drought, crop and vegetation loss, and then the establishment of a desert landscape like those described above. Drought alone does not cause desertification, but misuse of the land during drought greatly increases the chances of a stressed ecosystem reverting to desert. Desertification is associated with a number of other symptoms, including destruction of native and planted vegetation, accelerated and high rates of soil erosion, reduction of surface and groundwater resources, increased saltiness of remaining water supplies, and famine. Desertification can be accelerated by human-induced water use, population growth, and settlement in areas that do not have the water resources to sustain the exogenous population. In the following section, several examples of drought and of regions at various stages on the way to desertification are described.

DROUGHT DISASTERS

Drought is very different from normal desert processes. A drought is a prolonged reduction in the amount of rainfall for a region. It is one of the slowest of all major natural disasters to affect people, but it is also among the most severe, causing more deaths, famine, and displacement than most other more spectacular disasters.

Drought often presages the expansion of desert environments, and regions like Africa's sub-Saharan Sahel have experienced periods of drought, desert expansion, and desert contraction several times in the past few tens of thousands of years. At present, much of the Sahara is expanding southward, and peoples of the Sahel have suffered immensely, and there is no immediate relief in sight.

Droughts typically begin imperceptibly, with seasonal rains often not appearing on schedule. Farmers and herdsmen may be waiting for the rains to water their freshly planted fields and to water their flocks, but the rains do not appear. Local water sources such as stream, rivers, and springs may begin to dry up until only deep wells are still able to extract water out of groundwater aquifers. This is typically not enough to sustain crops and livestock, so they begin to be slaughtered or to die of starvation and dehydration. Crops don't grow, and natural vegetation begins to dry up and die. Brushfires often come next, wiping away the dry brush. Soon people start to become weak, and they can not manage to walk out of the affected areas, so they stay and the weak, elderly, and young of the population may die off. Famine and disease may follow, killing even more people.

One recent example of this is highlighted by the Sudan, where years of drought have exacerbated political and religious unrest, and where opposing parties raid Red Cross relief supplies, sabotage the other side's attempts at establishing aid and agriculture, and cause the Sudanese people to suffer. The Sudan is in the sub-Saharan Sahel, which is a large region between about 14° and 18° N latitude, characterized by scrubby grasslands and getting on average between fourteen and twenty-three inches of rain per year. In the late 1960s, this amount of rainfall had fallen to about half of its historical average, and the 25 million people of the Sahel began suffering. One of the unpleasant aspects of human nature is that slow-moving, long-lasting disasters like drought tend to bring out the worst in many people. War and corruption often strike drought-plagued regions once relief and foreign aid begin to bring outside food sources. This food may not be enough to feed the whole population, so factions break off and try to take care of their own people. By 1975, about 200,000 people had died, millions of herd animals were dead, and crops and the very structure of society in many Sahel countries were ruined. Children were born brain-damaged because of malnutrition and dehydration, and corruption had set in. Since then, the region has been plagued with continued (although more sporadic) drought, but the infrastructure of the region has not returned and the people continue to suffer. A new drought that was becoming severe in late 2002 has put 38 million people at risk of starvation and famine, with conditions in 2003 remarkably similar to those of the mid-1970's in the region.

Drought Caused by Changes in Global Atmospheric Circulation

Understanding of the causes of droughts and floods has advanced dramatically in the past decade. Global oceanic and atmospheric circulation patterns undergo frequent shifts that affect large parts of the globe, particularly those arid and semiarid parts affected by Hadley cell circulation. One of the better-known secondary variations in global circulation is known as the El Niño–Southern Oscillation (ENSO). It is now understood that fluctuations in global circulation can account for natural disasters including the Dust Bowl of the 1930s in the United States's Plains states. Similar global climate fluctuations may explain the drought, famine, and desertification of parts of the Sahel, and the great famines of Ethiopia and Sudan in the 1970s and 1980s. Much of Africa, including the Sahel region, has become increasingly dry and desert-like over the past hundred years or more, and any attempts to restart agriculture and repopulate regions evacuated during previous famines in this region may be fruitless and lead to further loss of life.

Hadley cells are the name given to the globe-encircling belts of air that rise along the equator, dropping moisture as they rise in the tropics (Figure 1.2). As the air moves away from the equator, it cools and becomes drier, then descends at 15–30° N and S latitude, where it either returns to the equator or moves poleward. The locations of the Hadley cells move north and south annually in response to the changing apparent seasonal movement of the sun. High-pressure systems form where the air descends, and stable clear skies and intense evaporation characterize these areas because the air is so dry. Another pair of major global circulation belts is formed as air cools at the poles and spreads toward the equator. Cold polar fronts form where the polar air mass meets the warmer air that has circulated around the Hadley cell from the tropics. In the belts between the polar front and the Hadley cells, strong westerly winds develop. The position of the polar front and extent of the west-moving wind is controlled by the position of the polar jet stream (formed in the upper troposphere), which is partly fixed in place in the northern hemisphere by the high Tibetan Plateau and the Rocky Mountains. Dips and bends in the jet stream path are known as Rossby waves, and these partly determine the location of high- and low-pressure systems. These Rossby waves tend to be semistable in different seasons and have predictable patterns for summer and winter. If the pattern of Rossby waves in the jet stream changes significantly for a season or longer, it may cause storm systems to track to different locations than normal, causing local droughts or floods. Changes to this global circulation may also change the locations of regional downwelling cold dry air. This can cause long-term drought and desertification. Such changes may persist for periods of several weeks, months, or years, and may explain several of the severe droughts that have affected Asia, Africa, North America, and elsewhere.

A secondary air circulation phenomenon, known as the El Niño–Southern Oscillation, can also have profound influences on the development of drought conditions and the desertification of stressed lands. Hadley cells migrate north and south with summer and winter, shifting the locations of the most intense heating. There are several zonal oceanic-atmospheric feedback systems that influence global climate, but the most influential is that of the Austro-Asian system. In normal northern hemisphere summers, the location of the most intense heating in Austral-Asia shifts from equatorial regions to the Indian subcontinent along with the start of the Indian monsoon. Air is drawn onto the subcontinent, where it rises and moves outward to Africa and the central Pacific. In northern hemisphere winters, the location of this intense heating shifts to Indonesia and Australia, where an intense low-pressure system develops over this mainly maritime region. Air is sucked in, moves upward, and flows back out at tropospheric levels to the east Pacific. High pressure develops off the coast of Peru in both situations, because cold upwelling water off the coast here causes the air to cool, inducing atmospheric downwelling. The pressure gradient set up thus causes easterly trade winds to blow from the coast of Peru across the Pacific to the region of heating, causing warm water to pile up in the Coral Sea off the northeast coast of Australia. This also causes sea level to be slightly depressed off the coast of Peru, and more cold water upwells from below to replace the lost water. This positive feedback mechanism is rather stable, as it enhances the global circulation because more cold water upwelling off Peru induces more atmospheric downwelling, and more warm water piling up in Indonesia and off the coast of Australia cause atmospheric upwelling in this region.

This stable linked atmospheric and oceanic circulation breaks down and becomes unstable every two to seven years, probably from some inherent chaotic behavior in the system. At these times, the Indonesian-Australian heating center migrates eastward, and the build-up of warm water in the western Pacific is no longer held back by winds blowing westward across the Pacific. This causes the elevated warm water mass to collapse and move eastward across the Pacific, where it typically appears off the coast of Peru by the end of December. The El Niño–Southern Oscillation events are when this warming is particularly strong, with temperatures increasing by 4–6°C and remaining high for several months. This phenomenon is also associated with a reversal of the atmospheric circulation around the Pacific such that the dry downwelling air is now located over Australia and Indonesia, and the warm upwelling air is located over the eastern Pacific and western South America.

The arrival of El Niño is not good news in Peru, since it causes the normally cold upwelling and nutrient-rich water to sink to great depths, and the fish either must migrate to better feeding locations or die. The fishing industry collapses at these times, as does the fertilizer industry that relies on the bird guano normally produced by birds that eat fish and anchovy and that also die during El Niño events. The normally cold dry air is replaced with warm moist air, and the normally dry or desert regions of coastal Peru receive torrential rains, with associated floods, landslides, death, and destruction (see Chapter 5). Shoreline erosion is accelerated in El Niño events, because the warm water mass that moved in from across the Pacific raises sea levels by 4–25 inches, enough to cause significant damage.

The end of ENSO events also leads to abnormal conditions in that they seem to turn on the "normal" type of circulation in a much stronger way than is normal. The cold upwelling water returns off Peru with such a ferocity that it may move northward, flooding a 1–2° band around the equator in the central Pacific ocean with water that is as cold as 20° C. This phenomenon is known as La Niña ("the girl" in Spanish).

The alternation between ENSO, La Niña, and normal ocean-atmospheric circulation has profound effects on global climate and the migration of different climate belts on yearly to decadal time scales, and it is thought to account for about a third of all the variability in global rainfall. ENSO events may cause flooding in the western Andes and southern California, and may cause a lack of rainfall in other parts of South America including Venezuela, northeastern Brazil, and southern Peru. It may change the climate that causes droughts in Africa, Indonesia, India, and Australia, and is thought to have caused the failure of the Indian monsoon in 1899 that resulted in regional famine with the deaths of millions of people. Recently, the seven-year cycle of floods on the Nile has been linked to ENSO events, and famine and desertification in the Sahel, Ethiopia, and Sudan can be attributed to these changes in global circulation as well.

Drought and Desertification in the Sahel and Sub-Saharan Africa

The Sahel region offers one of the world's most tragic examples of how poorly managed agricultural practices, when mixed with long-term drought conditions, can lead to disaster and permanent desertification. A similar lesson is found in the Rajputana desert of India, one of the cradles of civilization that has become desert because of poor land use coupled with natural drought cycles.

"Sahel" means boundary in Arabic, and the Sahel forms the southern boundary of the world's largest desert, the Sahara. It is home to about 25 million people, most of who are nomadic herders and subsistence farmers. In the summer months of June and July, heat normally causes air to rise and this is replaced by moist air from the Atlantic, which brings 14–23 inches of rain per year. In the Sahel, the normal northward movement of the wet intertropical convergence zone stopped during an ENSO event in 1968. Further climatic changes in the early 1970s led to only about half of the normal rain falling until 1975. The additional lack of moisture was brought on by complications from the temperature cycles of the northern and southern oceans becoming out of synchronicity at this time, and the region suffered long-term drought and permanent desertification.

As the rains continued to fail to come and the air masses continued to evaporate surface water, the soil moisture was drastically reduced, which further reduced evaporation and cloud cover. The vegetation soon died off, the soils became dry and hot, and near-surface temperatures further increased. Soon, the plants were gone, the soils were exposed to the wind, and the region became plagued with blowing dust and sand. Approximately 200,000 people died, and 12 million heads of livestock perished. Parts of the region were altered to desert with little chance of returning to their previous state.

The desertification of the Sahel was enhanced by the agricultural practices of the people of the region. Nomadic and marginal agriculture was strongly dependent on the monsoon, and when the rains didn't come for several years, the natural and planted crops died and many of the remaining plants were used for fires to offset the cost of fuel. This practice greatly accelerated the desertification process. The Sahara is now thought to be overtaking the Sahel by migrating southward at approximately three miles per year.

The cycle of drought and famine is repeating itself in the Sahel. In November of 2002, Meles Zenawi, the Prime Minister of Ethiopia warned that 15 million people in his country are in immediate danger of suffering from famine and starvation as a result of the recent drought. It is estimated that the number of people affected is two or three times as many as in the terrible drought disaster of 1984–1985. The scope of the disaster threat was widened merely a month later, with the U.N. World Food Program estimating in December of 2002 that at least 38 million people are at risk of starvation. The causes of the current drought are similar to earlier droughts. The short and long rains have failed, as a result of the changing global climate patterns. Making the situation worse is that Ethiopia is still paying back debts incurred during the last drought, costing about 10 percent of the country's revenues. The countries in the region have been suffering from regional conflicts for a number of years, exacerbated by drought and scarce food supplies. This history of strife has made it difficult to get food aid to places like landlocked Ethiopia; for instance, Eritrea, once part of Ethiopia, offered use of its port at Assab for relief shipments designated for Ethiopia, even though it is estimated that one third of its population of 3.5 million also face severe hunger. Ethiopia refused the offer, probably because of tensions remaining after 30 years of guerrilla war.

The U.S. Dust Bowl of the 1930s

Drought disasters are not limited to sub-Saharan Africa. Soon after the great stock market crash of October 1929, the central United States farmland suffered one of its worst drought disasters known, plunging the United States into the Great Depression of the 1930s. Changes in the upper-level atmospheric circulation patterns caused upper-level dry air to sink into the Great Plains region, and as the air sank it became warmer and drier. The air seemed to soak the moisture right out of the ground and the crops died, exposing the barren soils to the action of the wind. The winds blew across the Plains states, raising huge clouds of dust known as rollers that moved like thousands-of-feet-tall steamrollers across the plains. This dust permeated everything, filling homes, lungs, eyes, and every available space with the fine-grained airborne plague. The dust storms became so bad that they blocked out the sun, and they moved across east coast cities and even hampered shipping in the Atlantic when they covered ships.

Soon people began leaving the plains in the thousands. Many moved west to California, where the land was available but is now overcrowded and plagued with drought. The Plains states were left in a shambles as the Dust Bowl days continued into the late 1930s. Only afterward, with considerable study, have we been able to determine that much of the disaster could have been prevented. The weather conditions could not have been changed and the drought would have occurred, but the severity might have been lessened if the farmers in the region knew that the techniques they were using were actually contributing to the disaster. The farmers were digging deeply into the native soils, disrupting the root systems of existing plants, killing drought-resistant plants, and replacing these with higher-yield crops with shallow root systems. When the drought came, these crops died, and the bare soil was exposed and was removed by the strong winds. Even though this area had suffered droughts before, it was mostly not farmed at those times, and the drought-resistant plants native to the region were able to prevent the soil from being eroded by the winds. Now, modern farming techniques are employed in the region, and it should be able to sustain another drought similar to the 1930s without such a huge disaster.

DROUGHT AND WATER SHORTAGE BROUGHT ON BY POPULATION GROWTH

Drought can also be brought on by rapid increases in population, water use (and abuse), and migrations of people into desert or semiarid regions. Although these regions may never have been able to sustain large populations with their indigenous water supplies, settlement of places like southern California, the United States's desert southwest, and the rapid population expansion in the Middle East all offer examples of how drought-like water shortage conditions are experienced by the people living in these regions.

Drought, Politics, and the Middle East

Water shortage, or drought, coupled with rapid population growth provides for extreme volatility for any region. In the Middle East, water shortage issues are coupled with longstanding political and religious differences. The Middle East region, stretching from North Africa and the Arabian Peninsula, through the Levant to Turkey, and along the Tigris-Euphrates Valleys, has only three major river systems and a few smaller rivers (Figure 8.8). The population stands at about 160 million people. The Nile has an annual discharge of about 82 billion cubic yards, whereas the combined Tigris-Euphrates system has an annual discharge of 93 billion cubic yards. Some of the most serious water politics and drought issues in the Middle East arise from the four states (Jordan, Israel, Lebanon, and Syria) that share the relatively small amounts of water of the Jordan River, which has an annual discharge of less than 2 billion cubic yards. It has been estimated that with current water usage and population growth, many nations in this region have only 10–15 years left before the agriculture and eventual security of these nations will be seriously threatened.

The region is arid, receiving 1–8 inches of rain per year, and has many drought years with virtually no rain. The Middle East has a population growth rate of about 3.5 percent per year, one of the fastest in the world, and many countries in the region have inefficient agricultural practices that contribute to the growing problem of desertification in the region. Some of the problems include planting water-intensive crops, common flooding and furrow methods of irrigation, spraying methods of irrigation that lose much of the water to evaporation, and poor management of water and crop resources. These growing demands on the limited water supply, coupled with political strife resulting from shared usage of waterways that flow through multiple countries, has set up the region for a major confrontation over water rights. Many of the region's past and present leaders have warned that water issues may be the cause of the next major conflict in the Middle East—in the words of the late King Hussein of Jordan, water issues "could drive nations of the region to war" (Figure 8.8). As an illustration of this point, a large spring in southern Lebanon that feeds the Hasbani River, a tributary to the Jordan, has become a matter of international dispute. In late 2002 Lebabon tapped the spring and is pumping a large amount of the water to irrigate crops in Lebanon. Israel has objected, saying that the pumping violates international agreements on sharing the water, and has threatened to bomb the pumping stations if the situation is not resolved. Israel has brought the matter up with the United Nations and the United States, saying that if the United States wants Israel to stay out of any Perisan Gulf conflict if attacked, then the U.S. should first solve the problem of Hasbani springs.

Water use by individuals is by necessity much less in countries in the Middle East than in America and in other western countries. For instance, in the United States, every American has about 11,000 cubic yards of fresh water potential to use each year, whereas citizens of Iraq have about 6,000, Turkey 4,400, and Syria about 3,000. Along the Nile, Egyptians have about 1,200 cubic yards available for each citizen. In the Levant, Israelis have a freshwater potential of 500 cubic yards per person per year, and Jordanians have only 280 cubic yards per year.

The Nile, the second-longest river on Earth, forms the main water supply for nine north African nations and disputes have grown over how to share this water with growing demands. The Blue Nile flows out of the Ethiopian Highlands and meets the White Nile in the Sudan north of Khartoum, then flows through northern Sudan and into Egypt. The Nile is dammed at Aswan, forming Lake Nassir, then flows north through the fertile valley of Egypt to the Mediterranean.

The Nile is the only major river in Egypt, and nearly all of Egypt's population lives in the Nile Valley. About 3 percent of the nation's arable land stretches along the Nile Valley, but 80 percent of Egypt's water use goes to agriculture in the valley. The government has been attempting to improve agricultural and irrigation techniques, which in many places have not changed considerably for 5,000 years. If the Egyptians embraced widespread use of drip irrigation and other modern agricultural practices, then Egypt's demands on water could easily be reduced by 50 percent or more.

Egypt has initiated a massive construction and national reconstruction project, the aim of which is to establish a new second branch of the Nile River, extending

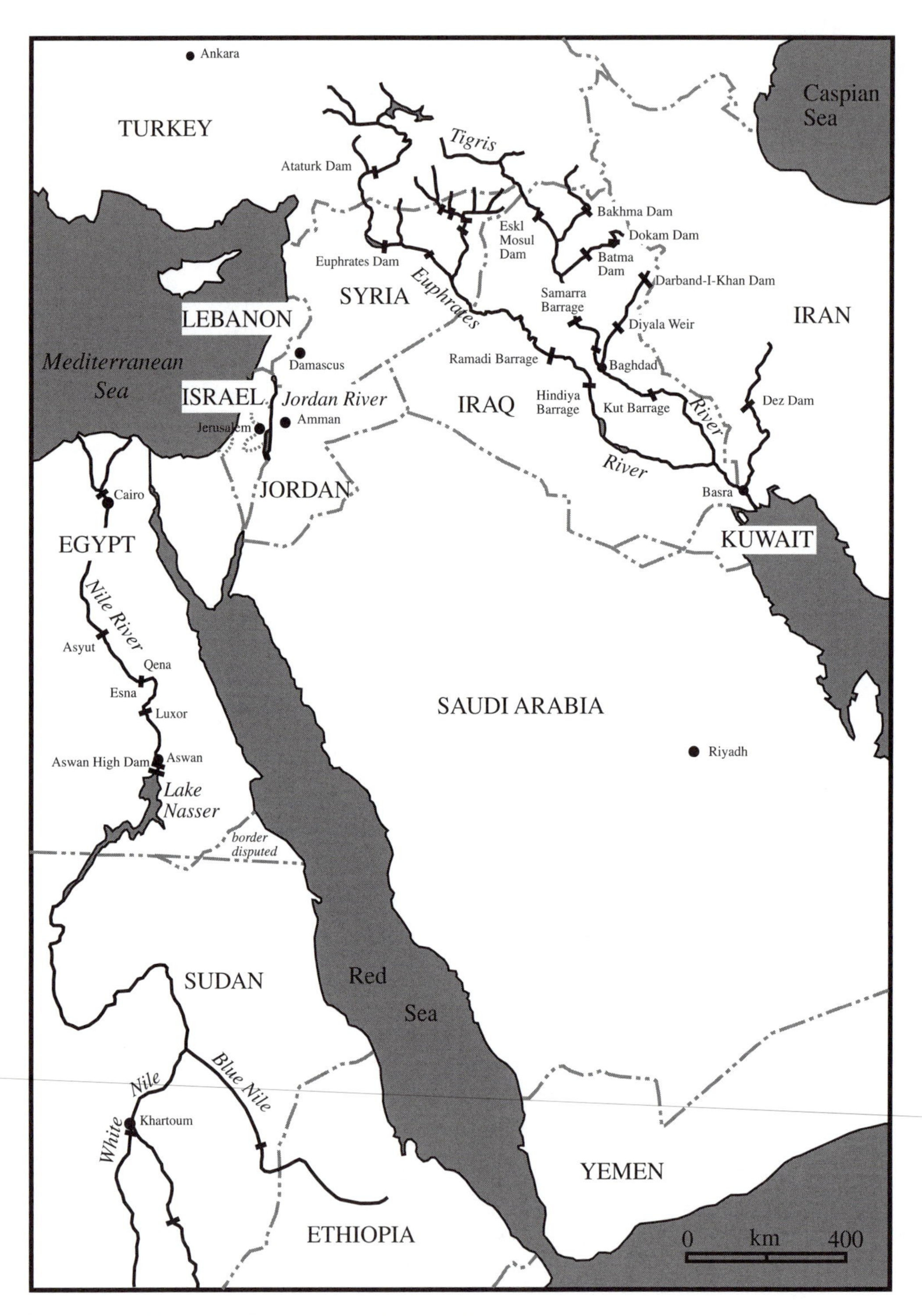

Figure 8.8. Map of the Middle East, showing the major rivers that supply most of the region's water. Much of the politics and tension in the region revolves around supplying a rapidly growing population with a small and dwindling source of fresh water.

from Lake Nassir in the south, across the scorching Western Desert, and emerging at the sea at Alexandria. This ambitious project starts in the Tushka Canal area, where water is drained from Lake Nassir and steered into a topographic depression that winds its way north through some of the hottest, driest desert landscape on Earth. The government plans to move thousands of farmers and industrialists from the familiar Nile Valley into this national frontier, hoping to alleviate overcrowding. Cairo's population of 13 million is increasing at a rate of nearly 1 million per year. If successful, this plan could reduce the water demands on the limited resources of the river.

There are many obstacles with this plan. Will people stay in a desert where temperatures regularly exceed 120°F? Will the water make it to Alexandria, having to flow through unsaturated sands, and through a region where the evaporation rate is 200 times greater than the precipitation rate? How will drifting sands and blowing dust affect plans for agriculture in the Western Desert? Much of the downriver part of the Nile is suffering from lower water and silt levels than needed to sustain agriculture or even the current land surface. So much water is used, diverted, or dammed upstream that parts of the Nile Delta have actually started to subside (sink) beneath sea level. These regions desperately need to receive the annual silt layer from the flooding Nile to rebuild the land surface and keep it from disappearing beneath the sea.

There are also political problems with establishing the New River through the Western Desert. Ethiopia contributes about 85 percent of the water to the Nile, yet it is experiencing severe drought and famine in the eastern part of the country. There is no infrastructure to get the water from the Nile to the thirsty lands and people to the east. Sudan and Egypt have longstanding disputes over water allotments, and Sudan is not happy that Egypt plans to establish a new river that will further their use of the water. Water is currently flowing out of Lake Nassir, filling up several small lake depressions to the west, and evaporating between the sands.

The Jordan River basin is host to some of the most severe drought and water shortage issues in the Middle East. Israel, Jordan, Syria, Lebanon, and the Palestinians share the Jordan River water (Figure 8.9), and this resource is much more limited than water along the Nile or in the Tigris-Euphrates system. The Jordan River is short (100 miles) and is made of three main tributaries, each with different characteristics. The Hasbani River has a source in the mountains of Lebanon and flows south to Lake Tiberias, and the Banias flows from Syria into the same lake. The smaller Dan River flows from Israel. The Jordan River then flows out of Lake Tiberias and is joined by water from the Yarmuk flowing out of Syria into the Dead Sea, where any unused water evaporates.

The Jordan River is the source for about 60 percent of the water used in Israel and 75 percent of the water used in Jordan. The other water used by these countries is largely from groundwater aquifers. Israel has almost exclusive use of the coastal aquifer along the Mediterranean shore, whereas disputes arise over use of aquifers from the West Bank and Golan Heights.

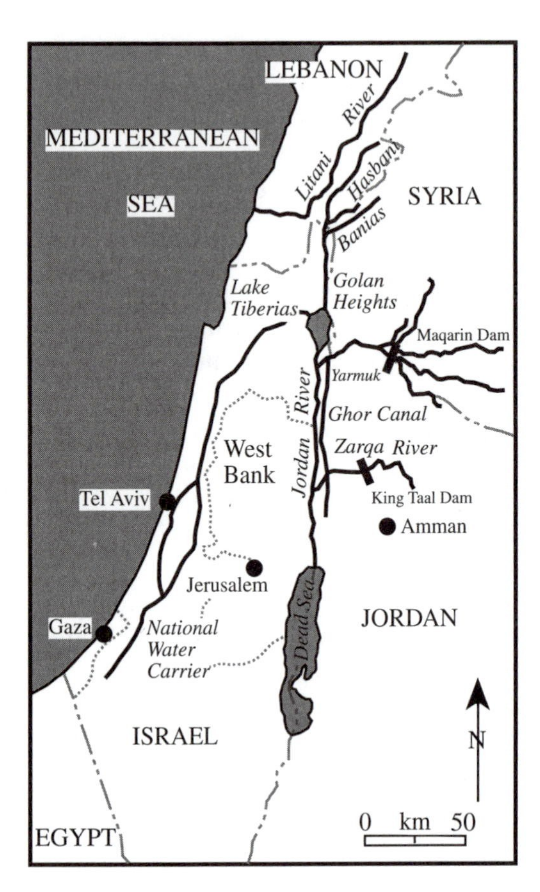

Figure 8.9. Detailed map of the Jordan River basin, showing how water divides form major political boundaries. The Jordan River has the smallest amount of water out of the major rivers in the Middle East, yet must supply millions of people with fresh water.

These areas are mountainous, get more rain and snowfall than the other parts of the region, and have some of the richest groundwater deposits in the region. Since the 1967 war, Israel has tapped the groundwater beneath the West Bank and now gets approximately 30–50 percent of its water supply from groundwater reserves beneath the mountains of the West Bank. The Palestinians get about 80 percent of their water from this mountain aquifer. A similar situation exists for the Golan Heights, though with lower amounts of reserves. These areas therefore have attained a new significance in terms of regional negotiations for peace.

The main problems of water use stem from the shortage of water compared to the population, effectively making drought conditions. The situation is not likely to get better given the alarming 3.5 percent annual population growth rate. Conservation efforts have only marginally improved the water use problem, and it is unlikely that there will be widespread rapid adoption of many of the drip-irrigation techniques used in Israel throughout the region. This is partly because it takes a larger initial investment in drip irrigation than in conventional furrow and flooding types of irrigation systems. Many of the farmers can not afford this investment, even if it would improve their long-term yields and decrease their use of water.

Sporadic droughts have made this situation worse in recent years, such that in 1999 Israel cut in half the amount of water it supplies to Jordan, and Jordan declared drought conditions and mandated water rationing. Jordan currently uses 73 percent of its water for irrigation, and if this number could be reduced by adoption of more efficient drip irrigation, the current situation would be largely in control.

One possible way to alleviate the problem of the drought and water shortage would be to explore for water in unconventional aquifer systems such as fractures or faults, which are plentiful in the region.

Many faults are porous and permeable structures that are several tens of meters wide and thousands of meters long and deep. They may be thought of as vertical aquifers, holding as much water as conventional aquifers (e.g., see Kusky and El-Baz, 1999). If these countries were to successfully explore for and exploit water in these structures, the water shortage and regional tensions might be reduced. This technique has proven effective in many other places in the Middle East, Africa, and elsewhere, and would probably work here as well. One exploration strategy used by several independent teams is to map the faults and fractures using satellite imagery, then do some further analysis in the field and computer modeling to determine which fracture systems might be more likely to yield significant groundwater resources.

Another set of problems plague the Tigris-Euphrates drainage basin and the countries that share water along their course (Figure 8.8). There are many political differences between the countries of Turkey, Syria, and Iraq, and the Kurdish people have been fighting for an independent homeland in this region for more than a decade. One of the underlying causes of dispute in this region is also the scarce water supply in a drought-plagued area. Turkey is in the midst of a massive dam construction campaign, with the largest dam being the Attaturk on the Euphrates. Overall, Turkey is spending an estimated $32 billion on twenty-two dams and nineteen hydroelectric plants. The aim is to increase the irrigated land in Turkey by 40 percent and to supply 25 percent of the nation's electricity through the hydroelectric plants. This system of dams also now allows Turkey to control the flow of the Tigris and the Euphrates, and, if it pleases, Turkey can virtually shut off the water supply to its downstream neighbors. At present, Turkey is supplying Syria and Iraq with what it considers to be a reasonable amount of water but what Syria and Iraq claim is inadequate. In response, Syria and Iraq have been supporting Kurdish separatist groups in their effort to keep pressure on Turkey to keep supplying them with additional water. Turkey is currently building a pipeline to bring water to drought-stricken Cyprus. Turkey and Israel are forging new partnerships and have been exploring ways to export water from Turkey and import it to Israel, which could help the drought in the Levant.

United States Desert Southwest

The history of development of the U.S. desert southwest was also crucially dependent on bringing water resources into this semiarid region. Much of California was regarded as worthless desert scrub land until huge water projects designed by the Bureau of Land Reclamation diverted rivers and resources from all over the West. In a brilliant exposé of much of the controversy and corruption associated with the diversion of water resources from Owens Valley, the Trinity River, the Colorado River, and many other western sources, M. Reisner (*Cadillac Desert,* 1986) paints a picture of development of the U.S. desert southwest that is ominous and has many parallels to ill-fated societies elsewhere in the history of the world. Can we continue to expand in the desert and demand more and more water resources from a depleting source? Are the soils becoming too salty to sustain agriculture? What is the future of the region if global climate continues to change and the deserts expand further?

RESOURCES AND ORGANIZATIONS

Print Resources

Abrahams, A. D., and Parsons, A. J. *Geomorphology of Desert Environments.* Chapman and Hall, 1994, 674 pp.

Al-Dabi, H. *Detection of Anthropogenic Changes in a Sand Dune Field of Northeastern Kuwait Using Remotely Sensed Imagery.* Master's thesis, Boston University, 1996, 75 pp.

Anderson, E. "Water Conflict in the Middle East—A New Initiative." *Janes Intelligence Review* 4, no. 5 (1992), 227–30.

Bagnold, R. A. *The Physics of Blown Sand and Desert Dunes.* London: Methuen, 1941, 65 pp.

Blackwell, Major James. *Thunder in the Desert: The Strategy and Tactics of the Persian Gulf War.* New York: Bantam, 1991, 252 pp.

Bryson, R., and Murray, T. *Climates of Hunger.* Canberra: Australian National University Press, 1977.

El-Baz, F., and Himida, I. *Sand Accumulations and Groundwater in the Sahara.* Cairo: UNESCO, International Geological Correlation's Program Project 391, Desert Research Center, 1996, 31 pp.

El-Baz, F., Kusky, T. M., Himida, I., and Abdel-Mogheeth, S., eds. *Ground Water Potential of the Sinai Peninsula, Egypt.* Cairo: Desert Research Center, 1998, 219 pp.

El-Baz, F., et al., eds. *Atlas of the State of Kuwait from Satellite Images.* Kuwait City: Kuwait Foundation for the Advancement of Science, 2000.

Freidman, N. *Desert Victory: The War for Kuwait.* Annapolis, Md.: Naval Institute Press, 1991, 435 pp.

Hack, J. T. "Dunes of the Western Navajo County." *Geographical Review* 31 (1941), 240–63.

Kusky, T. M., and El-Baz, F. "Structural and Tectonic Evolution of the Sinai Peninsula, Using Landsat Data: Implications for Ground Water Exploration." *Egyptian Journal of Remote Sensing* 1 (1999), 69–100.

McKee, E. D., ed. *A Study of Global Sand Seas.* U.S. Geological Survey Professional Paper 1052, 1979.

Reisner, M. *Cadillac Desert: The American West and Its Disappearing Water.* New York: Penguin, 1986, 582 pp.

Sorkhabi, R. "Water for Peace or Wars over Water: Hydropolitics in the Middle East." *The Professional Geologist* 30, no. 13 (1993), 8–9.

Starr, J. R. "Water Wars." *Foreign Affairs,* no. 82 (Spring 1991).

Starr, J. R., and Stoll, D. C., eds. *The Politics of Scarcity: Water in the Middle East.* Boulder, Colo.: Westview Press, 1988.

Walker, A. S. *Deserts: Geology and Resources.* U.S. Geological Survey, Publication 421–577, 1996, 60 pp.

Webster, D. "Alashan, China's Unknown Gobi." *National Geographic* (2002), 48–75.

Non-Print Resources

Web Sites

The Aircav Web site:
http://www.aircav.com/survival/asch13/asch13p02.html
This Web site describes how to survive in a desert.

Fact Sheets on the United Nations Convention to Combat Desertification:
www.unccd.int%2Fpublicinfo%2Ffactsheets%2Fmenu.php

NASA's Earth Observatory Web site:
http://www.earthobservatory.nasa.gov
This site covers desertification and drought.

U.S. Geological Survey's Web site on deserts:
http://pubs.usgs.gov/gip/deserts/contents/
This site contains a huge amount of information about deserts, desert geology, hazards, and resources.

University of Arizona College of Agriculture and Life Sciences Web site:
http://geography.about.com/gi/dynamic/offsite.htm?site=http%3A%2F%2Fag.arizona.edu
This Web site contains an online journal that focuses on desert regions. Current and past issues are available.

Organizations

Desert Research Center, Egypt
http://www.drc-egypt.com/researchdivision.html
The DRC includes scientific staff and experts in desert regions, and it supports studies of the Sahara and other deserts, primarily in Egypt. Some of the topics of interest include 1) exploring and evaluating groundwater aquifers in desert regions, 2) managing desert and newly reclaimed soils with respect to their agricultural use and development, 3) monitoring and analyzing desertification phenomena and recommending solutions to stop or minimize desertification processes with the aid of space maps taken by the DRC's Satellite Receiving Station, and 4) monitoring and recording sand dune movement and using various techniques to stabilize sand dunes.

Desert Research Institute
http://www.dri.edu/
The DRI is a part of the University of Nevada. The 400 members of the DRI staff are concerned with water resources and air quality, global climate change and the physics of the earth's turbulent atmosphere, humanity's historic struggle to adapt to harsh environments, and its urgent search today for the technology of the next century.

CHAPTER 9

Glaciers and Glaciation

Glaciers are any permanent body of ice (recrystallized snow) that shows evidence of gravitational movement. Glaciers are an integral part of the *cryosphere,* which is that portion of the planet where temperatures are so low that water exists primarily in the frozen state. Most glaciers are presently found in the polar regions and at high altitudes. However, at several times in Earth's history, glaciers have advanced deeply into mid-latitudes and the climate of the entire planet was different. Some models suggest that at one time the entire surface of the earth may have been covered in ice, a state referred to as the "snowball Earth."

Glaciers are dynamic systems, always moving under the influence of gravity and changing drastically in response to changing global climate systems. Thus, changes in glaciers may reflect coming changes in the environment. There are several types of glaciers. *Mountain glaciers* form in high elevations and are confined by surrounding topography, like valleys. These include cirque glaciers, valley glaciers, and fiord glaciers. *Piedmont glaciers* are fed by mountain glaciers, but terminate on open slopes beyond the mountains. Some piedmont and valley glaciers flow into open water, bays, or fiords and are known as *tidewater glaciers. Ice caps* form dome-shaped bodies of ice and snow over mountains and flow radially outward. *Ice sheets* are huge, continent-sized masses of ice that presently cover Greenland and Antarctica. These are the largest glaciers on Earth. Ice sheets contain about 95 percent of all the glacier ice on the planet. The polar ice sheet that covers Antarctica consists of two parts that meet along the Transantarctic Mountains. It shows ice shelves, which are thick glacial ice that floats on the sea. These form many icebergs by calving, which move northward into the shipping lanes of the southern hemisphere.

Polar glaciers form where the mean average temperature lies below freezing, and these glaciers have little or no seasonal melting because they are always below freezing. Other glaciers, called *temperate glaciers,* have seasonal melting periods during which the temperature throughout the glacier may be at the *pressure melting point* when the ice can melt at that pressure and both ice and water coexist. All glaciers form above the snow line, which is the lower limit at which snow remains year-round. It is at sea level in polar regions and at 5,000–6,000 feet at the equator (for example, Mount Kilamanjaro in Tanzania has glaciers, although these are melting rapidly).

Glaciers present two main categories of hazards. The first affect those who are working, living on, or living near glaciers or are transporting goods by sea, river, or land in glacially influenced areas. The second set of hazards is more global in nature and reflects climate change that brings on widespread glaciations. Glaciers also represent sensitive indicators of climate change and global warming, shrinking in times of warming and expanding in times of cooling. Glaciers may be thought of as the "canaries in the coal mine" for climate change.

The earth has experienced at least three major periods of long-term frigid climate and ice ages, interspersed with periods of warm climate. The earliest well-documented ice age is the period of the "snowball Earth" in the Late Proterozoic (about 600–800 million years ago), although there is evidence of several even earlier glaciations. The Late Paleozoic saw another ice age lasting about 100 million years, from 250 to 350 million years ago. The planet entered the present ice age about 55 million years ago. The underlying causes of these different glaciations are varied and include anomalies in the distribution of continents, oceans, and associated currents; variations in the amount of incoming solar radiation; and changes in the atmospheric balance between the amount of incoming and outgoing solar radiation.

FORMATION OF GLACIERS

Glaciers form mainly by the accumulation and compaction of snow, and are deformed by flow under the influence of gravity. When snow falls it is very porous, and with time the pore spaces close by precipitation and compaction. When snow first falls, it has a density of about one-tenth that of ice; after a year or more, the density is transitional between snow and ice, and it is called *firn.* After several years, the ice has a density of 0.9 gm/cm^3, and it flows under the force of gravity. At this point, glaciers are considered to be metamorphic rocks composed of the mineral ice.

The mass and volume of glaciers are constantly changing in response to the seasons and to global climate changes. The mass balance of a glacier is determined by the relative amounts of accumulation and ablation (mass loss through melting and evaporation or calving). Some years see a mass gain leading to a glacial advance, whereas some periods have a mass loss and a glacial retreat. The glacial front or terminus shows these effects.

Glaciers have two main zones that are best observed at the end of the summer ablation period. The *zone of accumulation* is found in the upper parts of the glacier and is still covered by the remnants of the previous winter's snow. The zone of ablation is below this and is characterized by older dirtier ice, from which the previous winter's snow has melted. An equilibrium line, marked by where the amount of new snow exactly equals the amount that melts that year, separates these two zones (Figure 9.1).

MOVEMENT OF GLACIERS

When glacial ice gets thick enough, it begins to flow and deform under the influence of gravity. The thickness of the ice must be great enough to overcome the internal forces that resist movement, and these forces depend on the temperature of the glacier. The thickness at which a glacier starts flowing also depends on how steep the slope it is on happens to be: Thin glaciers can move on steep slopes, whereas glaciers must become very thick to move across flat surfaces. The flow is by the process of *creep,* or the deformation of individual mineral grains. This creep leads to the preferential orientation of mineral (ice) grains forming *foliations and lineations,* much the same way as in other metamorphic rocks.

Some glaciers develop a layer of *meltwater* at their base, allowing *basal sliding* and *surging* to occur. Where glaciers flow over ridges, cliffs, or steep slopes, their upper surface fails by cracking, which forms large deep crevasses that can be up to 200 feet deep. A thin blanket of snow, making for very dangerous conditions, can cover these crevasses.

Ice in the central parts of valley glaciers moves faster than ice at the sides because of frictional drag against the valley walls on the side of the glacier. Similarly, a profile with depth into the glacier would show that they move the slowest along their bases and faster internally and along their upper surfaces. When a glacier surges, it may temporarily move as fast along its base as it does in the center and top. This is because during surges, the glacier is essentially riding on a cushion of meltwater along the glacial base, and frictional resistance is reduced during surge events. During meltwater-enhanced surges, glaciers may advance by as much as several kilometers in a year. Events like this may happen in response to climate changes.

Calving refers to a process in which icebergs break off from the fronts of tidewater glaciers or ice shelves. Typically, the glacier will crack with a loud noise that sounds like an explosion, and then a large chunk of ice will splash into the water, detaching from the glacier. Glaciers retreat rapidly by calving.

GLACIATION AND GLACIAL LANDFORMS

Glaciation is the modification of the land's surface by the action of glacial ice. When glaciers move over the land's surface, they plow up the soils, abrade and file down the bedrock, carry and transport the sedimentary load, steepen valleys, and then leave thick deposits of glacial debris during retreat.

In glaciated mountains, a distinctive suite of landforms forms from glacial action (Figure 9.2). Glacial *striations* are scratches on the surface of bedrock that were formed when the glacier dragged boulders across the bedrock surface. *Rouche mountainees* and other asymmetrical landforms form when the glacier plucks pieces of bedrock away from a surface and carries it away. The step faces in the direction of

Figure 9.1. Zones of Glaciers, from the Harding and Sargent Icefields, Kenai Peninsula, Alaska. (a) and (b) Zone of accumulation, showing thick snow accumulating in cirques, compacting to fern and ice, and flowing out into intermountain valleys. (c) View of Glaciers exiting from zone of accumulation, moving through zone of ablation, and forming tidewater glaciers in McCarty Fiord, Kenai Fiords National Park. (d) Transition from zone of accumulation (clean white snow in background) to zone of ablation (dirty snow with crevasses in foreground), with prominent medial moraine. (e) View "down-glacier" along medial moraine of Grewingk glacier, with Cook Inlet forearc basin in background. (f) Toe or "snout" of rapidly retreating Wosnesinski glacier, Kenai Peninsula. Note lack of vegetation in recently deglaciated U-shaped valley. Photos by T. Kusky

transport. *Cirques* are bowl-shaped hollows that open downstream and are bounded upstream by a steep wall. Frost wedging, glacial plucking, and abrasion all work to excavate cirques from previously rounded mountaintops. Many cirques contain small lakes called *tarns,* which are blocked by small ridges at the base of the cirque. Cirques continue to grow during glaciation, and where two cirques form on opposite sides of a mountain, a ridge known as an *arete* forms. Where three cirques meet, a steep-sided mountain known as a *horn* forms. The Matterhorn of the Swiss Alps is an example of a glacially carved horn.

Valleys that have been glaciated have a characteristic U-shaped profile with tributary streams entering above the base of the valley, often as waterfalls. In contrast, streams generate V-shaped valleys. *Fiords* are deeply indented glaciated valleys that are partly filled by the sea. In many places that were formerly overlain by glaciers, elongate streamlined forms known as drumlins occur. These are both depositional (composed of debris) and erosional (composed of bedrock) features.

GLACIAL TRANSPORT

Glaciers transport enormous amounts of rock debris, including some large boulders, gravel, sand, and fine silt. The glacier may carry this at its base, on its surface, or internally. Glacial deposits are characteristically poorly sorted or nonsorted, with large boulders next to fine silt. Most of a glacier's load is concentrated along its base and sides, because in these places plucking and abrasion are most effective.

Active ice deposits till as a variety of *moraines,* which are ridge-like accumulations of drift deposited on the margin of a glacier. A terminal moraine represents the farthest point of travel of the glacier's terminus. Glacial debris left on the sides of glaciers form lateral moraines; where two glaciers meet, their moraines merge and are known as a medial moraine (Figure 9.1).

Rock flour is a general name for the deposits at the base of glaciers, where they are produced by crushing and grinding by the glacier to make fine silt and sand. *Glacial drift* is a general term for all sediment deposited directly by glaciers or by glacial meltwater in streams, lakes, and the sea. *Till* is glacial drift that was deposited directly by the ice. It is a nonsorted random mixture of rock fragments. *Glacial marine drift* is sediment deposited on the seafloor from floating ice shelves or bergs. Glacial marine drift may include many isolated pebbles or boulders that were initially trapped in glaciers on land, then floated in icebergs that calved off from tidewater glaciers. These rocks melted out while over open water and fell into the sediment on the bottom of the sea. These isolated stones are called dropstones and are often one of the hallmark signs of ancient glaciations in rock layers that geologists find in the rock record. *Stratified drift* is deposited by meltwater and may include a range of sizes that are deposited in different fluvial or lacustrine environments.

Glacial *erratics* are glacially deposited rock fragments with lithologies different than those of underlying rocks. In many cases, the erratics are composed of rock types that do not occur in the area they are

Figure 9.2. Glacial landforms in recently deglaciated areas. (a) A drowned cirque and fiord, in which regional subsidence has caused previously glaciated areas above sea level to reside at and below sea level. Note ridge at edge of tarn lake, rounded valley, arete, and horn. (b) Cirque, showing tarn lake, and arete in background. (c) North side of range in Denali National Park, showing braided stream in wide glacial valley. Note U-shaped valleys that lead toward Mount McKinley, hidden from view in the clouds. Photos by T. Kusky

resting in, but are only found hundreds or even thousands of miles away. Many glacial erratics in the northern part of the United States can be shown to have come from parts of Canada. Some clever geologists have used glacial erratics to help them find mines or rare minerals that they have found in an isolated erratic; in other words, they have used their knowledge of glacial geology to trace the boulders back to their sources by following the orientation of glacial striation in underlying rocks. Recently, diamond mines were discovered in Nunavut, northern Canada, by tracing diamonds found in glacial till back to their source region.

Sediment deposited by streams washing out of glacial moraines is known as outwash and is typically deposited by braided streams (Figure 9.2c) (see also Chapter 6). Many of these form on broad plains known as outwash plains. When glaciers retreat, the load is diminished, and a series of outwash terraces may form.

HOW ARE GLACIAL HAZARDS STUDIED, AND WHO STUDIES THEM?

Glaciologists, who employ a wide variety of techniques depending on the goals of their studies, study glaciers. Some glaciologists are concerned with the movement of glaciers, and they may place stakes in various parts of the glacier and around its edges to measure its movement with time and to determine how much the glacier has advanced or retreated. Such measurements have improved in accuracy in recent years with the advent of Global Positioning System (GPS) technologies, with which subcentimeter displacements can now be measured.

Remote sensing technologies are also commonly employed for studies of glaciers. Time-series satellite images can show the position of glaciers at various times, so rates of movement can be calculated. Satellite radar and aircraft radar-altimeter data can be used to determine the thickness of snowfall and compare this information with rates of movement to determine the mass balance of glaciers. This information helps determine whether the glacier is experiencing net loss or gain of volume.

Other glaciologists are concerned with the physical conditions of deformation of the ice, temperature of the ice with depth, and how the ice may or may not be coupled to the underlying substratum. Also, many glacial studies are focused on using the isotopic, pollen, and other records in ice cores to determine the paleoclimate history of the past few tens of millions of years of Earth history. To accomplish this goal, glaciologists must drill and extract ice cores, and preserve them at subfreezing temperatures for measurement in the laboratory. The ages of the ice cores must be accurately determined, which in some cases can be done by counting down "annual rings" much like counting tree rings. Once the age of the ice layer is determined, glaciologists may analyze the ice, air bubbles trapped in the ice, or other trapped particles that reveal clues to climate history. Numerous ice cores from Greenland and Antarctica are currently being studied to help decipher the climate history of Earth for the past 100,000 years.

WHAT CAUSES GLACIATIONS?

In the last 2.5 billion years, several periods of glaciation have been identified, separated by periods of mild climate similar to that of today. Glaciations seem to form through a combination of several different factors. One of the variables is the amount of incoming solar radiation, and this changes in response to several astronomical effects. Another variable is the amount of heat that is retained by the atmosphere and ocean, or the balance between the incoming and outgoing heat. A third variable is the distribution of land masses on the planet. Shifting continents can influence the patterns of ocean circulation and heat distribution, and placing a large continent on one of the poles can cause ice to build up on that continent, increasing the amount of heat reflected back to space and lowering global temperatures in a positive feedback mechanism.

Changes in Amounts of Incoming Solar Radiation—Astronomical Effects

Astronomical effects influence the amount of incoming solar radiation. Minor variations in the path of the earth in its orbit around the sun and the inclination or tilt of its axis cause variations in the amount of solar energy reaching the top of the atmosphere. These variations are thought to be responsible for the advance and retreat of the northern and southern hemisphere ice sheets in the past few million years. In the past 2 million years alone, the earth has seen the ice sheets advance and retreat approximately twenty times! The climate record, as deduced from ice-core records from Greenland and isotopic tracer studies from deep ocean, lake, and cave sediments, suggests that the ice builds up gradually over periods of about 100,000 years, then retreats rapidly over a period of decades to a few thousand years. These patterns result from the cumulative effects of different astronomical phenomena.

Several movements are involved. The earth rotates around the sun following an elliptical orbit, and the shape of this elliptical orbit is known as its eccentricity. The *eccentricity* changes cyclically with time over a period of 100,000 years, alternately bringing the earth closer to and farther from the sun in summer and winter. This 100,000-year cycle is about the same as the general pattern of glaciers advancing and retreating every 100,000 years in the past 2 million years, suggesting that this is the main cause of variations within the present day ice age.

The earth's axis is presently tilting by 23.5° away from the orbital plane, and the tilt varies between 21.5° and 24.5°. The tilt changes by plus or minus 1.5° from a tilt of 23° every 41,000 years. When the tilt is greater, there is greater seasonal variation in temperature.

Wobble of the rotational axis describes a motion much like a top rapidly spinning and rotating with a wobbling motion, such that the direction of tilt toward or away from the sun changes even though the tilt amount stays the same. This wobbling phenomenon is known as precession of the equinoxes and it has the effect of placing different hemispheres closest to the sun in different seasons. Presently, the precession of the equinoxes is such that the earth is closest to the sun during the northern hemisphere's winter. This precession changes with a double cycle, with periodicities of 23,000 years and 19,000 years.

Because each of these astronomical factors act on different time scales, they interact in a complicated way known as Milankovitch cycles, after the Yugoslavian scientist, Milutin Milankovitch, who first analyzed them in the 1920s. Using the power of understanding these cycles, we can make predictions of where the earth's climate is heading, whether we are heading into a warming or cooling period, and whether we need to plan for sea level rise, desertification, glaciation, sea level drops, floods, or droughts.

Changes in Amount of Heat Lost to Space—Atmospheric Effects

Global climate represents a balance between the amount of solar radiation received and the amount of this energy that is retained in a given area. The planet receives about 2.4 times as much heat in the equatorial regions than in the polar regions. The atmosphere and oceans respond to this unequal heating by setting up currents and circulation systems that redistribute the heat more equally. These circulation patterns are in turn affected by the ever-changing pattern of the distribution of continents, oceans, and mountain ranges.

The amounts and types of gases in the atmosphere can modify the amount of incoming solar radiation. For instance, cloud cover can cause much of the incoming solar radiation to be reflected back to space before being trapped by the lower atmosphere. On the other hand, certain types of gases (known as greenhouse gases) allow incoming short wavelength solar radiation to enter the atmosphere, but trap this radiation when it tries to escape in its longer wavelength reflected form. This causes a buildup of heat in the atmosphere and can lead to a global warming known as the greenhouse effect.

The amount of heat trapped in the atmosphere by greenhouse gases has varied greatly over Earth's history. One of the most important greenhouse gases is carbon dioxide (CO_2). Plants, which release O2 to the atmosphere, now take up CO_2 by photosynthesis. In the early part of Earth's history in the Precambrian, before plants covered the land surface, photosynthesis did not remove CO_2 from the atmosphere, with the result that CO_2 levels were much higher than at present. Atmospheric CO_2 is also presently taken up by marine organisms that remove it from the ocean's surface water (which is in equilibrium with the atmosphere) and use the CO_2 along with calcium to form their shells and mineralized tissue. These organisms make $CaCO_3$ (calcite), which is the main component of limestone, a rock composed largely of the dead remains of marine organisms. Approximately 99 percent of the planet's CO_2 is presently removed from the atmosphere and ocean system because it has been locked up in rock deposits of limestone on the continents and on the sea floor. If this amount of CO_2 were released back into the atmosphere, the global temperature would increase dramatically. In the early Precambrian, when this CO_2 was free in the atmosphere, global temperatures averaged about 550°F (290°C).

The atmosphere redistributes heat

quickly by forming and redistributing clouds and uncondensed water vapor around the planet along atmospheric circulation cells. Oceans are able to hold and redistribute more heat because of the greater amount of water in the oceans, but they redistribute this heat more slowly than the atmosphere. Surface currents are formed in response to wind patterns, but deep ocean currents that move more of the planet's heat follow courses that are more related to the bathymetry (topography of the sea floor) and the spinning of the earth than they are related to surface winds.

The balance of incoming and outgoing heat from the earth has determined the overall temperature of the planet through time. Examination of the geological record has enabled paleoclimatologists to reconstruct periods when the earth had glacial periods, hot and dry periods, hot and wet periods, and cold and dry periods. In most cases, the earth has responded to these changes by expanding and contracting its climate belts. Warm periods see an expansion of the warm subtropical belts to high latitudes, and cold periods see an expansion of the frigid cold climates of the poles to low latitudes. What is next?

THE GLACIAL AGES

At several times in Earth's history, large portions of the earth's surface have been covered with huge ice sheets. About 10,000 years ago, all of Canada, much of the northern United States, and most of Europe were covered with ice sheets, as was about 30 percent of the world's land mass. These ice sheets lowered sea level by about 320 feet, exposing the continental shelves and leaving the locations of cities including New York, Washington, and Boston 100 miles from the sea!

Glaciations have happened frequently in the past 55 million years and could come again at almost any time. In the late 1700s and early 1800s, Europe experienced a "little ice age" when many glaciers advanced out of the Alps and destroyed many small villages. Ice ages have occurred at several other times in the ancient geologic past, including in the Late Paleozoic (about 250–350 million years ago), Silurian (435 million years ago), and Late Proterozoic (about 600–800 million years ago). During parts of the Late Proterozoic glaciation, it is possible that the entire surface of the earth had a temperature below freezing and was covered by ice. What causes these glaciations, and what drives the planet in and out of these frozen climate states?

In the late Proterozoic, the earth experienced one of the most profound ice ages in the history of the planet. Isotopic records and geologic evidence suggests that the entire earth's surface was frozen, though some workers dispute the evidence and claim that there would be no way for the earth to recover from such a frozen state. In any case, it is clear that in the Late Proterozoic, during the formation of the supercontinent of Gondwana, the earth experienced one of the most intense glaciations ever with the lowest average global temperatures in known Earth history.

One of the longest-lasting glacial periods was the Late Paleozoic ice age that lasted about 100 million years, indicating a long-term underlying cause of global cooling. Of the variables that operate on these long-time scales, it appears that the distribution and orientation of continents seems to have caused the Late Paleozoic glaciation. The Late Paleozoic saw the amalgamation of the planet's landmasses into the supercontinent of Pangea. The southern part of Pangea, known as Gondwana, consisted of present-day Africa, South America, Antarctica, India, and Australia. During the drift of the continents in the Late Paleozoic, Gondwana slowly moved across the South Pole, and huge ice caps formed on these southern continents during their passage over the pole. The global climate was overall much colder with the subtropical belts becoming very condensed and the polar and subpolar belts expanding to low latitudes.

It seems that during all major glaciations, there was a continent situated over one of the poles. We now have Antarctica over the South Pole, and this continent has huge ice sheets on it. When continents rest over a polar region, they accumulate huge amounts of snow that gets converted into ice sheets that are several kilometers thick, which reflect more solar radiation back to space and lower global sea water temperatures and sea levels.

Another factor that helps initiate glaciations is to have continents distributed in a roughly north-south orientation across equatorial regions. Equatorial waters receive more solar heating than polar waters. Continents block and modify the simple east-to-west circulation of the oceans that is induced by the spinning of the planet. When continents are present on or near the equator, they divert warm water currents to high latitudes, bringing warm water to these latitudes. Since warm water evaporates much more effectively than cold water, having warm water move to high latitudes promotes evaporation, cloud formation, and precipitation. In cold high-latitude regions, the precipitation falls as snow, which persists and builds up glacial ice.

The Late Paleozoic glaciation ended when the supercontinent of Pangea began breaking apart, suggesting a further link between tectonics and climate. It may be that the smaller land masses could not divert the warm water to the poles any more, or perhaps enhanced volcanism associated with the breakup caused additional greenhouse gases to build up in the atmosphere, raising global temperatures.

The planet began to enter a new glacial period about 55 million years ago, following a 10-million-year-long period of globally elevated temperatures and expansion of the warm subtropical belts into the subarctic. This Late Paleocene (see Figure 1.3) *global hot house* saw the oceans and atmosphere holding more heat than at any other time in Earth's history, but temperatures at the equator were not particularly elevated. Instead, the heat was distributed more evenly around the planet such that there were probably fewer violent storms (with a small temperature gradient between low and high latitudes) and overall more moisture in the atmosphere. It is thought that the planet was so abnormally warm during this time because of several factors, including a distribution of continents that saw the equatorial region free of continents. This allowed the oceans to heat up more efficiently, raising global temperatures. The oceans warmed so much that the deep ocean circulation changed, and the deep currents that are normally cold became warm. This melted frozen gases known as *methane hydrates* that accumulate on the sea floor, releasing huge amounts of methane to the atmosphere. Methane is a greenhouse gas, and its increased abundance in the atmosphere trapped solar radiation in the atmosphere, contributing to global warming. In addition, this time saw vast outpourings of mafic lavas in the North Atlantic Ocean realm, and these volcanic eruptions were probably accompanied by the release of large amounts of CO_2, which would have increased the greenhouse gases in the atmosphere and further warmed the planet. The global warming during the Late Paleocene was so extreme that about 50 percent of all the single-celled organisms living in the deep ocean became extinct.

After the Late Paleocene hot house, the earth entered a long-term cooling trend that we are currently still in, despite the present warming of the past century. This current ice age was marked by the growth of Antarctic glaciers starting about 36 million years ago until about 14 million years ago, when the Antarctic ice sheet covered most of the continent with several miles of ice. At this time, global temperatures had cooled so much that many of the mountains in the northern hemisphere were covered with mountain and piedmont glaciers similar to those in southern Alaska today. The ice age continued to intensify until, at 3 million years ago, extensive ice sheets covered the northern hemisphere. North America was covered with an ice sheet that extended from northern Canada to the Rocky Mountains; across the Dakotas, Wisconsin, Pennsylvania, and New York; and on the continental shelf. At the peak of the glaciation 18,000–20,000 years ago, about 27 percent of the land's surface was covered with ice. Midlatitude storm systems were displaced to the south and desert basins of the U.S. southwest, Africa, and the Mediterranean received abundant rainfall and hosted many lakes. Sea level was lowered by 425 feet to make the ice that covered the continents, so most of the world's continental shelves were exposed and eroded.

The causes of the Late Cenozoic glaciation are surprisingly not well-known but seem related to Antarctica coming to rest over the South Pole and other plate tectonic motions that have continued to separate the once contiguous land masses of Gondwana and that have changed global circulation patterns in the process. Two of the important events seem to be the closing of the Mediterranean Ocean around 23 million years ago and the closing of the Isthmus of Panama 3 million years ago. These tectonic movements restricted the east-to-west flow of equatorial waters, causing the warm water to move to higher latitudes where evaporation promotes snowfall. An additional effect seems to be related to uplift of some high mountain ranges including the Tibetan plateau, which has changed the pattern of the air circulation associated with the Indian monsoon.

The closure of the Isthmus of Panama is closely correlated with the advance of northern hemisphere ice sheets, suggesting a causal link. This thin strip of land has drastically altered the global ocean circulation such that there is no longer an effective communication between Pacific and Atlantic Ocean waters, and it diverts warm currents to near-polar latitudes in the

North Atlantic, enhancing snowfall and northern hemisphere glaciation. Since 3 million years ago, the ice sheets in the northern hemisphere have alternately advanced and retreated, apparently in response to variations in the earth's orbit around the sun and other astronomical effects. These variations change the amount of incoming solar radiation on time scales of thousands to hundreds of thousands of years. Together with the other longer-term effects of shifting continents, changing global circulation patterns, and abundance of greenhouse gases in the atmosphere, most variations in global climate can be approximately explained. This knowledge may help predict where the climate is heading in the future and may help model and mitigate the effects of human-induced changes to the atmospheric greenhouse gases. If we are heading into another warm phase and the existing ice on the planet melts, sea level will quickly rise by 210 feet, inundating many of the world's cities and farmlands. Alternately, if we enter a new ice sheet stage, sea levels will be lowered, and the planet's climate zones will be displaced to more equatorial regions.

TYPES OF GLACIAL HAZARDS

Glaciers present several near-field hazards to those who live near or travel on the ice. Most people do not experience these hazards unless they are on the glacier.

Crevasses

Crevasses are extremely hazardous and may be hidden under fresh layers of snow or wind-blown drifts. They form most typically over bedrock ridges where the upper surface of the glacier must bend in extension, opening crevasses (Figure 9.1). Crevasses may be long and narrow but up to several hundred feet deep. There have been a number of accidents on glaciers when hikers, explorers, or natives of glaciated lands have fallen into crevasses. The result is unpleasant and most typically deadly. As a person falls into a crevasse, they get wedged tightly into the bottom of the crevasse, often injured but still alive. The person's body heat slowly melts an envelope around the person, and they sink slightly deeper into the crevasse. Gradually, the person gets squeezed into a smaller and smaller place, and either suffocates from constriction or freezes from hypothermia. Glaciers are typically in remote locations, so death usually results before help can arrive.

Calving

Tidewater glaciers and ice shelves are prone to spectacular calving events, during which huge pieces of the front of a glacier break off and plunge into the water (Figure 9.3). Although calving is one of the more spectacular events offered by nature, it may also be hazardous. Many a small boat, kayak, and sightseer have been plunged into icy water when they have gotten too close to the front of a tidewater glacier. These glacial fronts are extremely unstable and may calve at any instant, sending hundreds of thousands of tons of ice plunging into the water in one thunderous crash. This generates large waves, often many tens of feet high, which are capable of capsizing even moderate-sized boats.

Figure 9.3. (a) McCarty Glacier, Alaska, is a valley glacier fed by many smaller mountain glaciers. It is a tidewater glacier that is rapidly retreating by calving. (b) McCarty glacier used to fill Northwestern Lagoon past the island visible in the background less than a century ago but has retreated to its present position by calving. Photos by T. Kusky

Avalanches

Glaciers and recently glaciated valleys have abundant very steep surfaces that are prone to avalanches. Active glaciers are continuously plucking material away from the bases of mountain valleys, causing higher material to avalanche onto the glacier where it gets carried away. Many glaciers are so covered with avalanche debris that they look like dirt-filled valleys. Snow also avalanches onto the main glacial surfaces, adding to the material that may get converted to ice. Recently glaciated valleys have many avalanches, particularly during wet periods and during shaking by earthquakes.

Icebergs and Sea Ice

Ice that has broken off an ice cap or polar sea, or that has calved off a glacier and is floating in open water, is known as sea ice. Sea ice presents a serious hazard to ocean traffic and shipping lanes and has sunk numerous vessels, including the famous sinking of the *Titanic* in 1912 that killed 1,503 people.

There are four main categories of sea ice. The first comes from ice that formed on polar seas in the Arctic Ocean and around Antarctica (Figure 9.4). The ice that forms in these regions is typically about three to four meters thick. Antarctica becomes completely surrounded by this sea ice every winter, and the Arctic Ocean is typically about 70 percent covered in the winter. During summer, many passages open up in this sea ice, but during the winter they re-close, forming pressure ridges of ice that may be up to tens of meters high. Recent observations suggest that the sea ice in the Arctic Ocean is thinning dramatically and rapidly and may soon disappear altogether. The ice cap over the Arctic Ocean rotates clockwise in response to the spinning of the earth. This spinning is analogous to putting an ice cube in a glass and slowly turning the glass. The ice cube will rotate more slowly than the glass because it is decoupled from the edge of the glass. About one-third of the ice is removed every year by the East Greenland current. This ice then moves south and becomes a hazard to shipping in the North Atlantic.

Icebergs from sea ice float on the surface, but between 81 and 89 percent of the ice will be submerged. The exact level that sea ice floats in the water depends on the exact density of the ice, as determined by the total amount of air bubbles trapped in the ice and how much salt got trapped in the ice during freezing.

A second group of sea ice forms as pack ice in the Gulf of St. Lawrence along the southeast coast of Canada; in the Bering, Beaufort, and Baltic Seas; in the Seas of Japan and Okhotsk; and around Antarctica. Pack ice builds up especially along the western sides of ocean basins, where cold currents are more common on the west sides of the oceans. Occasionally, during cold summers, pack ice may persist throughout the summer.

Several scenarios suggest that new ice ages may begin with pack ice that persists through many summers, gradually growing and extending to lower latitudes. Other models and data show that pack ice varies dramatically with a four- or five-year cycle, perhaps related to sunspot activity and the El Niño–Southern Oscillation (ENSO).

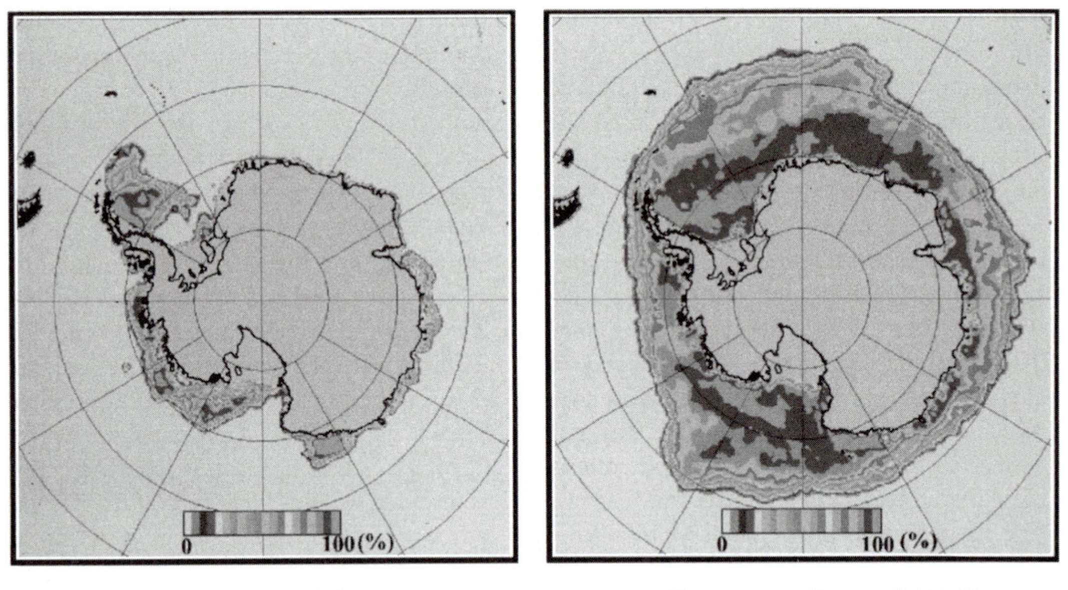

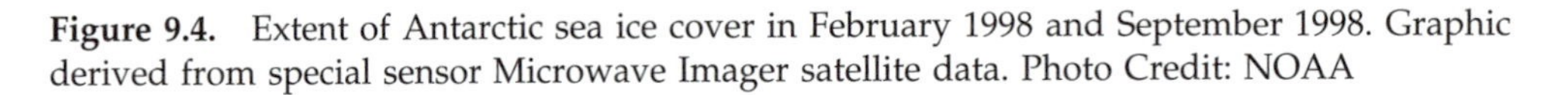

Figure 9.4. Extent of Antarctic sea ice cover in February 1998 and September 1998. Graphic derived from special sensor Microwave Imager satellite data. Photo Credit: NOAA

Pack ice presents hazards when it gets so extensive that it effectively blocks shipping lanes, or when leads (channels) into the ice open and close, forming pressure ridges that become too thick to penetrate with ice breakers. Ships attempting to navigate through pack ice have become crushed when leads close and the ships are trapped. Pack ice has terminated or resulted in disaster for many expeditions to polar seas, most notably Franklin's expedition in the Canadian Arctic and Scott's expedition to Antarctica. Pack ice also breaks up to form many small icebergs, but because these are not as thick as icebergs of other origins, they do not present as significant a hazard to shipping.

Pack ice also presents hazards when it drifts into shore, usually during spring breakup. With significant winds, pack ice can pile up on flat shorelines and accumulate in stacks up to fifty feet high. The force of the ice is tremendous and is enough to crush shoreline wharves, docks, buildings, and boats. Pack ice that is blown ashore also commonly pushes up high piles of gravel and boulders that may be thirty-five feet high in places. These ridges are common around many of the Canadian Arctic islands and mainland. Ice that forms initially attached to the shore presents another type of hazard. If it breaks free and moves away from shore, it may carry with it significant quantities of shore sediment, causing rapid erosion of beaches and shore environments.

Pack ice also forms on many high-latitude lakes, and the freeze-thaw cycle causes cracking of the lake ice. When lake water rises to fill the cracks, the ice cover on the lake expands and pushes over the shoreline, resulting in damage to any structures built along the shore. This is a common problem on many lakes in northern climates and leads to widespread damage to docks and other lakeside structures.

An unusual pack ice disaster has been occurring in northern Quebec, Canada, along the Ungava Peninsula on the east side of Hudson Bay. A series of dams has been built in Canada along rivers that flow into Hudson Bay, and these dams are used to generate clean hydroelectric energy. The problem is that these dammed rivers have annual spring floods, which before the dams were built would flush the pack ice out of Hudson Bay. Since the dams have been built, the annual spring floods are diminished, resulting in the pack ice remaining on Hudson Bay through the short summer. This situation has drastically changed the summer season on the Ungava Peninsula: As the warm summer winds blow across the ice they pick up cool moist air, and cold fogs now blow across the Ungava all summer. These fogs have drastically changed the local climate and have hindered the growth and development of the region.

Icebergs present the greatest danger to shipping. In the northern hemisphere, most icebergs calve off glaciers in Greenland or Baffin Island, Canada, then move south through the Davis Strait between Greenland and Baffin Island into shipping lanes in the North Atlantic off Newfoundland. Some icebergs calve off glaciers adjacent to the Barents Sea, and others come from glaciers in Alaska and British Columbia. In the southern hemisphere, most icebergs come from Antarctica, though some come from Patagonia in southern Argentina.

Once in the ocean, icebergs drift with ocean currents, but because of the Coriolis force they are deflected to the right in the northern hemisphere and to the left in the southern hemisphere. Most icebergs are 100–300 feet high and up to about 2,000 feet in length. However, in March 2000, a huge iceberg broke off the Ross Ice Shelf in Antarctica, and this berg was roughly the size of the state of Delaware. It had an area of 4,500 square miles and stuck 205 feet out of the water. Icebergs in the northern hemisphere pose a greater threat to shipping, as those from Antarctica are too remote and rarely enter shipping lanes. Ship collisions with icebergs have resulted in numerous maritime disasters, especially in the North Atlantic on the rich fishing grounds of the Grand Banks off the coast of Newfoundland.

Icebergs are now tracked by satellite, and ships are updated with their positions so that they can avoid any collisions that could prove fatal. Radio transmitters are placed on larger icebergs to more closely monitor their locations, and many ships now carry more sophisticated radar and navigational equipment that helps them track the positions of large icebergs.

Icebergs also pose a serious threat to oil drilling platforms and sea floor pipelines in high-latitude seas. Some precautions have been taken such as building seawalls around or near shore platforms, but not enough planning has gone into preventing an iceberg colliding with and damaging an oil platform, or from one being dragged across the sea floor and rupturing a pipeline.

SEA LEVEL CHANGES

Global sea levels are currently rising as a result of the melting of the Greenland and Antarctica ice sheets and the thermal expansion of the world's ocean waters due to global warming. We are presently in an interglacial stage of an ice age: Sea levels have risen nearly 400 feet since the last glacial maximum 20,000 years ago and about six inches in the past 100 years. The rate of sea level rise seems to be accelerating and may presently be as much as an inch every 8–10 years. If all the ice on both ice sheets were to melt, global sea levels would rise by 230 feet (sixty-six meters), inundating most of the world's major cities and submerging large parts of the continents under shallow seas. The coastal regions of the world are densely populated and are experiencing rapid population growth. Approximately 100 million people presently live within one meter of the present-day sea level. If sea level were to rise rapidly and significantly, the civilized world would experience an economic and social disaster of a magnitude not yet experienced by them. Many areas would become permanently flooded or subject to inundation by storms, beach erosion would be accelerated, and water tables would rise.

The Greenland and Antarctic ice sheets have some significant differences that cause them to respond differently to changes in air and water temperatures. The Antarctic ice sheet is about ten times as large as the Greenland ice sheet, and since it sits on the South Pole, Antarctica dominates its own climate. The surrounding ocean is cold even during summer, and much of Antarctica is a cold desert with low precipitation rates and high evaporation potential. Most meltwater in Antarctica seeps into underlying snow and simply refreezes with little runoff into the sea. Antarctica hosts several large ice shelves that are fed by glaciers moving at rates of up to a thousand feet per year. Most ice loss in Antarctica is accomplished through calving and basal melting of the ice shelves at rates of 10–15 inches per year.

In contrast, Greenland's climate is influenced by warm North Atlantic currents and its proximity to other land masses. Climate data measured from ice cores taken from the top of the Greenland ice cap show that temperatures have varied significantly in cycles of years to decades. Greenland also experiences significant summer melting and abundant snowfall, it has few ice shelves, and its glaciers move quickly at rates of up to miles per year. These fast-moving glaciers are able to drain a large amount of ice from Greenland in relatively short amounts of time.

The Greenland ice sheet is thinning rapidly along its edges, losing an average of 15–20 feet in the past decade. In addition, tidewater glaciers and the small ice shelves in Greenland are melting faster than the Antarctic ice sheets, with rates of melting between 25–65 feet per year. About half of the ice lost from Greenland is through surface melting that runs off into the sea. The other half of ice loss is through calving of outlet glaciers and melting along the tidewater glaciers and ice shelf bases.

These differences between the Greenland

and Antarctic ice sheets lead them to play different roles in global sea level rise. Greenland contributes more to the rapid short-term fluctuations in sea level, responding to short-term changes in climate. In contrast, most of the world's water available for raising sea level is locked up in the slowly changing Antarctic ice sheet. Antarctica, therefore, contributes more to the gradual, long-term sea level rise.

What is causing the rapid melting of the polar ice caps? Most data suggest that the current melting is largely the result of the gradual warming of the planet in the past 100 years through the effects of greenhouse warming. Greenhouse gases have been increasing at a rate of more than 0.2 percent per year, and global temperatures are rising accordingly. The most significant contributor to the greenhouse gas buildup is CO_2, which is produced mainly by the burning of fossil fuels. Other gases that contribute to greenhouse warming include carbon monoxide, nitrogen oxides, methane (CH_4), ozone (O_3), and chlorofluorocarbons. Methane is produced by gas from grazing animals and termites, nitrogen oxides are increasing because of the increased use of fertilizers and automobiles, and chlorofluorocarbons are increasing from the release from aerosols and refrigerants. Together, the greenhouse gases have the effect of allowing shorter wavelength incoming solar radiation to penetrate the gas in the upper atmosphere but then trapping the solar radiation after it is re-emitted from the earth in a longer wavelength form. The trapped radiation causes the atmosphere to heat up, leading to greenhouse warming. Other factors also influence greenhouse warming and cooling, including the abundance of volcanic ash in the atmosphere and solar luminosity variations, as evidenced by sun spot variations.

Measuring global (also called *eustatic*) sea level rise and fall is difficult because many factors influence the relative height of the sea along any coastline. These vertical motions of continents are called epeirogenic movements and may be related to plate tectonics, to rebound from being buried by glaciers, or to changes in the amount of heat added to the base of the continent by mantle convection. Continents may rise or sink vertically, causing apparent sea level change, but these sea level changes are relatively slow compared to changes induced by global warming and glacial melting. Slow, long-term sea level changes can also be induced by changes in the amount of sea floor volcanism associated with sea floor spreading. At some times in Earth's history, sea floor spreading was particularly vigorous, and the increased volume of volcanoes and the mid-ocean ridge system caused global sea levels to rise.

Steady winds and currents can mass water against a particular coastline, causing a local and temporary sea level rise. Such a phenomenon is associated with the ENSO, causing sea levels to rise by 4–8 inches in the Austral-Asia region. When the warm water moves east in an ENSO event, sea levels may rise 4–20 inches across much of the North and South American coastlines. Other atmospheric phenomena can also change sea level by centimeters to meters locally on short time scales. Changes in atmospheric pressure, salinity of sea waters, coastal upwelling, onshore winds, and storm surges all cause short-term fluctuations along segments of coastline. Global or local warming of waters can cause them to expand slightly, causing a local sea level rise. It is even thought that the extraction and use of groundwater and its subsequent release into the sea might be causing sea level rise of about 1.3 mm per year. Seasonal changes in river discharge can temporarily change sea levels along some coastlines, especially where winter cooling locks up large amounts of snow that melt in the spring.

It is clear that attempts to estimate eustatic sea level changes must be able to average out the numerous local and tectonic effects in order to arrive at a globally meaningful estimate of sea level change. Most coastlines seem to be dominated by local fluctuations that are larger in magnitude than any global sea level rise. Recently, scientists have employed satellite radar technology to precisely measure sea surface height and to document annual changes in sea level. Radar altimetry is able to map sea surface elevations to the sub-inch scale globally, providing an unprecedented level of understanding of sea surface topography. Satellite techniques support the concept that global sea levels are rising at about .01 inches per year.

What effect will rising sea levels have on the world's cities and low-lying areas? Are we going to enter a new glacial period, and, if so, what will be the consequences?

RESOURCES AND ORGANIZATIONS

Print Resources

Alley, R. B., and Bender, M. L. "Greenland Ice Cores: Frozen in Time." *Scientific American* (February 1998).

Dawson, A. G. *Ice Age Earth.* London: Routledge, 1992, 293 pp.

Douglas, B., Kearney, M., and Leatherman, S. *Sea Level Rise: History and Consequences.* International Geophysics Series, vol. 75. San Diego: Academic Press, 2000, 232 pp.

Erickson, J. *Glacial Geology: How Ice Shapes the Land.* Changing Earth Series. New York: Facts on File, 1996.

Molina, B. "Modern Surge of Glacier Comes to an End." *EOS* 75, no. 47 (November 22, 1994).

Schneider, D. "The Rising Seas." *Scientific American* (March 1997).

Stone, G. "Exploring Antarctica's Islands of Ice." *National Geographic* (December 2001), 36–51.

Thomas, R. "Remote Sensing Reveals Shrinking Greenland Ice Sheet." *EOS* 82, no. 34 (2001), 369–73.

Non-Print Resources

Videos

Glaciers, Emecorp, 1989, 15 mins.

Award-winning photography shows glacial erosion in action! Advancing ice masses carve U-shaped valleys and leave a trail of lakes and erratics, while melting glaciers deposit kames, drumlins, and outwash plains. Covers alpine and continental glaciation. Animation sequences help explain key processes.

Glaciers: Earth Explored Series, Public Broadcasting System, 1985, 28 mins.

Web Sites

NASA Goddard Space Flight Center's Scientific Visualization Studio

http://webserv.gsfc.nasa.gov/SVS/stories/greenland/

This site offers many extremely impressive views and animations of Greenland's receding glaciers.

Polar Pointers

http://www-bprc.mps.ohio-state.edu/cgi-bin/genpp.cgi

This Web site has many links to interesting sites about glaciers, glacial hazards, and arctic themes.

Organizations

Arctic Climate System Study Mission
http://acsys.npolar
The scientific goal of the Arctic Climate System Study (ACSYS) project is to ascertain the role of the Arctic in global climate by attempting to find answers to the following related questions: (1) What are the global consequences of natural or manmade changes in the Arctic climate system? and (2) Is the Arctic climate system as sensitive to increased greenhouse gases as climate models suggest?

National Aeronautics and Space Administration (NASA) Program for Arctic Regional Climate Assessment
Researches the Greenland ice sheet by measuring changes in ice sheet volume, uses satellite and aircraft data to monitor glaciers, and conducts field programs to measure in situ properties of glaciers.

National Science Foundation Polar Research Program
http:www.nsf.gov/home/polar/
NSF states the mission of the Polar Research Program as follows: "The earth's polar regions offer compelling scientific opportunities, but their isolation and their extreme climates challenge the achievement of these opportunities. The Foundation's programs for support of research in the Antarctic and the Arctic acknowledge the need to understand the relationships of these regions with global processes and the need to understand the regions as unique entities. NSF's polar programs, most of which are supported through the Office of Polar Programs, thus provide support for investigations in a range of scientific disciplines."

NOAA Paleoclimatology Program
National Geophysical Data Center
http://www.ngdc.noaa.gov/paleo/
The center is the central location for paleoclimate data, research, and education. The program's mission, as stated on its Web page, is to "help the World share scientific data and information related to climate system variability and predictability. Our mission is to ensure the international paleoclimate research community meets the scientific goals of programs including IPCC, IGBP PAGES, WCRP CLIVAR, and NOAA's Climate and Global Change Program."

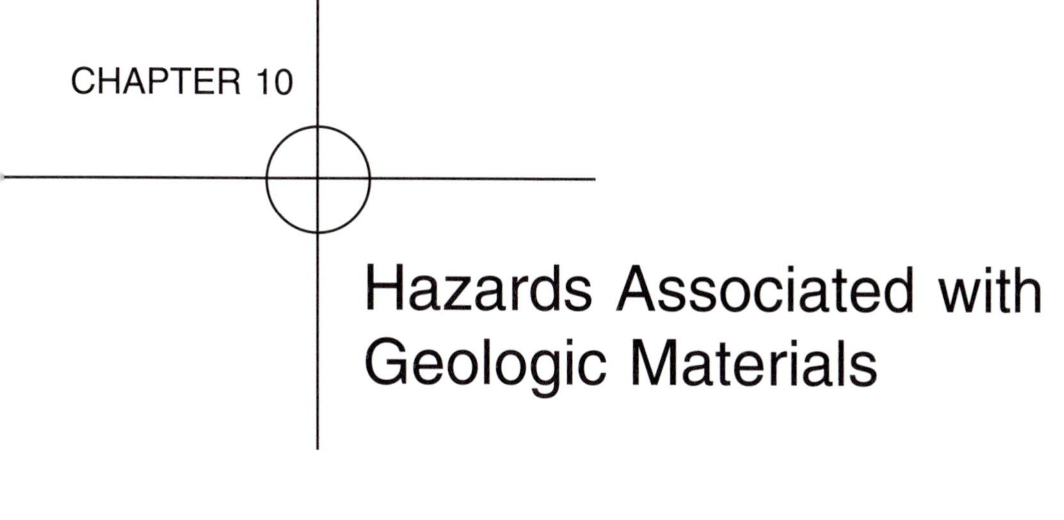

Hazards Associated with Geologic Materials

INTRODUCTION

Some materials are simply dangerous. A game that was popular in the 1970s involved tossing a make-believe 8-ball time bomb from person to person, and the point was that you didn't want to be caught holding the ball when it exploded. Similarly, there are some natural geologic materials that you would not want to be caught holding, ever. Natural geologic processes have concentrated some of the more hazardous elements and compounds in some locations and left them virtually absent in others. Many of these hazards from geologic materials are silent yet deadly. Poisonous gases creep into homes, dissolved chemicals make their way into groundwater wells, and radioactive particles are constantly shooting through our bodies. Scientists are only beginning to understand what causes hazardous elements to be concentrated in some cases and what needs to be done to minimize the hazards associated with geologic materials.

Uranium and plutonium occur naturally and look like many other rocks. Would you know if you were holding an ounce of plutonium? You would find out soon. Luckily, naturally concentrated radioactive minerals are rare, and you are not likely to pick up a hazardous sample on a roadside foray. However, uranium and other radioactive minerals have been concentrated naturally in a few cases, even leading to natural nuclear explosions in the West African nation of Gabon.

Toxic metals, such as selenium, zinc, arsenic, lead, and other elements, may be concentrated in some locations and make their way through soils to drinking water supplies, food supplies, and even into the air we breathe. Exposure to toxic levels of these and other trace elements is a serious global health issue with billions of people exposed to potentially harmful material hazards. It is important to understand where these and other elements are naturally concentrated as well as what the effects of removing them from the environment, such as during lead-mining operations, may have on human health. An emerging field of science is *medical geology,* which deals with the effects of geologic materials and processes on human, plant, and animal health (Finkelman et al., 2001). Although the effects of airborne and trace quantities of some hazardous materials are only recently widely appreciated, some of the harmful effects of certain minerals have been known for thousands of years. Aristotle noted lead poisoning in some lead miners, and Hippocrates described relationships between environmental hazards and human health.

Asbestos is presently being removed from thousands of buildings because of the perceived threat it poses to those people exposed to breathing its deadly fibers. Asbestos is a fire retardant and was widely used as an additive to insulation materials until it was determined that breathing asbestos fibers could cause cancer. Asbestos, however, is relatively inert and does not tend to leap into the air system as many would fear. If left alone, asbestos poses little threat to those walking past insulated pipes and the like. However, if asbestos becomes airborne as fine-grained dust particles, it can be deadly. Other dusts can carry pathogens, toxins, soil fungi, heavy metals, and other harmful elements that can adversely affect human health.

Other materials are hazardous in different ways: For instance, there are many clays that expand by 400 percent when water is added to their environment. The expansion of these clays is powerful and has the force to slowly crumble bridges, foundations, and tall buildings. Damage from expanding clays is one of the most costly of all natural hazards in the United States, causing billions of dollars of damage every year.

Invisible and odorless poisonous gases that are emitted from underlying rocks, minerals, and soils invade many homes and businesses in the United States and abroad. The most common hazardous gas of this class is radon, which can cause lung cancer and other ailments. Millions of homes in the United States alone have radon problems, and the populace is beginning to understand and mitigate this problem.

Other people are drinking their demise. On the Indo-Gangetic plain, south of the Himalayan Mountains in India and Bangladesh, many groundwater wells are contaminated with natural arsenic. Twenty-five to seventy-five million people are drinking this water, becoming contaminated, and suffering from the awful effects of arsenic poisoning. Much of this suffering is preventable. It is a simple matter to get water from nearby, uncontaminated wells. However, many of the local people do not understand the danger and can not afford the energy to walk the extra five kilometers to get the water from a clean well. Monitoring groundwater arsenic levels could prevent much of this arsenic poisoning, and this is being done by several UN organizations with limited success. Getting drinking water in many parts of the world is not as simple as turning on the faucet in the United States. Somebody, usually the women, must carry vessels from the home to the well, fill the vessels, and carry them back to the home. Some wells are simple and easily accessed; others involve long climbs down treacherous paths in narrow caves that lead to the groundwater level.

Some natural and human-induced processes related to geologic materials are hazardous on both local and global scales. For instance, there are thousands of underground fires burning in underground coal seams, especially in China. These fires are releasing tons of hazardous gases to the atmosphere and may be contributing to global warming.

TYPES OF MATERIAL HAZARDS

Natural processes in many places on the planet concentrate potentially hazardous geologic materials. The hazards to people that are posed by these elements depend on the way that the people interact with their environment, which can be very different between different cultures. Primitive cultures that live off the land are more susceptible to hazards and disease associated with contaminated or poor water quality, toxic elements in plants harvested from contaminated soils, and insect- and animal-borne diseases associated with unsanitary environments. In contrast, more developed societies are more likely to be affected by air pollution, different types of water pollution, and indoor pollution such as radon exposure. Some diseases reflect a complex interaction between humans, insects or animals, climate, and the natural concentration of some elements in the environment. For instance, *schistosomiasis*-bearing snails are abundant in parts of Africa and Asia where natural waters are rich in calcium, but in similar climates in South America, schistosomiasis is rare. It is thought that this difference is because the waters in South America are calcium poor, and the disease-bearing snails need the calcium to build their shells.

All life that we know is built from a few basic elements, including hydrogen, carbon, nitrogen, oxygen, phosphorous, sulfur, chlorine, sodium, magnesium, potassium, and calcium. Some other elements are important for life, as they play vital roles in controlling how tissues and organs work. Trace element metals are present in very dilute quantities in our bodies, and some that are known to be important for life functions include fluorine, chromium, manganese, iron, cobalt, copper, zinc, selenium, molybdenum, and iodine. Other elements accumulate in tissue as it ages, but their function, and whether they are beneficial or detrimental, is yet to be determined. These *age elements* include nickel, arsenic, aluminum, and barium.

The distribution of elements in the natural environment is complex and may be changed by many different processes. Geologic processes such as volcanism may concentrate certain elements in some locations even to ore-grade or unhealthy levels. When these igneous rocks are weathered, the concentrations of specific elements may be increased or decreased in the soil horizon, depending on the element, climate, and other factors. After this, biologic processes may further concentrate elements. Together, processes of leaching and accumulation of elements during soil formation, biological concentration, and many other processes may concentrate or disperse elements that may be harmful to humans.

Hazardous Elements, Minerals, and Materials

There are more than 100 naturally occurring elements, many of which are toxic to humans in high doses. The same elements may be beneficial or even necessary in small dilute doses and pose little or no threat in intermediate concentrations. Most elements show similar toxicity effects on people, although not all are necessary in small doses nor are they all toxic in high doses. Understanding the effects of trace elements in the environment on human health is the huge and rapidly growing field of medical geology and is beyond the scope of this book. However, a few specific examples are instructory.

Some minerals are hazardous when exposed in the natural environment or when extracted in mining operations. In particular, selenium, asbestos, silica, coal dust, and lead can be harmful when inhaled or when present in high concentrations in the environment. These are discussed next, along with the beneficial mineral iodine.

Iodine Iodine occurs naturally in the geologic environment and is released from rocks by weathering. It is readily soluble in water, so most iodine makes its way to the sea after it is leached from bedrock or soil. A deficiency of iodine in the body can lead to several adverse health effects including thyroid disease and goiter.

There is a strong correlation between the geography of occurrence of thyroid disease and a deficiency of iodine in the environment. Much of the northern half of the conterminous United States has soils with low iodine contents, and it is the same region that yields most of the thyroid disease cases in the United States.

Selenium Selenium is one of the most toxic elements known in the environment. Like most elements, selenium is needed in small concentrations for normal biological functions. Concentrations of 0.04 to 0.1 parts per million are healthy, but any concentrations of over 4 parts per million are toxic.

Selenium is produced naturally by volcanic activity, and it is usually ejected as small particles that fall out near the volcano, resulting in a concentration near volcanic vents. Selenium in natural soils ranges from 0.1 parts per million to more than 12,000 parts per million in organic-rich soils. Selenium exists in insoluble form in acidic soils and in soluble form in alkaline soils. Selenium may also be concentrated by biological activity. Some plants take up soluble selenium and concentrate it in their structures. The efficiency of this process is dependent on what form (soluble or insoluble) selenium exists in the environment. Selenium is also concentrated in human tissue to about 1,000 times the background level in fresh water. It is also concentrated by up to 2,000 times the natural background level in marine fish.

The concentration of selenium in biologic material has persisted back through geological time, and many coals and fossils fuels are also rich in selenium. Burning coal releases large amounts of selenium to the atmosphere, and this selenium then rains down on the landscape.

Asbestos Asbestos was widely used as a flame retardant in buildings through the mid-1970s, and it is present in millions of buildings in the United States. It was also used in vinyl flooring, ceiling tiles, and roofing material. It is no longer used in construction since it was recognized that asbestos might cause certain types of diseases, including asbestosis (pneumoconiosis), a chronic lung disease. Asbestos particles get lodged in the lungs, and the lung tissue hardens around the particles, decreasing lung capacity. This decreased lung capacity causes the heart to work harder, leading to heart failure and death. Virtually all deaths from asbestosis can be attributed to long-term exposure to asbestos dust in the workplace before environmental regulations governing asbestos were put in place. A less common disease associated with asbestos is mesothelioma, a rare cancer of the lung and stomach linings. Asbestos has become one of the most devastating occupational hazards in U.S. history, costing billions of dollars for cleaning up asbestos in schools, offices, homes, and other buildings. Approximately $3 billion a year is currently spent on asbestos removal in the United States.

Asbestos is actually a group of six related minerals, all with similar physical and chemical properties. Asbestos includes minerals from the amphibole and serpentine groups that are long and needle-shaped, making it easy for them to get lodged in people's lungs. The Occupational Safety and Health Administration (OSHA) defined asbestos as having dimensions of greater than 5 micrometers (0.002 inches) long, with a length to width ratio of at least 3:1. The minerals in the amphibole group included in this definition are grunerite (known also as amosite), reibeckite (crocidolite), anthophyllite, tremolite, and actinolite, and the serpentine group mineral that fits the definition is chrysotile. Almost all of the asbestos used in the United States is chrysotile (known as white asbestos), and about 5 percent of the asbestos used was crocidolite (blue asbestos) and amosite (brown asbestos). There is currently considerable debate among geologists, policy-makers, and medical officials on the relative threats from different kinds of asbestos.

In 1972, OSHA and the U.S. government began regulating the acceptable levels of asbestos fibers in the workplace. The Environmental Protection Agency (EPA) agreed and declared asbestos a Class A carcinogen. The EPA composed the Asbestos Hazard Emergency Response Act, which was signed by President Reagan in 1986. OSHA gradually lowered the acceptable limits from a preregulated estimate of greater than 1,600 fibers per cubic centimeter (4,000 fibers per cubic inch) to 1.6 fibers per cubic centimeter (4 particles per cubic inch) in 1992. Responding to public fears about asbestosis, Congress passed a law requiring that any asbestos-bearing material that appeared to be visibly deteriorating must be removed and replaced by non-asbestos-bearing material. This remarkable ruling has resulted in billions of dollars being spent on asbestos removal, which in many cases may have been unnecessary. The asbestos can only be harmful if it is an airborne particle, and only long-term exposure to high concentrations leads to disease. In some cases, it is estimated that the processes of removing the asbestos resulted in the inside air becoming more hazardous than before removal, as the remediation can cause many small particles to become airborne and fall as dust throughout the building.

Asbestos fibers in the environment have led to some serious environmental disasters, as the hazards were not appreciated during early mining operations before the late 1960s. One of the worst cases occurred in the town of Wittenoom, Australia. Crocidolite was mined in Wittenoom for twenty-three years between 1943 and 1966, and the mining was largely unregulated. Asbestos dust filled the air of the mine and the town, and the 20,000 people who lived in Wittenoom breathed the fibers in high concentrations daily. More than 10 percent or 2,300 people who lived in Wittenoom have since died of asbestosis, and the Australian government has condemned the town and is in the process of burying it in deep pits to rid the environment of the hazard.

In the United States, W.R. Grace and Co. in Libby, Montana, afflicted hundreds of people with asbestos-related diseases through mining operations. Vermiculite was mined at Libby from 1963 to 1990 and shipped to Minneapolis to make insulation products, but the vermiculite was mixed with the tremolite (amphibole) variety of asbestos. In 1990, the EPA tested residents of Libby and found that 18 percent of residents who had been there for at least six months had various stages of asbestosis and that 49 percent of the W.R. Grace mine employees had asbestosis. The mine was closed down, and Libby is now being considered a potential superfund site by the EPA. The problem was not limited to Libby, however; twenty-four workers at the processing plant in Minneapolis have since died from asbestosis, and one resident who lived near the factory has also died. The EPA and Minnesota's Department of Health are currently assessing the level of exposure to other nearby residents.

Silica and Coal Dust Other minerals can be hazardous if they are made into small airborne particles that can become lodged in the lungs. As with asbestos, both silica and coal mining operations release into the air large amounts of small dust particles, which are also known respectively as quartz dust and coal dust. Workers exposed to these types of dust are at risk for diseases that are broadly similar to asbestosis.

Quartz dust is commonly produced during rock drilling and sandblasting operations. These practices produce airborne particles of various sizes, the largest of which are naturally filtered by hair and mucous membranes during inhalation. However, some of the smallest particles can work their way deep into the lungs and get lodged in the air sacs of the lungs, where they can do great harm. When small particles get trapped in the air sac, the lungs react by producing fibrolitic nodules and scar tissue around the trapped particles, reducing lung capacity in a disease called silicosis. This disease is easily preventable by simply wearing a respiratory mask when exposed to silica fibers, although this is not yet a common practice.

Coal dust has presented a long-term health problem in the United States and elsewhere in the world, with underground coal miners being at high risk for developing this disease. Mining operations inevitably release many fine particles of coal into the air. These particles also may get lodged in the lungs, resulting in a myriad of diseases including chronic bronchitis and emphysema, which are collectively known as black lung disease. The longer a miner works underground, the greater the risk of developing black lung disease. Miners who work underground for less than ten years have about a 10 percent chance of developing these symptoms, whereas miners who have worked underground for more than forty years have a 60 percent chance of developing black lung disease.

Lead Lead is a metalliferous element used primarily for pipes, solder, batteries, bullets, pigments, radioactivity shields, and wheel weights. Lead is a known environmental hazard, and ingestion of large amounts of lead can lead to developmental problems in children, retardation, brain damage, and birth defects. It also may lead to kidney failure, multiple sclerosis, and brain cancer. Some researchers speculate that the fall of the Roman Empire was partly caused by lead poisoning. The Romans drank a lot of wine, and lead was concentrated at several different steps in the process they used to make the wine. The upper class also drank from lead cups and had water pumped into their homes in lead pipes. It is thought that lead poisoning contributed to brain damage, retardation, and the high incidence of birth defects among the Romans. These ideas are supported by the high contents of lead measured in the remains of some exhumed Roman burials. Remarkably, the lead content of ice cores from Greenland representing the Roman Empire period (500 B.C.–300 A.D.) also preserve about four times the normal level of lead, reflecting the increased mining and use of lead by the Romans.

Lead is present in the natural environment in several different forms. Galena is the most common ore mineral, forming shiny cubes with a silvery "lead" color. Lead is not generally hazardous in its natural mineral form, but becomes hazardous when mined and released from smelters as particulates, when leached from pipes or other fixtures, or when released into the air from automobile fumes. These processes can lead to high concentrations of native lead in soils, streams, and rivers. Lead may then be taken up by plants or aquatic organisms and thus enter the food chain where it can do great damage. Lead paint is also a great hazard in many homes in the United States, as lead was used as a paint additive until the 1970s. Paint in many older homes is peeling and ingested by infants, and paint along window frames is turned into airborne dust when windows are opened and closed. Environmental regulations in many states now require the removal of lead paint from homes upon the sale or leasing of properties.

The largest lead smelter in the United States, in Herculaneum, Missouri, brings an example of the legacy of lead mining. Herculaneum is located about thirty miles south of St. Louis in the heart of the nation's largest lead deposits and has been the site of mining operations for generations. The problem in Herculaneum is that the town's smelter releases thirty-four tons of emissions per year (reduced from 800 tons per year a generation ago), including fine-grained lead dust. This rains down on the local community, and the local street dirt has been tested as containing 30 percent lead (Wilgoren, 2002). Signs on the streets in town warn children not to play in the streets, curbs, or sidewalks, and parents are vigilant in attempting to keep the dust off toys, shoes, and out of the food and water supply. All their efforts are not enough, though, and the State of Missouri is now replacing the soil on 535 properties contaminated by lead. Many of the children and adults in the town are suffering the effects of lead poisoning, with retardation, stunted growth, hearing loss, and clusters of brain cancer and multiple sclerosis in town. One-quarter of all the children in the town tested positive for lead poisoning in 2001. Lead contamination has long been suspected in Herculaneum, but it wasn't until 2002 that the federal government stepped in. The Environmental Protection Agency has recently (January 2002) initiated a large-scale relocation program, initially moving 100 families with young children or pregnant women to safer locations. This may only be the beginning of the end, as Congressman Dick Gephardt, the former House Democratic leader, has suggested that the government shut down the Doe Run Lead Smelter and perhaps relocate the 2,800 families remaining in the town of Herculaneum.

Soil Hazards: Expansive Clays, Quicksand, and Dissolved Pollutants

Sometimes the very soil on which we stand can prove hazardous. There are many different types of soils (see Chapter 5), and most of them are relatively inert. However, other soils expand and contract dramatically with changes in seasonal moisture content, causing damage to structures built on them. Other soils become liquid like quicksand when agitated and may swallow people and structures. Still other soils contain high concentrations of contaminated water, radioactive gases, and other elements that can prove hazardous. In this section, we discuss hazardous aspects of soil geology.

Soil can be defined as that part of the regolith that can support rooted plants. Soil is formed by a combination of physical and chemical weathering and by organic decay. The organic component is important, because organic decay produces nutrients to support plant life. Soils can be thought of as complex ecosystems, and every cubic-meter may contain millions of living organisms. Soils are important parts of our environment and determine land use patterns, types of agriculture that can be supported, which areas have groundwater aquifers, and which are aquicludes that stop the movement of water. Soils also play a vital role in locating sites of waste disposal, buildings, and other structures. The material properties of soils are very variable and control hillslope stability, landslide potential, and groundshaking potential during earthquakes.

The amount of water saturation in soils is extremely variable both from place to place and in time at any given place. Water saturation of soils is important for understanding soil hazards, since water saturation largely controls the mechanical properties of soils. The influence of water on the mechanical properties of soil can be appreciated by considering attempts to build two sandcastles: one in dry sand and one in wet sand. Both sands may be exactly the same in all aspects except for water content. Attempts to build sandcastles in the wet sand are successful—the sand sticks together and walls, towers, and tunnels can be constructed. However, attempts to build sandcastles in dry sand will be unfruitful. Walls will slide and remain at the angle of repose (an unsatisfactory 22° for sandcastles), towers can not be constructed, and tunnels are hopeless. The simple addition of water changes the mechanical properties so much that the castle can be constructed in wet sand, whereas constructing the castle in dry sand is not viable.

The effect of water may be the opposite in soils with different structure, particularly those with more clay minerals. This is exemplified by clay-rich road surfaces that are common in many desert environments. When these are dry, driving on them is simple and stable, though dusty. However, if it rains the clay becomes wet and slippery, and the road may become nearly impossible to drive on. Likewise, water in many soils percolates through the pores and lubricates grain boundaries, making the entire soil unstable.

Engineers determine the potential hazards of soils by evaluating several different properties of the soil. The most significant variables are the size of the particles in the soil, the types of solid particles, and how much water is present. Porosity and permeability are important, as is the overall mixture of solid, liquid, and gas.

The *strength* of a soil is a function of frictional forces, determined by the soil's particle size, shape, density, surface roughness, and several other factors. A soil's cohesion, or how well the particles stick together, is a function of the water content of the soil as well as the strength of the molecular and electrostatic forces.

The *sensitivity* of a soil is a measure of how the strength changes with shaking or with other disturbances such as those associated with excavation or construction. Sensitivity is dependent on soil type and particularly on clay content. Soils with high clay contents may lose up to 75 percent of their strength with shaking, whereas sand- and gravel-rich soils tend to be less sensitive and retain more of their strength during shaking.

The *compressibility* of a soil is determined by many things, foremost of which is the amount of fine-grained and organic materials present in the soil. Many of these materials are prone to settling when they are built on.

The *corrosiveness* of a soil is a measure of its ability to corrode or chemically decompose buried objects, such as pipes, wires, tanks, and posts. The composition of the soil and composition of the buried material, as well as the amount of water percolating through the soil, determine a soil's corrosiveness.

The *shrink/swell potential* of a soil is extremely important to know for construction to begin in an area. The shrink/swell potential is a measure of a soil's ability to add or lose water at a molecular level. Some particular types of clays are said to be *expansive,* and they add layers of water molecules between the plates of clay minerals (made of silica, aluminum, and oxygen), loosely bonding the water in the mineral structure. Damage from shrinking and swelling clays in soils is the most costly of all natural hazards and disasters, costing more than $2 billion a year in the United States alone. Most expansive clays are rich in montmorillonite, a clay mineral that can expand up to fifteen times its normal dry size. Most soils do not expand more than 25–50 percent of their dry volume, but it must remembered that an expansion of 3 percent is considered hazardous.

Damage from shrinking/swelling soils is mostly to bridges, foundations, and roadways, all of which may crack and move during expansion. Regions with pronounced wet and dry seasons tend to have a greater problem with expansive clays than regions with more uniform precipitation distributed throughout the year. This is because the changes in the adhered water content of the clays change less in regions where the soil moisture remains more constant. Some damage from shrinking and swelling soils can be limited, especially around homes. Trees that are growing near foundations can cause soil shrinkage during dry seasons and expansion in wet seasons. These dangers of shrinking/swelling soils can be avoided by not planting trees too closely to homes and other structures. Local topography and drainage details also influence the site-specific shrink/swell potential. Buildings should not be placed in areas with poor drainage as water may accumulate there and lead to increased soil expansion. Local drainage may be modified to allow runoff away from building sites, reducing the hazards associated with soil expansion.

In general, soils that are rich in clay minerals and organic material tend to have low strength, low permeability, high compressibility, and the greatest shrink/swell potential. These types of soils should be avoided for construction projects when possible. If they can not be avoided, steps must be taken to accommodate these undesirable traits into the building construction. Sand- and gravel-rich soils pose much less danger than clay and organic rich soils. These soils are well-drained and strong, have low compressibility and sensitivity, and have low shrink/swell potential.

HAZARDS ASSOCIATED WITH GROUND WATER CONTAMINATION

Fresh water is one of the most important resources in the world. Wars have been and will be fought over the ability to obtain fresh water, and water rights are hot political issues in places where it is scarce, like the American West and the Middle East (see Chapter 8). Since we live in a finite world with a finite amount of fresh water, and since the global population is growing rapidly, it is likely that fresh water will become an increasingly important topic for generations to come.

Less than 1 percent of the planet's water is *groundwater,* which may be defined as all the water contained within spaces in bedrock, soil, and regolith. However, the volume of groundwater is thirty-five times the volume of freshwater in lakes and streams. Americans and people of other nations have come to realize that groundwater is a vital resource for their nation's survival and are only recently beginning to appreciate that much of the world's groundwater resources have become contaminated by natural and human-aided processes. Approximately 40 percent of drinking water in the United States comes from groundwater reservoirs. About 80 billion gallons of groundwater are pumped out of these reservoirs every day in the United States.

Where does this water come from? It comes from rainfall, where it seeps into the ground and slowly makes its way downhill toward the sea. There is water everywhere beneath the ground surface, and most of this occurs within 2,500 feet (750 meters) of the surface. The volume of groundwater is estimated to be equivalent to a layer that is 180 feet (fifty-five meters) thick and is spread evenly over the earth's land surface.

The distribution of water in the ground can be divided into the *unsaturated* and the *saturated* zones. The top of the water table is defined as the upper surface of the saturated zone. Below this surface, all openings are filled with water.

Movement of Groundwater

Most of the water under the ground doesn't just sit there; it is constantly in motion, although rates are typically only centimeters per day. The rates of movement are controlled by the amount of open space in the bedrock or regolith and how the spaces are connected.

Porosity is the percentage of total volume of a body that consists of open spaces. Sands and gravels typically have about 20 percent open spaces, whereas clays have about 50 percent. The sizes and shapes of grains determine their porosity, which is also influenced by how much they are compacted, cemented together, or deformed.

In contrast, *permeability* is a body's capacity to transmit fluids or to allow the fluids to move through its open pore spaces. Permeability is not directly related to porosity, because if all the pore spaces in a body are isolated, then it may have high porosity, but the water may be trapped and unable to move through the body. Permeability is also affected by *molecular attraction,* the force that makes thin films of water stick to things instead of being forced to the ground by gravity. If the pore spaces in a material are very small, as in a clay, then the force of molecular attraction is strong enough to stop the water from flowing through the body. When the pores are large, the water in the center of the pores is free to move.

After a rainfall, much of the water stays near the surface, because clay in the near-surface horizons of the soil retains much water because of its molecular attraction. This forms a layer of soil moisture in many regions and is able to sustain seasonal plant growth.

Some of this near-surface water evaporates, and plants use some of the near-surface water. Other water runs directly off into streams. The remaining water seeps into the saturated zone or into the water table. Once in the saturated zone it moves by percolation, gradually and slowly, from high areas to low areas under the influence of gravity. These lowest areas are usually lakes or streams. Many streams form where the water table intersects the surface of the land (Figure 10.1).

Once in the water table, the paths that individual particles follow vary and the transit time from surface to stream may vary from days to thousands of years along a single hillside! Water can flow upward because of high pressure at depth and low pressure in stream.

The Groundwater System

Groundwater is best thought of as a system of many different parts. Some of these act as conduits and reservoirs, and others act as off-ramps and on-ramps into the groundwater system. *Recharge areas* are where water enters the groundwater system, and *discharge areas* are where water leaves the groundwater system. In humid

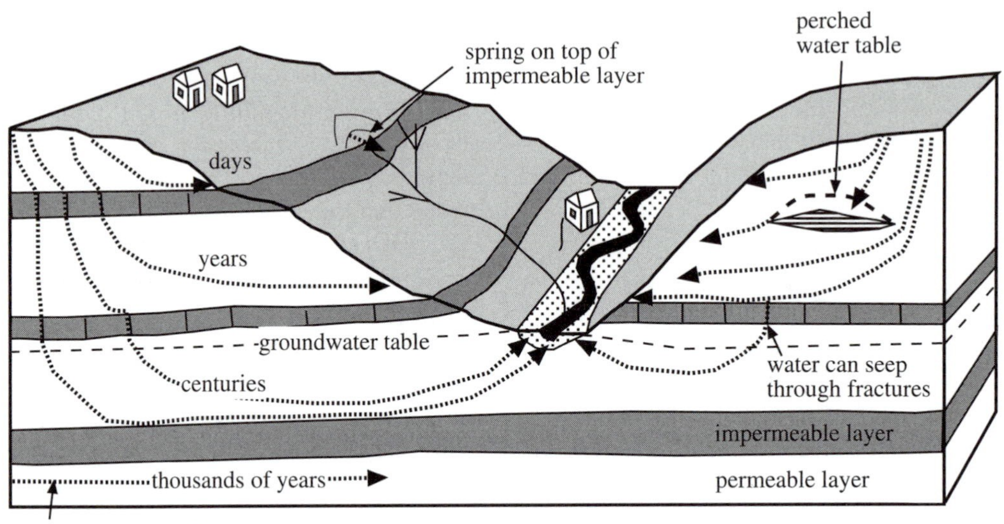

Figure 10.1. Schematic view of the groundwater system, showing how rainwater and surface runoff may make its way to the stream system, or seep into the groundwater system. Water in the ground may take days to thousands of years to move across a hillside, depending on the depth and route it takes. Some water may be trapped above impermeable layers forming a perched aquifer, whereas other water molecules may make it to the saturated zone, and become part of the main water table.

climates, recharge areas encompass nearly the land's entire surface except for streams and floodplains, whereas in desert climates, recharge areas consist mostly of the mountains and alluvial fans. Discharge areas consist mostly of streams and lakes.

The level of the water table changes with different amounts of precipitation. In humid regions, it reflects the topographic variation, whereas in dry times or places it tends to flatten out to the level of the streams of lakes. Water flows faster when the slope is greatest, so groundwater flows faster during wet times. The fastest rate of groundwater flow yet observed in the United States is 800 feet per year (250 meters/year).

Aquifers include any body of permeable rock or regolith saturated with water through which groundwater moves. Gravels and sandstone make good aquifers, but clay is so impermeable that it makes bad aquifers or even aquicludes. Other fractured rock bodies also make good aquifers.

Springs are places where groundwater flows out at the ground surface. Springs can form where the ground surface intersects the water table or at a vertical or horizontal change in permeability, such as where water in gravels on a hillslope overlay a clay unit and the water flows out on the hill along the gravel/clay boundary.

Water Wells and Use of Ground Water

Most wells fill with water simply because they intersect the water table. However, the rocks below the surface are not always homogeneous, which can result in a complex type of water table known as a perched water table (Figure 10.1). These result from discontinuous bodies in the subsurface, which create bodies of water at elevations higher than the main water table.

Confined Aquifers In many regions, a permeable layer, typically a sandstone, is confined between two impermeable beds, creating a confined aquifer. In these systems, water only enters the system in a small recharge area, and if this is in the mountains, then the aquifer may be under considerable pressure. This is known as an *artesian system.* Water that escapes the system from the fracture or well reflects the pressure difference between the elevation of the source area and the discharge area (hydraulic gradient), and it rises above the aquifer as an *artesian spring* or *artesian well.* Some of these wells have made fountains that have spewed water 200 feet (sixty meters) high!

One example of an artesian system is found in Florida, where water enters in the recharge area and is released near Miami about 19,000 years later.

Groundwater Dissolution

Groundwater also reacts chemically with the surrounding rocks; it may deposit minerals and cement together as grains, causing a reduction in porosity and permeability, or form features like stalactites and stalagmites in caves. In other cases, particularly when water moves through limestone, it may dissolve the rock, forming caves and underground tunnels. Where these dissolution cavities intersect the surface of the earth, they form *sinkholes,* which are discussed in Chapter 11.

Groundwater Contamination

Natural groundwater is typically rich in dissolved elements and compounds derived from the soil, regolith, and bedrock that the water has migrated through. Some of these dissolved elements and compounds are poisonous, whereas others are tolerable in small concentrations but harmful in high concentrations. Groundwater is also increasingly becoming contaminated by human and industrial waste, and the overuse of groundwater resources has caused groundwater levels to drop and led to other problems, especially along coastlines. Seawater may move in to replace depleted fresh water, and the ground surface may subside when the water is removed from the pore spaces in aquifers.

The United States Public Health Service has established limits on the concentrations of dissolved substances (called total dissolved solids, or t.d.s.) in natural water that is used for domestic and other uses. Table 10.1 lists these standards for the United States. It should be emphasized that many other countries, particularly those with chronic water shortages, have much more lenient standards. Sweet water is preferred for domestic use and has less than 500 mil-

Table 10.1
Drinking Water Standards for the United States

Water Classification	Total Dissolved Solids (T.D.S.)
Sweet	< 500 mg/L
Fresh	500–1,000 mg/L
Slightly Saline	1,000–3,000 mg/L
Moderately Saline	3,000–10,000 mg/L
Very Saline	10,000–35,000 mg/L
Brine	$> 35,000$ mg/L

ligrams (mg) of total dissolved solids per liter (L) of water. Fresh and slightly saline water, with t.d.s. of 1,000–3,000 mg/L, is suitable for use by livestock and for irrigation. Water with higher concentrations is unfit for humans or livestock. Irrigation of fields using waters with high concentrations of t.d.s. is also not recommended as the water will evaporate but leave the dissolved salts and minerals behind, degrading and eventually destroying the productivity of the land.

The quality of groundwater can be reduced or considered contaminated by either a high amount of total dissolved solids or by the introduction of a specific toxic element. Most of the total dissolved solids in groundwaters are salts that have been derived from dissolution of the local bedrock or from soils derived from the bedrock. Salts may also seep into groundwater supplies from the sea along coastlines, particularly if the water is being pumped out for use. In these cases, seawater often moves in to replace the depleted fresh water.

Dissolved salts in groundwater commonly include the bicarbonate (HCO_3) and sulfate (SO_4) ions, which are often attached to other ions. Dissolved calcium (Ca) and magnesium (Mg) ions can cause the water to become "hard." Hard water is defined as containing more than 120 parts per million dissolved calcium and magnesium. Hard water is difficult to lather with soap, and it forms a crusty mineralization buildup on faucets and pipes. Adding sodium (Na) in a water softener can soften hard water, but people with heart problems or those who are on a low-salt diet should not do this. Hard water is common in areas where the groundwater has moved through limestone or dolostone rocks, which contain high concentrations of calcium- and magnesium-rich rocks that are easily dissolved by groundwater.

Groundwater may have many other contaminants, some natural and others that are the result of human activity. We are concerned here primarily with natural contaminants; the human pollutants including animal and human waste, pesticides, industrial solvents, petroleum products, and other chemicals are a serious problem in many areas.

Groundwater contamination, whether natural or human-induced, is a serious problem because of the importance of the limited water supply. Pollutants in the groundwater system do not simply wash away with the next rain, as many dissolved toxins in the surface water system do. Groundwater pollutants typically have a *residence time,* or average length of time that it remains in the system, of hundreds or thousands of years. Many groundwater systems are capable of cleaning themselves from natural biological contaminants using bacteria, but other chemical contaminants have longer residence times.

Arsenic in Groundwater Arsenic poisoning leads to a variety of horrific diseases, including hyperpigmentation (abundance of red freckles), hyperkeratosis (scaly lesions on the skin), cancerous lesions on the skin, and squamous cell carcinoma (a type of skin cancer). Arsenic may be introduced into the food chain and body in several ways. In Guizhou Province, China, villagers dry their chili peppers indoors over coal fires. Unfortunately, the coal is rich in arsenic (containing up to 35,000 parts per million arsenic; Finkelman et al., 2001), and much of this arsenic is transferred to the chili peppers during the drying process. Thousands of the local villagers are now suffering arsenic poisoning, with cancers and other forms of the disease ruining families and entire villages.

Most naturally occurring arsenic is introduced into the food chain through drinking contaminated groundwater. Arsenic in groundwater is commonly formed by the dissolution of minerals from weathered rocks and soils. In Bangladesh and West Bengal, India, 25–75 million people are at risk for arsenosis because of high concentrations of natural arsenic on groundwater.

Since 1975, the maximum allowable level of arsenic in drinking water in the United States has been 50 parts per billion. The EPA has been considering adopting new standards on the allowable levels of arsenic in drinking water. Scientists from the National Academy recommend that the allowable levels of arsenic be lowered to 10 parts per billion, yet many administration officials (including the president) want the limit to be raised and to be higher than that even in some third world countries like Bangladesh. The EPA suggested that the new standard of 10 parts per billion go into effect in January 2001, but this level was overruled by the Bush administration. The issue is cost: The EPA estimates that it would cost businesses and taxpayers $181 million per year to bring arsenic levels to the proposed 10 parts per billion level, although some private foundations suggest that this estimate is too low by a factor of three. They estimate that the cost would be passed on to the consumer and residential water bills would quadruple. The EPA estimates that the health benefits from such a lowering of arsenic levels would prevent between seven and thirty-three deaths from arsenic-related bladder and lung cancer per year.

These issues reflect a delicate and difficult choice for the government. The EPA tries to "maximize health reduction benefits at a cost that is justified by the benefits." How much should be spent to save seven to thirty-three lives per year? Would the money be better spent elsewhere?

Arsenic is not concentrated evenly in the groundwater system of the United States or anywhere else in the world. The U.S. Geological Survey issued a series of maps in 2000 showing the concentration of arsenic in tens of thousands of groundwater wells in the United States. Arsenic is concentrated most in the southwestern part of the United States, with a few peaks elsewhere such as southern Texas, parts of Montana (due to mining operations), and in parts of the upper Plains states. Perhaps a remediation plan that attacks the highest concentrations of arsenic would be the most cost-effective plan that has the highest health benefit.

Contamination by Sewage A major problem in groundwater contamination is sewage. If chloroform bacteria get into the groundwater, the aquifer is ruined, and care must be taken and samples analyzed before the water is used for drinking. In many cases, sand filtering can remove bacteria, and aquifers contaminated by chloroform bacteria and other human waste can be cleaned more easily than aquifers contaminated by many other elemental and mineral toxins.

Although serious, detailed discussion of groundwater contamination by human waste is beyond the scope of this book; the reader is referred to the sources listed at the end of the chapter for more detailed accounts.

Hazards of Radon Gas

Radon is a poisonous gas that is produced as a product of radioactive decay product of the uranium decay series. Radon is a heavy gas, and it presents a serious indoor hazard in every part of the country. It tends to accumulate in poorly ventilated basements and well-insulated homes that are built on specific types of soil or bedrock that are rich in uranium minerals. Radon is known to cause lung cancer and, since it is an odorless colorless gas, it can go unnoticed in homes for years. However, the hazard of radon is easily mitigated and homes can be made safe once the hazard is identified.

Uranium is a radioactive mineral that spontaneously decays to lighter "daughter" elements by losing high-energy particles at a predictable rate known as a half-life. The half-life specifically measures how long it takes for half of the original or parent element to decay to the daughter element. Uranium decays to radium through a long series of steps with a cumulative half-life of 4.4 billion years. During these steps, intermediate daughter products are produced, and high-energy particles including *alpha particles,* consisting of two protons and two neutrons, are released. This produces heat. The daughter mineral radium is itself radioactive, and it decays with a half-life of 1,620 years by losing an alpha particle, thus forming the heavy gas radon. Since radon is a gas, it escapes out of the minerals and ground and makes its way to the atmosphere where it is dispersed, unless it gets trapped in people's homes. If it gets trapped, it can be inhaled and do damage. Radon is a radioactive gas, and it decays with a half-life of 3.8 days, producing daughter products of polonium, bismuth, and lead. If this decay occurs while the gas is in someone's lungs, then the solid daughter products become lodged in their lungs, which is how the damage from radon is initiated. Most of the health risks from radon are associated with the daughter product polonium, which is easily lodged in lung tissue. Polonium is radioactive, and its decay and emission of high-energy particles in the lungs can damage lung tissue eventually causing lung cancer.

There is a huge variation in the concentration of radon between geographic regions and in specific places in those regions. There is also a great variation in the concentration of the gas at different levels in the soil, home, and atmosphere. This variation is related to the concentration and type of radioactive elements present at a location. Radioactivity is measured in a unit known as a *picocurie* (pCi), which is approximately equal to the amount of radiation produced by the decay of two atoms per minute.

Soils have gases trapped between the individual grains that make up the soil, and these soil gases have typical radon levels of 20 pCi per liter to 100,000 pCi per liter, with most soils in the United States falling in the range of 200–2,000 pCi/L. Radon can also be dissolved in groundwater with typical levels falling between 100–2 million pCi/Liter. Outdoor air typically has 0.1–20 pCi/Liter, and radon inside people's homes ranges from 1–3,000 pCi/Liter, with 0.2 pCi/Liter being typical.

Why is there such a large variation in radon levels, and how can homeowners, water users, and others know which air is safe to breathe and which water is safe to drink? There are many natural geologic variations that lead to the complex distribution of hazardous radon, and these are examined below.

Formation and Movement of Radon Gas One of the main variables controlling the concentration of radon at any site is the initial concentration of the parent element uranium in the underlying bedrock and soil. If the underlying materials have high concentrations of uranium, it is more likely that homes built in the area may have high concentrations of radon. Most natural geologic materials contain a small amount of uranium, typically about 1–3 parts per million (ppm). The concentration of uranium is typically about the same in soils derived from a rock as in the original source rock. However, some rock (and soil) types have much higher initial concentrations of uranium, ranging up to and above 100 ppm. Some of the rocks that have the highest uranium contents include some granites, some types of volcanic rocks (especially the rhyolites), phosphate-bearing sedimentary rocks, and the metamorphosed equivalents of all of these rocks.

As the uranium in the soil gradually decays, it leaves its daughter product, radium, in concentrations proportional to the initial concentration of uranium. The radium then decays by forcefully ejecting an alpha particle from its nucleus. This ejection is an important step in the formation of radon, since every action has a reaction. In this case, the reaction is the recoil of the nucleus of the newly formed radon. Most radon remains trapped in minerals once it forms. However, if the decay of radium happens near the surface of a mineral, and if the recoil of the new nucleus of radon is away from the center of the grain, the radon gas may escape the bondage of the mineral. It will then be free to move in the intergranular space between minerals, soil, or cracks in the bedrock, or become absorbed in groundwater between the mineral grains. Less than half (10–50 percent) of the radon produced by decay of radium actually escapes the host mineral. The rest is trapped inside where it eventually decays, leaving the solid daughter products behind as impurities in the mineral.

Once the radon is free in the open or water-filled pore spaces of the soil or bedrock, it may move rather quickly. The exact rate of movement is critical to whether or not the radon enters homes, because radon does not stay around for very long with a half-life of only 3.8 days. The rates at which radon moves through a typical soil depend on how much pore space there is in the soil (or rock), how connected these pore spaces are, and the exact geometry and size of the openings. Radon moves quickly through very *porous* and *permeable* soils such as sand and gravel, but moves very slowly through less permeable materials such as clay. Radon also moves very quickly through fractured material, whether it is bedrock, clay, or concrete.

Considering how the rates of radon movement are influenced by the geometry of pore spaces in a soil or bedrock underlying a home, and how the initial concentration of uranium in the bedrock determines the amount of radon available to move, it becomes apparent that there should be a large variation in the concentration of radon from place to place. Homes built on dry permeable soils can accumulate radon quickly because it can migrate through the soil quickly. Conversely, homes built on impermeable soils and bedrock are unlikely to concentrate radon beyond their natural background levels.

Radon becomes hazardous when it enters homes and becomes trapped in poorly ventilated or well-insulated areas. Radon moves up through the soil and moves toward places with greater permeability. Home foundations are often built with a very porous and permeable gravel envelope surrounding the foundation to allow for water drainage. This also has the effect of focusing radon movement and bringing it close to the foundation, where the radon may enter through small cracks in the concrete, seams, spaces around pipes, sumps, and other openings, as well as through the concrete that may be moderately porous. Most modern homes intake less than 1 percent of their air from the soil. However, some homes, particularly older homes with cracked or poorly sealed foundations, low air pressure, and other entry points for radon, may intake as much as 20 percent of their internal air from the soil. These homes tend to have the highest concentrations of radon.

Radon can also enter the home and body through the groundwater. Homes that rely on well water may be taking in water with high concentrations of dissolved radon. This radon can then be ingested, or it can be released from the water by agitation in the home. Radon is released from high-radon water by simple activities such as taking showers, washing dishes, or running faucets. Radon can also come from some municipal water supplies, such as those supplied by small towns that rely on well fields that take the groundwater and distribute it to homes without providing a reservoir for the water to linger in while the radon decays to the atmosphere. Most larger cities, however, rely on reservoirs and surface water supplies, where the radon has had a chance to escape before being used by unsuspecting homeowners.

Radon Hazard Mapping A greater understanding of the radon hazard risk in an area can be obtained through mapping the potential radon concentrations in an area. This can be done at many scales of observation. Radon concentrations can also be measured locally to know what kinds of mitigation are necessary to reduce the health risks posed by this poisonous gas.

The broadest sense of risk can be obtained by examining regional geologic maps and determining whether or not an area is located above potential high-uranium content rocks such as granites, shales, and rhyolites. These maps are available through the U.S. Geological Survey and many state geological surveys. The U.S. Department of Energy has flown airplanes with radiation detectors across the country and produced maps that show the measured surface radioactivity on a regional scale. These maps give a very good indication of the amount of background uranium concentration in an area and thus are related to the potential risk for radon gas.

More detailed information is needed by local governments, businesses, and homeowners to assess whether or not they need to invest in radon remediation equipment. Geologists and environmental scientists are able to measure local soil radon gas levels using a variety of techniques, typically involving placing a pipe into the ground and sucking out the soil air for measurement. Other devices may be buried in the soil to more passively measure the formation of the damage produced by alpha particle emission. Using such information, the radon concentrations in certain soil types can be established. This information can be integrated with soil characteristic maps produced by the U.S. Department of Agriculture and by state and county officials to make more regional maps of potential radon hazards and risks.

Most homeowners must resort to private measurements of radon concentrations in their homes by using commercial devices that detect radon or measure the damage from alpha particle emission. The measurement of radon levels in homes has become a standard part of home sales transactions, so more data and awareness of the problem has risen in the past ten years. If your home or business does have a radon problem, an engineer or contractor can simply and cheaply (typically less than $1,000 for an average home) design and build a ventilation system that can remove the harmful radon gas, making the air safe to breathe.

Hazards of Natural (and Induced) Underground Fires

In many places around the world, fires have been burning in underground deposits of coal and other flammable rock material for hundreds or even thousands of years. Fires in underground coal seams have been ignited spontaneously by lightning, surface fires, and spontaneous combustion for millions of years. These fires scorch the overlying surface material, turning it into a barren brittle material known as *clinker.* More recently, the incidence of fires has increased dramatically because of mining operations and other human activities. Thousands of underground fires are currently burning on the planet, with most concentrated in Asia, where it is estimated that approximately 20 percent of China's annual coal production is burned in underground fires (Revkin, 2002).

Underground fires present numerous hazards. First, they release poisonous gases through cracks, fissures, and other openings to the surface. These fumes kill vegetation and pollute the air, and are contributing a large volume of greenhouse gases to the atmosphere, thus contributing to global warming. It is estimated that millions of tons of carbon dioxide alone are being added to the atmosphere by underground fires. The volume of carbon dioxide produced annually by underground coal fires in China is approximately equal to that produced by all the cars and small trucks in the United States (Revkin, 2002).

Once the underground fires burn through an area, the coal seam is gone and leaves an unsupported roof over the former seam. This causes instability, and the roof may collapse into the space formerly occupied by the coal seam. Such collapse often propagates to the surface, where sinkholes and other collapse structures may swallow formerly productive land (see Chapter 11). In cases where the fires are burning closer to the surface, subsidence may occur as the fire burns, turning the land into a sunken moonscape of clinker.

Ignition of underground fires may occur several different ways. First, if a flammable rock such as coal is exposed at the surface, it may be ignited by surface fires or lightning. As the fire burns, it can propagate underground and burn for tens, hundreds, or thousands of years until the fuel is used up. One underground coal fire in Australia, known as Burning Mountain, has been burning for at least 2,000 years. Fires can also ignite spontaneously. Minerals in the coal such as pyrite release small amounts of heat when they are exposed to oxygen. If the coal is in an enclosed area, such as a mine or natural cavity, then the heat can build up and eventually ignite the coal. Underground coal and peat can also ignite from forest fires or lightning strikes, such as those that plagued the Indonesian region in 1997 after years of drought. Smoke from the Indonesian fires and burning underground fuel made its way thousands of kilometers across the ocean to Australia and Pacific island regions. Underground fires sparked by surface fires in Indonesia are burning to this day.

Humans have ignited other underground fires. For instance, an underground fire in a coal seam in Centralia, Pennsylvania, has been burning since 1961 and has destroyed much of the town over the path of the fire. The fire was ignited when trash dumped into an abandoned mine shaft caught fire and ignited the coal seam exposed on the mine wall. The fire propagated through the existing tunnels that provide the perfect mix of fuel, air, and heat. As the fire moves through the tunnels and outward along the coal seam, the overlying land smolders and turns into clinker. The U.S. federal government has had to buy out and evict most of Centralia's residents, at a cost of $40 million to taxpayers.

It is extremely difficult to extinguish underground fires. Several different methods have been tried with limited success. First, barriers can be constructed in tunnels to attempt to block the movement of the fire, but in many cases the fires can burn right around the barriers. Tunnels can be filled with solid or foam material, but if the fire has moved away from the tunnels into the coal seam, this will also be ineffective. Noncombustible gases can be pumped into the caves or mines to suffocate the fire by depriving it of fuel. This is also difficult, as many of the mines are very porous, and the noncombustible gases can escape outward while oxygen can move inward. In a few examples, the surface area over burning underground fires has been dammed and flooded, causing the water to seep into the ground and extinguishing the fires. However, in most cases underground fires continue to burn and nearby residents must adapt to the situation. Many coal seams in China and India are actively mined as they burn, and residents live nearby breathing the toxic fumes.

Living next to burning underground coal fires is extremely hazardous, as exemplified by the Jharia mine in India, which has been on fire since 1916. This long-burning fire caused a mine wall to collapse in 1995, releasing surface water into the mine. The sudden influx of water caused a steam explosion, killing sixty miners. This disaster did not hinder further operations in the mine, however. The population around the mines has more than doubled in the past twenty years, putting even more people at risk.

RESOURCES AND ORGANIZATIONS

Print Resources

Alley, W. M., Reilly, T. E., and Franke, O. L. *Sustainability of Ground-Water Resources.* U.S. Geological Survey Circular 1186, 1999, 79 pp.

"Asbestos: Try Not to Panic." *Consumer Reports* (July 1995), 468–69.

Birkland, P. W. *Soils and Geomorphology.* New York: Oxford University Press, 1984.

Clark, J. W., Viessman, W., Jr., and Hammer, M. J. *Water Supply and Pollution Control.* New York: Harper and Row, 1977, 857 pp.

Cothern, C. R., and Smith, J. E., Jr. *Environmental Radon.* New York: Plenum Publishing Corp., 1987, 378 pp.

Finkelman, R. B., Skinner, H. C., Plumlee, G. S., and Bunnell, J. E. "Medical Geology," *Geotimes* (November 2001), 20–23.

Gates, A. E., and Gundersen, L.C.S., eds. "Geologic Controls on Radon." Geological Society of America Special Paper 271, 1992, 88 pp.

Goldsmith, D. F. "Health Effects of Silica Dust Exposure." *Silica: Physical Behavior, Geochemistry and Materials Applications, Reviews in Mineralogy* 29 (1994), 545–606.

Hirshleiffer, J., De haven, J. C., and Milliman, J. W. *Water Supply: Economics, Technology, and Policy.* Chicago: The University of Chicago Press, 1960, 378 pp.

Mutamansky, J. M. "The War on Black Lung." *Earth and Mineral Sciences* 59 (1990), 6–10.

Nazaroff, W. W., and Nero, A. V., Jr., eds. *Radon and Its Decay Products in Indoor Air.* New York: John Wiley and Sons, 1988, 518 pp.

Otton, J. K., Gundersen, L.C.S., and Schumann, R. R. *The Geology of Radon.* U.S. Geological Survey, General Interest Publication 1993–0-356–733, 1993, 29 pp.

Revkin, A. C. "Sunken Fires Menace Land and Climate." *New York Times,* January 15, 2002, sec. D1–4.

Ross, M. "The Health Effects of Mineral Dusts," in *The Environmental Geochemistry of Mineral Deposits, Part A: Processes, Techniques, and Health Issues.* Society of Economic Geologists, Reviews in Economic Geology, 6A, 1999, pp. 339–56.

Solley, W. B., Pierce, R. R., and Perlman, H. A. "Estimated Use of Water in the United States in 1995." U.S. Geological Survey Circular 1200, 1988, 71 pp.

U.S. Environmental Protection Agency. *Home Buyer's and Seller's Guide to Radon.* EPA 402-R-93–003, 1993, 32 pp.

———. *Consumer's Guide to Radon Reduction: How to Reduce Radon Levels in Your Home.* EPA 402-K92–003, 1992, 17 pp.

U.S. Environmental Protection Agency and Centers for Disease Control. *A Citizen's Guide to Radon: The Guide to Protecting Yourself and Your Family from Radon.* 2nd ed. EPA 402-K92–001, 1992, 15 pp.

West, T. R. *Geology Applied to Engineering.* Englewood Cliffs, N.J.: Prentice Hall, 1995, 560 pp.

Wilgoren, J. "100 Families Leaving Tainted Town for Cleanup." *New York Times,* January 19, 2002, sec. 1, A11.

Non-Print Resources

1999 risk assessment for radon in drinking water:
www.nap.edu/catalog/6287.html?onpi_newsdoc091598

The American Lung Association Web site on asbestos:
www.lungusa.org/air/envasbestos.html

Cost/benefit analysis of arsenic standards for drinking water:
www.epa.gov/safewater/arsenic/html

The Environmental Protection Agency (EPA) map of radon zones:
www.epa.gov/iaq/radon/
The map assigns each of the 3,141 counties in the United States to one of three potential hazard levels. The map is an average for each county and does not yield the indoor radon level in any particular home. It assigns high risk (red zone 1) to counties that have predicted average radon levels of greater than 4 picocuries per liter (pCi/L), moderate risk to zone 2 (orange) counties with predicted indoor radon levels of 2–4 pCi/L, and low risk to zone 3 (yellow) counties with less than 2 pCi/L for predicted indoor radon levels.

International Agency for Research on Cancer:
www.iarc.fr
Site contains information on other mineral hazards.

The National Science Foundation Web site on arsenic and other hazards in drinking water:
www.nsf.org

The Occupational Safety and Health Administration (OSHA) Web site on asbestos:
www.osha-slc.gov/SLTC/asbestos/

U.S. Department of Labor Occupational Safety and Health Administration:
www.osha-slc.gov/SLTC/silicacrystalline/
Site contains information on silica dust.

The U.S. Geological Survey maps of arsenic:
http://co.water.usgs.gov/trace/arsenic/

The World Health Organization Web site on arsenic in groundwater:
www.who.int/inf-fs/en/fact210.html

Organizations

Environmental Protection Agency
1200 Pennsylvania Avenue, NW
Washington, DC, 20460
http://www.epa.gov/
The E.P.A.'s mission is to protect human health and to safeguard the natural environment—air, water, and land—upon which life depends. For 30 years, the EPA has been working for a cleaner, healthier environment for the American people.

National Science Foundation
4201 Wilson Boulevard
Arlington, Virginia 22230
703–292–5111
http://www/nsf.gov
The N.S.F. funds basic research including that about material hazards.

Occupational Safety and Health Administration
U.S. Department of Labor
Occupational Safety & Health Administration
200 Constitution Avenue, NW
Washington, DC 20210
www.osha.gov
Occupational Safety and Health Administration. The mission of the Occupational Safety and Health Administration (OSHA) is to save lives, prevent injuries and protect the health of America's workers. OSHA monitors and regulates contaminants in the workplace, and improves

safety standards by reducing risk and exposure to many hazardous materials.

United States Geological Survey
U.S. Department of the Interior
USGS National Center,
12201 Sunrise Valley Drive
Reston, VA 20192
703–648–4000
http://www.usgs.gov/

The U.S. Geological Survey is the primary organization charged with monitoring the nation's groundwater supply. They also are responsible for mapping soils and soil hazards, and for determining risks from other natural hazardous substances such as arsenic and elemental and mineral anomalies. The U.S. Geological Survey has created maps showing potential hazards from arsenic in groundwater, as well as many other similar thematic maps.

CHAPTER 11

Natural Geologic Subsidence Hazards

INTRODUCTION

Natural geologic subsidence is the sinking of land relative to sea level or some other uniform surface. Subsidence may be a gradual, barely perceptible process, or it may occur as a catastrophic collapse of the surface. Subsidence occurs naturally along some coastlines, and it occurs in areas where groundwater has dissolved cave systems in rocks such as limestone. It may occur on a regional scale, affecting an entire coastline, or it may be local in scale, such as when a sinkhole suddenly opens and collapses in the middle of a neighborhood. Other subsidence events reflect the interaction of humans with the environment and include ground surface subsidence as a result of mining excavations, groundwater and petroleum extraction, and several other processes. Compaction is a related phenomenon in which the pore spaces of a material are gradually reduced, condensing the material and causing the surface to subside. Subsidence and compaction do not typically result in death or even injury, but they do cost Americans alone tens of millions of dollars per year. The main hazard of subsidence and compaction is damage to property.

Subsidence and compaction directly affect millions of people. Residents of New Orleans live below sea level and are constantly struggling with the consequences of living on a slowly subsiding delta. Coastal residents in the Netherlands have constructed massive dike systems to try to keep the North Sea out of their slowly subsiding land. The city of Venice, Italy, has dealt with subsidence in a uniquely charming way, drawing tourists from around the world. Millions of people live below the high tide level in Tokyo. The coastline of Texas along the Gulf of Mexico is slowly subsiding, placing residents of Baytown and other Houston suburbs close to sea level and in danger of hurricane-induced storm surges and other more frequent flooding events. In Florida, sinkholes have episodically opened up and swallowed homes and businesses, particularly during times of drought. What are the causes of this sinking of the land, and where is subsidence most and least likely to occur?

The driving force of subsidence is gravity, with the style and amount of subsidence controlled by the physical properties of the soil, regolith, and bedrock underlying the area that is subsiding. Subsidence does not require a transporting medium, but it is aided by other processes such as groundwater dissolution that can remove mineral material and carry it away in solution, creating underground caverns that are prone to collapse.

Natural subsidence has many causes. Dissolution of limestone by underground streams and water systems is one of the most common, creating large open spaces that collapse under the influence of gravity. Groundwater dissolution results in the formation of sinkholes—large, generally circular depressions that are caused by collapse of the surface into underground open spaces.

Earthquakes may raise or lower the land suddenly, as in the case of the 1964 Alaskan earthquake when tens of thousands of square miles suddenly sank or rose three to five feet, causing massive disruption to coastal communities and ecosystems. Earthquake-induced ground shaking can also cause liquefaction and compaction of unconsolidated surface sediments, also leading to subsidence. Regional lowering of the land surface by liquefaction and compaction is known from the massive 1811 and 1812 earthquakes in New Madrid, Missouri, and from many other examples.

Volcanic activity can cause subsidence, as when underground magma chambers empty out during an eruption. In this case, subsidence is often the lesser of many hazards that local residents need to fear. Subsidence may also occur on lava flows when lava empties out of tubes or underground chambers.

Some natural subsidence on the regional scale is associated with continental scale tectonic processes. The weight of sediments deposited along continental shelves can cause the entire continental margin to sink, causing coastal subsidence and a landward migration of the shoreline. Tectonic processes associated with extension, continental rifting, strike-slip faulting, and even collision can cause local or regional subsidence, sometimes at rates of several inches per year.

TYPES OF SURFACE SUBSIDENCE AND COLLAPSE

Some subsidence occurs because of processes that happen at depths of thousands of feet beneath the surface, which is referred to as *deep subsidence.* Other subsidence is caused by shallow near-surface processes and is known as *shallow subsidence. Tectonic subsidence* is a result of the movement of the plates on a lithospheric scale, whereas *human-induced subsidence* refers to cases where the activities of people, such as extraction of fluids from depth, have resulted in lowering of the land surface.

Compaction-related subsidence may be defined as the slow sinking of the ground surface because of reduced pore space, lowered pore pressure, and other processes that cause the regolith to become more condensed and occupy a smaller volume. Most subsidence and compaction mechanisms are slow and result in gradual sinking of the land's surface, whereas sometimes the process may occur catastrophically and is known as *collapse.*

Carbonate Dissolution and Sinkholes

The formation and collapse of sinkholes is one of the more dramatic deep subsidence mechanisms. Sinkholes form over rock that is readily dissolved by chemical weathering and groundwater dissolution, and these rocks are most typically limestones. In the United States, limestone dissolution and sinkhole formation is an important and dangerous process in Florida, much of the Appalachian Mountains, large parts of the Midwest, scattered areas in and east of the Rocky Mountains, and especially Missouri and northern Arkansas. Areas that are affected by groundwater dissolution, cave complexes, and sinkhole development are known as karst terrains. Globally, several regions are known for spectacular karst systems, including the cave systems of the Caucasus, southern Arabia including Oman and Yemen, Borneo, and the mature highly eroded karst terrain of southern China's Kwangsi Province.

The formation of caves and sinkholes in karst regions begins with a process of dissolution. Rainwater that filters through soil and rock may work its way into natural fractures or breaks in the rock, and chemical reactions that remove ions from the limestone slowly dissolve and carry away in solution parts of the limestone. Fractures are gradually enlarged, and new passageways are created by groundwater flowing in underground stream networks through the rock. Dissolution of rocks is most effective if the rocks are limestone and if the water is slightly acidic (acid rain greatly helps cave formation). Carbonic acid (H_2CO_3) in rainwater reacts with the limestone, rapidly (at typical rates of a few millimeters per thousand years) creating open spaces, cave and tunnel systems, and interconnected underground stream networks (Figure 11.1).

When the initial openings become wider, they are known as caves. Many caves are small pockets along enlarged or widened cavities, whereas others are huge open underground spaces. The largest cave in the world is the Sarawak Chamber in Borneo. This cave has a volume of 65 million cubic feet. The Majlis Al Jinn (Khoshilat Maqandeli) Cave in Oman is the second-largest cave known in the world and is big enough to hold several of the Sultan of Oman's royal palaces with a 747 flying overhead. Its main chamber is over 13 million cubic feet in volume, larger than the biggest pyramid at Giza. Other large caves include the world's third, fourth, and fifth largest caves: the Belize Chamber, Salle de la Verna, and the largest "Big Room" of Carlsbad Caverns, New Mexico. Each of these has a volume of at least 3 million cubic feet. The Big Room is a large chamber that is 4,000 feet (1,200 meters) long, 625 feet (190 meters) wide, and 325 feet (100 meters) high. Some caves form networks of linked passages that extend for many miles in length. For instance, Mammoth Cave in Kentucky has at least 300 miles of interconnected passageways. While the caves are forming, water flows through these passageways in underground stream networks.

In many parts of the world, the forma-

Figure 11.1. Photograph showing formation of a sinkhole at Winter Park, Florida, in 1981. Note the buildings and cars that plunged into the hole as it formed. Photo credit: U.S. Geological Survey.

tion of underground cave systems has led to parts of the surface collapsing into the caverns and tunnels, forming a distinctive type of topography known as karst topography. Karst is named after the Kars limestone plateau region in Serbia (the northwest part of the former Yugoslavia), where karst is especially well developed. Karst topography may take on many forms in different stages of landscape evolution but typically begins with the formation of circular pits on the surface known as sinkholes. These form when the roof of an underground cave or chamber suddenly collapses, bringing everything on the surface suddenly down into the depths of the cave. Striking examples of sinkhole formation surprised residents of the Orlando region in Florida in 1981 (mentioned earlier), when a series of sinkholes swallowed many businesses and homes with little warning. In this and many other examples, sinkhole formation is initiated after a prolonged drought or drop in the groundwater levels. This drains the water out of underground cave networks, leaving the roofs of chambers unsupported and making them prone to collapse.

The sudden formation of sinkholes in the Orlando area is best illustrated by the formation of the Winter Park sinkhole on May 8, 1981. The first sign that trouble was brewing was provided by the unusual spectacle of a tree suddenly disappearing into the ground at 7:00 P.M. as if being sucked in by some unseen force. Residents were worried, and rightfully so. Within ten hours, a huge sinkhole nearly 100 feet across and more than 100 feet deep had formed. It continued to grow, swallowing six commercial buildings, a home, two streets, six Porsches, and the municipal swimming pool, causing more than $2 million in damage. The sinkhole has since been converted into a municipal park and lake. More than 1,000 sinkholes have formed in parts of southern Florida in recent years.

Sinkhole topography is found in many parts of the world, including Florida, Indiana, Missouri, Pennsylvania, and Tennessee in the United States; the Karst region of Serbia; the Salalah region of Arabia; southern China; and many other places where the ground is underlain by limestone.

Sinkholes have many different forms. Some are funnel-shaped with boulders and unconsolidated sediment along their bottoms; others are steep-walled pipe-like features that have dry or water-filled bottoms. Some sinkholes in southern Oman are up to 900-feet-deep pipes, with caves at their bottoms where residents would get their drinking water until recently when wells were drilled. Villagers, mostly women, would have to climb down precarious vertical walls and then back out carrying vessels of water. The bottoms of some of these sinkholes are littered with bones, some dating back thousands of years, of water carriers who slipped on their route. Some of the caves are decorated with prehistoric cave art, showing that these sinkholes were used as water sources for thousands or tens of thousands of years (Figure 11.2).

Sinkhole formation is intricately linked to the lowering of the water table, as exemplified by the Winter Park example. When water fills the underground caves and passages, it slowly dissolves the walls, floor, and roof of the chambers, carrying the limestone away in solution. When the water table is lowered by drought, by overpumping of groundwater by people, or by other mechanisms, the roofs of the caves may no longer be supported and they may catastrophically collapse into the chambers, forming a sinkhole on the surface. In Florida, many of the sinkholes formed because officials lowered the water table level to drain parts of the Everglades to make more land available for development. This ill-fated decision was rethought and attempts have been made to restore the water table, but in many cases it was too late and the damage was done.

Many sinkholes form suddenly and catastrophically, with the roof of an underground void suddenly collapsing and dropping all of the surface material into the hole. Other sinkholes form more gradually, with the slow movement of loose unconsolidated material into the underground stream network eventually leading to the formation of a surface depression that may continue to grow into a sinkhole.

The pattern of surface subsidence resulting from sinkhole collapse depends on the initial size of the cave that collapses, the depth to the cavity, and the strength of the overlying rock. Big caves that collapse can cause a greater surface effect. For a collapse structure at depth to propagate to the surface, blocks must fall off the roof and into the cavern. The blocks fall by breaking along fractures and falling by the force of gravity. If the overlying material is weak,

Figure 11.2. Teyq sinkhole in southern Oman's Salalah region, which is the largest known sinkhole in the world. (a) Shows limestone on far wall of sinkhole dissected by many caves and karst features. (b) View looking down into the sinkhole, showing two dry streams (wadis) that drain into a cave (called a sparrow hole). When it rains in this region, the sinkhole may rapidly fill partly with water, some of which shoots out of the caves in the background of (a). All the water drains into the active sparrow hole, which functions like a drain in a bathtub. The water that enters here makes its way through an extensive cave system and exits at the base of a cliff about 20 kilometers (12 miles) away, near the sea. Photos by T. Kusky

the fractures will propagate outward, forming a cone-shaped depression with its apex in the original collapse structure. In contrast, if the overlying material is strong, the fractures will propagate vertically upward, resulting in a pipe-like collapse structure.

When the roof material collapses into the cavern, blocks of wall rock accumulate on the cavern floor. There is abundant pore space between these blocks so that the collapsed blocks take up a larger volume than they did when they were attached to the walls. In this way, the underground collapsed cavern may become completely filled with blocks of the roof and walls before any effect migrates to the surface. If enough pore space is created, then almost no subsidence may occur along the surface. In contrast, if the cavity collapses near the surface, then a collapse pit will eventually form on the surface.

It may take years or decades for a deep collapse structure to migrate from the depth where it initiates to the surface. The first signs of a collapse structure migrating to the surface may be tensional cracks in the soil, bedrock, or building foundations, which are formed as material pulls away from unaffected areas as it subsides. Circular areas of tensional cracks may enclose an area of contractional buckling in the center of the incipient collapse structure, as bending in the center of the collapsing zone forces material together.

After sinkholes form, they may take on several different morphological characteristics. Solution sinkholes are saucer-shaped depressions formed by the dissolution of surface limestone and have a thin cover of soil or loose sediment. These grow slowly and present few hazards, since they are forming on the surface and are not connected to underground stream or collapse structures. Cover-subsidence sinkholes form where the loose surface sediments move slowly downward to fill a growing solution sinkhole. Cover-collapse sinkholes form where a thick section of sediment overlies a large solution cavity at depth, and the cavity is capped by an impermeable layer such as clay or shale. A perched water table develops over the aquiclude. Eventually, the collapse cavity becomes so large that the shale or clay aquiclude unit collapses into the cavern, and the remaining overburden rapidly sinks into the cavern, much like sand sinking in an hourglass. These are some of the most dangerous sinkholes since they form rapidly and may be quite large. Collapse sinkholes are simpler but still dangerous. They form where the strong layers on the surface collapse directly into the cavity, forming steep-walled sinkholes.

Sinkhole topography may continue to mature into a situation in which many of the sinkholes have merged into elongate valleys and the former surface is found as flat areas on surrounding hills. Even this mature landscape may continue to evolve until tall steep-walled karst towers reach to the former land surface and a new surface has formed at the level of the former cave floor. The Cantonese region of southern China's Kwangsi Province best shows this type of karst tower terrain (Figure 11.3).

Human-Induced Subsidence

Several types of human activity can result in the formation of sinkholes or cause

Figure 11.3. The Stone Forest, a karst formation in Yunnan Province. ©The Purcell Team/CORBIS

other surface subsidence phenomena. Withdrawals of fluids from underground aquifers, depletion of the source of replenishment to these aquifers, and collapse of underground mines can all cause surface subsidence. In addition, vibrations from drilling, construction, or blasting can trigger collapse events, and the extra load of buildings over unknown deep collapse structures can cause them to propagate to the surface, forming a sinkhole. These processes reflect geologic hazards caused by human interaction with the natural geologic environment.

Mine Collapse Mining activities may mimic the formation of natural caves, since mining operations remove material from depth and leave roof materials partly unsupported. There are many examples of mines of different types that have collapsed, resulting in surface subsidence, sinkhole formation, and even the catastrophic draining of large lakes.

The mining of salt has created sinkholes and surface subsidence problems in a number of cases. Salt is mined in several different ways, including the digging of tunnels from which the salt is excavated and by injecting water into a salt deposit, removing the salt-saturated water and drying it to remove the salt for use. This second technique is called solution mining and has less control over where the salt is removed than the classical excavation-style mining. Shallow sinkholes and surface subsidence are typical and are expected around solution mining operations; for instance, the salt-mining operations near Hutchinson, Kansas, resulted in the formation of dozens of sinkholes, including one nearly 1,000 feet across that partly swallowed the salt processing plant.

One of the most spectacular of all salt-mining subsidence incidents occurred on November 21, 1980, on Lake Peigneur, Louisiana. The center of Lake Peigneur is occupied by a salt dome that forms Jefferson Island. This salt dome was mined extensively with many underground tunnels excavated to remove the salt. Southern Louisiana is also an oil-rich region, and on the ill-fated day in 1980, an oil-drilling rig accidentally drilled a hole into one of the mine shafts. Water began swirling into the hole, and the roof of the mine collapsed, setting the entire lake into a giant spinning whirlpool that quickly drained into the deep mine as if it were bathwater escaping down the drain. The oil drilling rig, ten barges, and a tugboat were sucked into the mine shafts, and much of the surrounding land collapsed into the collapse structure, destroying a home and other properties. After the mine was filled with water, the lake gradually filled again, but the damage was done (Figure 11.4).

Figure 11.4. The ruins of a building stand in the water of Lake Peigneur on Jefferson Island, evidence of the disaster that occurred in 1980. ©Philip Gould/CORBIS

Other types of mining have resulted in surface subsidence and collapse, including coal mining in the Appalachians and the Rocky Mountains. Most coal mine–related subsidence occurs where relatively shallow flat-lying coal seams have been mined, and the mine roofs collapse. The fractures and collapse structures eventually migrate to the surface, leading to elongate trains of sinkholes and other collapse structures.

Groundwater Extraction The extraction of groundwater, oil, gas, or other fluids from underground reservoirs can cause significant subsidence of the land's surface. In some cases, the removal of underground water is natural. During times of severe drought, soil moisture may decrease dramatically and drought-resistant plants with deep root systems can draw water from great depths, reaching many tens of meters in some cases. In most cases, however, subsidence caused by deep fluid extraction is caused by human activity.

This deep subsidence mechanism operates because the fluids that are extracted help support the weight of the overlying regolith. The weight of the overlying material places the fluids under significant pressure, known as hydrostatic pressure, which keeps the pressure between individual grains in the regolith at a minimum. This in turn helps prevent the grains from becoming closely packed or compacted. If the fluids are removed, the pressure between individual grains increases and the grains become more closely packed and compacted, occupying less space than before the fluid was extracted. This can cause the surface to subside. A small amount of this subsidence may be temporary or recoverable, but generally, once surface subsidence related to fluid extraction occurs, it is nonrecoverable. When this process occurs on the scale of a reservoir or entire basin, the effect can be subsidence of a relatively large area. Subsidence associated with underground fluid extraction is usually gradual but still costs millions of dollars in damage every year in the United States.

The amount of surface subsidence is related to the amount of fluid withdrawn from the ground and also to the compressibility of the layer from which the fluid has been removed. If water is removed from cracks in a solid igneous, metamorphic, or sedimentary rock, then the strength of the rock around the cracks will be great enough to support the overlying material and no surface subsidence is likely to occur. In contrast, if fluids are removed from a compressible layer such as sand, shale, or clay, then significant surface subsidence may result from fluid extraction. Clay and shale have a greater porosity and compressibility than sand, so extraction of water from clay-rich sediments results in greater subsidence than the same amount of fluid withdrawn from a sandy layer.

One of the most common causes of fluid extraction–related subsidence is the overpumping of groundwater from aquifers. If many wells are pumping water from the same aquifer, the cones of depression (see Chapter 10) surrounding each well begin to merge, lowering the regional groundwater level. Lowering of the groundwater table can lead to gradual, irreversible subsidence.

Surface subsidence associated with groundwater extraction is a serious problem in many parts of the southwestern United States. Many cities such as Tucson, Phoenix, Los Angeles, Salt Lake City, Las Vegas, and San Diego rely heavily on groundwater pumped from compressible layers in underground aquifers.

The San Joaquin Valley of California offers a dramatic example of the effects of groundwater extraction. Extraction of groundwater for irrigation over a period of fifty years has resulted in nearly thirty feet of surface subsidence. Parts of the Tucson Basin in Arizona are presently subsiding at an accelerating rate, and many investigators fear that the increasing rate of subsidence reflects a transition from temporary recoverable subsidence to a permanent compaction of the water-bearing layers at depth.

The world's most famous subsiding city is Venice, Italy. Venice is sinking at a rate of about one foot per century, and much of the city is below sea level or just above sea level and is prone to floods from storm surges and astronomical high tides in the Adriatic Sea (Figure 11.5). The city has subsided more than ten feet since it was founded near sea level. These aqua altas ("high waters" in Italian) flood streets as far as the famous Piazza San Marco. Venice has been subsiding for a combination of reasons, including compaction of the coastal muds that the city was built on. One of the main causes of the sinking of Venice has been groundwater extraction. Nearly 20,000 groundwater wells pumped water from compressible sediment beneath the city, with the result being that the city sunk into the empty space created by the withdrawal of water. The Italian government has now built an aqueduct system to bring drinking water to residents and has closed most of its 20,000 wells. This action has slowed the subsidence of the city, but it is still sinking, and this action may be too little, too late to spare Venice from the future effects of storm surges and astronomical high tides.

Mexico City is also plagued with subsidence problems caused by groundwater extraction. Mexico City is built on a several-thousand-foot-thick sequence of sedimentary and volcanic rocks, including a large dried lake bed on the surface. Most of the groundwater is extracted from the upper 200 feet of these sediments. Parts of Mexico City have subsided dramatically, whereas others have not. The northeast part of the city has subsided about twenty feet. Many of the subsidence patterns in Mexico City can be related to the underlying geology. In places like the northeast part of the city that are underlain by loose compressible sediments, the subsidence has been large. In other places underlain by volcanic rocks, the subsidence has been minor.

The extraction of oil, natural gas, and other fluids from the earth also may result in surface subsidence. In the United States, subsidence related to petroleum extraction is a large problem in Texas, Louisiana, and parts of California. One of the worst cases of oil field subsidence is that of Long Beach, California, where the ground surface has subsided thirty feet in response to extraction of underground oil. There are approximately 2,000 oil wells in Long Beach, pumping oil from beneath the city. Much of Long Beach's coastal area subsided below sea level, forcing the city to construct a series of dikes to keep the water out. When the subsidence problem was recognized and understood, the city began a program of reinjecting water into the oil field to replace the extracted fluids and to prevent further subsidence. This reinjection program was initiated in 1958, and since then the subsidence has stopped, but the land surface can not be pumped up again to its former levels.

Pumping of oil from an oil field west of Marina del Rey in Los Angeles, along the Newport-Inglewood Fault, resulted in subsidence beneath the Baldwin Hills Dam and Reservoir, leading to the dam's catastrophic failure on December 14, 1963. Oil extraction from the Inglewood oil field resulted in subsidence-related slip on a fault beneath the dam and reservoir, which was enough to initiate a crack in the dam foundation. The crack was quickly expanded by pressure from the water in the reservoir, which led to the dam's catastrophic failure

Figure 11.5. Chairs and tables from a bar float in a flooded St. Mark's Square in Venice Italy, Saturday, Nov. 16, 2002, after the morning's high tide didn't retract due to strong winds. According to Venice authorities, the water level, which reached 147 cm (58 inches), was the fifth highest level since the major flood in 1966 of 194 cm (78 inches). Venice has sunk more than 10 feet since the city was founded, largely because of groundwater extraction and compaction of coastal muds. If global sea level rises, many of the world's coastal cities may eventually have problems similar to those of Venice. AP Wide World Photos

at 3:38 P.M. on December 14, 1963. Sixty-five million gallons of water were suddenly released, destroying dozens of homes, killing five people, and causing $12 million in damage (Figure 11.6).

Subsidence Caused by Drought and Surface Water Overuse Drought and overuse of surface water can also lead to subsidence by depriving the groundwater system of the water that would normally replenish the aquifer. If the drought or overuse of the surface water persists for years, then the sediments in the aquifer may begin to compact, leading to surface subsidence. One of the more dramatic examples of the effects of surface water use leading to subsidence is provided by the water problem along the Jordan River on the Israeli-Jordanian border. Water in the Jordan is needed by the thirsty population in this dry area, and the greatly decreased flow of the Jordan caused by this water used has caused subsidence of the Dead Sea by about 1.5–2.0 inches per year. This subsidence caused by lack of groundwater replenishment is in addition to the tectonic subsidence in the

Figure 11.6. Baldwin Hills Dam disaster. Dam collapsed in California when a fracture related to ground subsidence propagated through the dam, causing it to collapse. This photo, made less than two minutes after the Baldwin Hill Dam burst, shows water thundering down the hillside into the heavily populated area. December 16, 1963. ©Bettmann/Corbis

Dead Sea pull-apart basin. The total subsidence is now about twenty feet since 1998, putting the Dead Sea a startling 1,400 feet below sea level.

Tectonic Subsidence

Plate tectonics is associated with subsidence of many types and scales, particularly on or near plate boundaries. Plate tectonics is associated with the large-scale vertical motions that uplift entire mountain ranges, drop basins to lower elevations, and form elongate kilometer-deep depressions in the earth's surface known as rifts. Plate tectonics also causes the broad flat coastal plains and passive margins (see Chapter 1) to slowly subside relative to sea level, causing the sea to move slowly into the continents. More local-scale folding and faulting can cause areas of the land surface to rise or sink, although at rates that rarely exceed one centimeter per year.

Extensional plate boundaries are naturally associated with subsidence, since these boundaries occur in places where the crust is being pulled apart, thinning, and sinking relative to sea level. Places where the continental crust has ruptured and is extending are known as continental rifts. In the United States, the Rio Grande rift in New Mexico represents a place where the crust has begun to rupture, and it is subsiding relative to surrounding mountain ranges. In this area, the actual subsidence does not present much of a hazard, since the land is not near the sea, and a large region is subsiding. The net effect is that the valley floor is slightly lower in elevation every year than it was the year before. The rifting and subsidence is sometimes associated with faulting when the basin floor suddenly drops, and the earthquakes are associated with their own sets of hazards. Rifting in the Rio Grande is also associated with the rise of a large body of magma beneath Soccoro, and if this magma body has an eruption it is likely to be catastrophic.

Large areas of the basin and range province of the southwestern United States are also subsiding. The region was topographically uplifted millions of years ago, and tectonic stresses are now pulling the entire region apart, causing locally rapid subsidence in the basins between ranges. Again, the main hazards from this type of subsidence are mainly associated with the earthquakes that sometimes accommodate the extension and subsidence.

The world's most extensive continental rift province is found in East Africa. An elongate subsiding rift depression extends from Ethiopia and Somalia in the north and south through Kenya, Uganda, Rwanda, Burundi, and Tanzania, then swings back toward the coast through Malawi and Mozambique (Figure 11.7). The East African rift system contains the oldest hominid fossils and is also host to areas of rapid land surface subsidence. Earthquakes are common, as are volcanic eruptions such as the catastrophic eruption of Nyiragongo, Congo, in January 2002. Lava flows from Nyiragongo covered large parts of the town of Goma, forcing residents to flee to neighboring Rwanda.

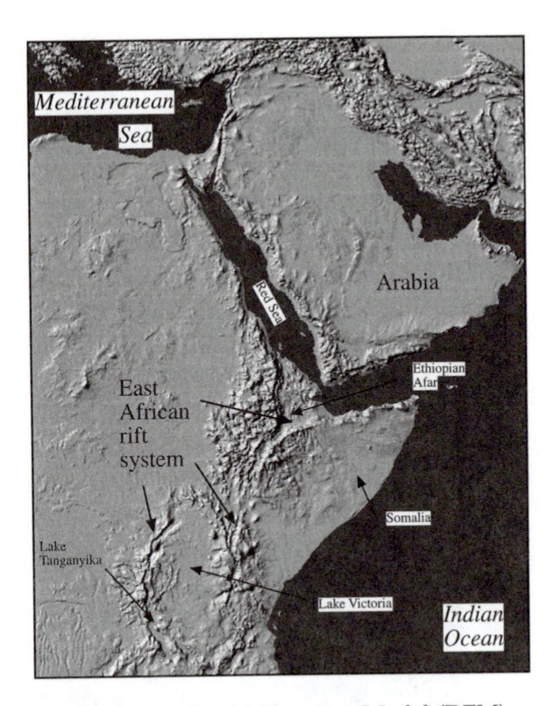

Figure 11.7. Digital Elevation Model (DEM) of the East African rift system, showing the deep depressions occupied by lakes. The northern part of the system breaks into three branches at the Afar depression—the Red Sea, the Gulf of Aden, and the East African rift system. Elevation data provided by the U.S. Geological Survey.

Subsidence in the East African rift system has formed a series of very deep elongate lakes, including Lakes Edward, Albert, Kivu, Malawi, and Tanganyika (Figure 11.7). These lakes sit on narrow basin floors that are bounded on their east and west sides by steep rift escarpments. The shoulders of the rifts slope away from the center of the rift, so sediments carried by streams do not enter the rift but are carried away from it. This allows the rift lakes to become very deep without being filled by sediments. It also means that additional subsidence can cause parts of the rift floor to subside well below sea level, such as Lake Abe in the Awash depression in the Afar rift. This lake and several other areas near Djibouti rest hundreds of feet below sea level. These lakes, by virtue of being so deep, become stratified with respect to dissolved oxygen, methane, and other gases. Methane is locally extracted from these lakes for fuel, although periodic overturning of the lakes' waters can lead to hazardous release of gases.

Transform plate boundaries, where one plate slides past another, can also be sites of hazardous subsidence. The strike-slip faults that comprise transform plate boundaries are rarely perfectly straight. Places where the faults bend may be sites of uplift of mountains or rapid subsidence of narrow elongate basins. The orientation of the bend in the fault system determines whether the bend is associated with contraction and the formation of mountains, or extension, subsidence, and the formation of the elongate basins known as pull-apart basins. Pull-apart basins typically subside quickly, have steep escarpments marked by active faults on at least two sides, and may have volcanic activity. Some of the topographically lowest places on Earth are in pull-apart basins, including the Salton Sea in California and the Dead Sea along the border between Israel and Jordan. The hazards in pull-apart basins are very much like those in continental rifts.

Convergent plate boundaries are known for tectonic uplift, although they may also be associated with regional subsidence. When a mountain range is pushed along a fault on top of a plate boundary, the underlying plate may subside rapidly. In most situations, erosion of the overriding mountain range sheds enormous amounts of loose sediment onto the underriding plate, so the land surface does not actually subside, although any particular rock layer will be buried and subside rapidly.

Subsidence from Earthquake Ground Displacements

Sometimes individual large earthquakes may displace the land surface vertically, resulting in subsidence or uplift. One of the largest and best-documented cases of earthquake-induced subsidence resulted from the March 27, 1964, magnitude 9.2 earthquake in southern Alaska. This earthquake tilted a huge (approximately 200,000 square kilometers) area of the earth's crust. Significant changes in ground level were recorded along the coast for more than 1,000 kilometers, including uplifts of up to eleven meters, subsidence of up to two meters, and lateral shifts of several to tens of meters. Much of the area that subsided was along Cook Inlet, north to Anchorage, Valdez, and south to Kodiak Island. Towns that were built around docks prior to the earthquake were suddenly located below the high tide mark, and entire towns had to move to higher ground. Forests that subsided found their root systems suddenly inundated by salt water, leading to the death of the forests. Populated areas located at previously safe distances from the high tide (and storm) line became prone to flooding and storm surges and had to be relocated.

Compaction-Related Subsidence on Deltas and Passive Margin Coastal Areas

Subsidence related to compaction and removal of water from sediments deposited on continental margin deltas, in lake beds, and in other wetlands poses a serious problem to residents trying to cope with the hazards of life at sea level. Deltas are especially prone to subsidence because the sediments that are deposited on deltas are very water-rich, and the weight of overlying new sediments compacts existing material, forcing the water out of pore spaces. Deltas are also constructed along continental shelves that are prone to regional-scale tectonic subsidence and are subject to additional subsidence forced by the weight of the sedimentary burden deposited on the entire margin. Continental margin deltas are rarely more than a few feet above sea level, so they are prone to the effects of tides, storm surges, river floods, and other coastal disasters. Any decrease in the sediment supply to keep the land at sea level has serious ramifications, subjecting the area to subsidence below sea level.

Some of the world's thickest sedimentary deposits are formed in deltas on the continental shelves, and these are of considerable economic importance because they also host the world's largest petroleum reserves. The continental shelves are divided into many different sedimentary environments. Beaches contain the coarse fraction of material deposited at the ocean front by rivers and sea cliff erosion. Quartz is typically very abundant, because of its resistance to weathering and its abundance in the crust. Beach sands tend to be well-rounded, as is anything else such as beach glass, because of the continuous abrasion caused by the waves dragging the particles back and forth. Many of the sediments transported by rivers are deposited in estuaries, which are semi-enclosed bodies of water near the coast in which fresh water and seawater mix. Near-shore sediments deposited in estuaries include thick layers of mud, sand, and silt. Many estuaries are slowly subsiding, and they get filled with thick sedimentary deposits. Deltas are formed where streams and rivers meet the ocean, and they drop their loads because of the reduced flow velocity. Deltas are complex sedimentary systems with coarse stream channels, fine-grained inter-channel sediments, and a gradation seaward to deep water deposits of silt and mud.

All the sediments deposited in the coastal environments tend to be water-rich when deposited and thus subject to water loss and compaction. Subsidence poses the greatest hazard on deltas, since these sediments tend to be thickest of all deposited on continental shelves. They are typically fine-grained muds and shales that suffer the greatest water loss and compaction. Unfortunately, deltas are also the sites of some of the world's largest cities, since they offer great river ports. New Orleans, Shanghai, and many other major cities have been built on delta deposits and have

Table 11.1
Subsidence Statistics for the Ten Worst-Case Coastal Cities

City	Maximum Subsidence (m)	Area Affected (km²)	Tectonic Environment
Los Angeles (Long Beach), California	9.0	50	Oilfield subsidence
Tokyo, Japan	4.5	3,000	Delta
San Jose, California	3.9	800	Delta
Osaka, Japan	3.0	500	Delta
Houston, Texas	2.7	12,100	Oilfield and coastal marsh
Shanghai, China	2.63	121	Delta
Niigata, Japan	2.5	8,300	Delta
Nagoya, Japan	2.37	1,300	Delta
New Orleans, Louisiana	2.0	175	Delta
Taipei, Taiwan	1.9	130	Coastal plain, Delta

subsided several meters since they were first built (Table 11.1). Many other cities built on these very compactible shelf sediments are also experiencing dangerous amounts of subsidence. What are the consequences of this subsidence for people who live in these cities, and how will they be affected by increased rates of subsidence caused by damming rivers that trap replenishing sediments upstream? How will these cities fare with current sea level rise, estimated to be occurring at a rate of an inch every ten years with more than six inches of rise in the past century? Whatever the response, it will be costly. Some urban and government planners estimate that protecting the populace from sea level rise on subsiding coasts will be the costliest endeavor ever undertaken by humans.

What is the fate of these and other coastal cities that are plagued with natural and human-induced subsidence in a time of global sea level rise? The natural subsidence in these cities is accelerated by human activities. First of all, the construction of tall heavy buildings on loose, compactible, water-rich sediments forces water out of the pore spaces of the sediment underlying each building, causing that building to subside. The weight of cities has a cumulative effect, and big cities built on deltas and other compactible sediment cause a regional flow of water out of underlying sediments, leading to subsidence of the city as a whole.

New Orleans has one of the worst subsidence problems of coastal cities in the United States. Its rate and total amount of subsidence are not the highest (see Table 11.1), but since nearly half of the city is at or below sea level, any additional subsidence will put the city dangerously far below sea level. Already, the Mississippi River is higher than downtown streets, and ships float by at the second-story level of buildings. Dikes keep the river at bay and keep storm surges from inundating the city. Additional subsidence will make these measures impractical. New Orleans, Houston, and other coastal cities have been accelerating their own sinking by withdrawing groundwater and oil from compactible sediments beneath the cities. They are literally pulling the ground out from under their own feet.

The combined effects of natural and human-induced subsidence, plus global sea level rise, have resulted in increased urban flooding of many cities and greater destruction during storms. Storm barriers have been built in some cases, but this is only the beginning. Thousands of kilometers of barriers will need to be built to protect these cities unless billions of people are willing to relocate to inland areas, which is an unlikely prospect.

What can be done to reduce the risks from coastal subsidence? First, a more intelligent regulation of groundwater extraction from coastal aquifers and of oil from coastal regions must be enforced. If oil is pumped out of an oil reservoir, then water should be pumped back in to prevent subsidence. Sea level is rising partly from natural astronomical effects and partly from human-induced changes to the atmosphere. It is not too early to start planning for sea level rises of a few feet. Seawalls should be designed and tested before construction on massive scales. Businesses and individuals should consider moving many operations inland to higher ground.

SUBSIDENCE HAZARD STATISTICS

What Are the Costs of Subsidence Hazards?

Mitigation of Property Damage from Subsidence and Compaction Most of the hazards and damage from subsidence, compaction, and collapse is to buildings constructed on material that is differentially sinking. Buildings are forced to either accommodate the changing volume of their foundations or to crack and crumble.

The most severe compaction is on soils and regolith made of organic and clay-rich soils, such as those on deltas, wetlands, and in artificially filled areas including reclaimed landfills. As building sites become scarce, it is becoming increasingly common to fill in marshes and wetlands and to reclaim old landfills for construction sites. When buildings are constructed on the organic-rich, clay-rich, poorly compacted soils, the weight of the artificial fill and the buildings may be enough to force water out of the underlying pore spaces, causing compaction and subsidence. The hazards from this type of compaction are most severe if parts of the surface compact and subside more than other nearby parts. This differential subsidence can result from uneven thickness of the compacting layer or from buildings that have unequal weights in different parts of the structure, exerting different stresses on underlying soils and thus causing differential compaction. Differential subsidence causes cracks in foundations and walls, causes pipelines and plumbing systems to be disrupted, and it can disrupt the regular flow of groundwater by changing the local slopes of underground units.

Conducting geological surveys before construction and adhering to proper construction techniques can reduce some of the damage from compaction and subsidence. All structures built on compactible regolith should be supported by pilings driven through the compressible layer to a noncompressible, competent layer such as bedrock, sand, or gravel. Pipelines should be constructed with flexible joints to accommodate differential compaction, and bridges should have expansion joints to accommodate differential movement between support pillars.

Some types of soils and regolith have proven particularly prone to compaction and subsidence, and even more stringent measures are needed to mitigate hazards of construction. Shrink-swell clays, discussed in Chapter 10, are notorious for compacting in dry seasons, causing millions of dollars of damage to foundations, bridges, and pipelines. In many cases, the best solution is to totally remove the shrink/swell clays from the site and replace them with suitably compacted nonswelling material such as sand and gravel. This technique is particularly expensive and is not often done. In most cases, reinforcing concrete with extra rebarb (steel rods) so that foundations and walls do not pull apart as the clay dries and shrinks mitigates the hazard, and buildings are placed on deep pilings that extend past the expansive clay.

Organic-rich soils are also particularly hazardous because of their enormous compaction potential. Organic-rich soils are formed in wetlands such as swamps and are common in coastal marshes and deltas. These organic-rich soils are very water-rich when they accumulate, and when the wetland or region containing the soil is drained, the soil compacts and air enters the pore spaces and begins to oxidize the organic material into soil. This oxidation results in the decomposition of the organic material, and decomposed organic matter occupies less space than the original material, resulting in compaction of the soil. This type of compaction of organic-rich soil is a major problem in old lake beds, deltas, wetlands, and coastal wetlands.

Buildings constructed on organic-rich soils should be built on pilings that extend below the organic-rich layer, if possible. On deltas, this is not generally possible since the organic-rich layers may extend for several miles below the surface. Homes and businesses built on pilings on reclaimed wetlands and coastal marshes often appear to be emerging, or rising, every year out of the ground on long concrete pilings. The ground is really sinking faster than the buildings, giving rise to this illusion. Continued compaction and subsidence of the organic-rich soils may cause the buildings to collapse off their pilings. New material should be regularly added beneath these structures to prevent them from toppling off unsupported pilings.

Detection of Incipient Sinkholes Some of the damage from sinkhole formation could be avoided if the location and general time of sinkhole formation could be predicted. At present, it may be possible to recognize places where sinkholes may be forming by monitoring for the formation of shallow depressions and extensional cracks on the surface, particularly circular depressions. Building foundations can be examined regularly for new cracks, and distances between slabs on bridges with expansion joints can be monitored to check for expansion related to collapse. Other remote sensing and geophysical methods may prove useful for monitoring sinkhole formation, particularly if the formation of a collapse structure is suspected. Shallow seismic waves can detect open spaces, and ground-penetrating radar can be used to map the bedrock surface and to look for collapse structures beneath soils. In some cases, it may be worthwhile to drill shallow test holes to determine if there is an open cavity at depth that is propagating toward the surface.

RESOURCES AND ORGANIZATIONS

Print Resources

Beck, B. F. *Engineering and Environmental Implications of Sinkholes and Karst*. Rotterdam: Balkema, 1989.

Dolan, R., and Grant Goodell, H. "Sinking Cities." *American Scientist* 74, no. 1 (1986), 38–47.

Drew, D. *Karst Processes and Landforms*. New York: MacMillan Education Press, 1985, 63 pp.

Ford, D., and Williams, P. *Karst Geomorphology and Hydrology*. London: Unwin-Hyman, 1989, 601 pp.

Holzer, T. L., ed. "Man-Induced Land Subsidence." Geological Society of America, *Reviews in Engineering Geology* 6, 1984.

Jennings, J. N. *Karst Geomorphology*. Oxford: Basil Blackwell, 1985.

White, W. B. *Geomorphology and Hydrology of Karst Terrains*. Oxford: Oxford University Press, 1988, 464 pp.

Non-Print Resources

Karst Web site:
http://hum.amu.edu.pl/sgp/spec/linkk.html
Links to sites about caves and karst systems; arranged by continent and region.

Karst Waters Institute Web site:
http://www.karstwaters.org/
Offers excellent descriptions of karst phenomenon, hazards, waters, and researchers in the field. Also posts a lexicon (glossary) of karst terminology.

Speleological Web site links:
http://hum.amu.edu.pl/sgp/spec/links.html
Large collection of links to speleological sites; ordered by continent.

U.S. Global Change Research Office Web site:
http://www.gcrio.org/geo/karst.html
Describes karst activity, its significance and causes, types of monitoring, and hazards.

Organizations

The Karst Waters Institute
P.O. Box 537
Charles Town, WV 25414
(304) 725–1211
http://www.karstwaters.org/
"The Karst Waters Institute (KWI) is a 501 (c)(3) non-profit institution whose mission is to improve the fundamental understanding of karst water systems through sound scientific research and the education of professionals and the public. The institute is governed by a Board of Directors and does not have or issue memberships. Institute activities include the initiation, coordination, and conduct of research, the sponsorship of conferences and workshops, and occasional publication of scientific works. KWI supports these activities by acting as a coordinating agency for funding and personnel, but does not supply direct funding or grants to individual researchers. As one way of increasing public awareness of karst and cave protection, the Institute publishes a list of the Top 10 endangered karst environments in the world. The third annual list is now available."

Solution Mining Research Institute
3336 Lone Hill Lane
Encinitas, California 92024–7262
858–759–7532
Fax: 858–759–7542
E-mail: smri@solutionmining.org
Web site: www.solutionmining.org
"The Solution Mining Research Institute (SMRI) is a private organization interested in the production of salt brine and the utilization of the resulting caverns for the storage of oil, gas, chemicals, compressed air and waste, and other

mining activities. SMRI is a worldwide organization with more than seventy members in Asia, Europe and North and South America. Participation by operators, researchers, suppliers, consultants, educators and government regulators is encouraged. Activities include: Sponsoring and funding research, Holding technical meetings, Conducting classes, Maintaining a library, and Participating in the development of government regulations."

United States Environmental Protection Agency
Office of Research and Development
Washington D.C. 20460.
http://www.epa.gov
The EPA works with other government agencies and private organizations to monitor subsidence and sinkhole hazards.

CHAPTER 12

Hazards of Sudden Catastrophic Geologic Events

INTRODUCTION

Some of the least likely events can cause the greatest amount of destruction. It is now recognized that an impact with a space object, probably a meteorite, caused the extinction of the dinosaurs and 65 percent of the other species on the planet at the end of the Cretaceous Period 66 millions years ago. The crater resulting from this meteor's impact is apparently preserved at Chicxulub on Mexico's Yucatán Peninsula, and the impact occurred at a time when the world's biosphere was already stressed, probably by massive amounts of volcanism and sea level fall. The volcanic fields from this time are preserved as vast lava plains in western India known as the Deccan traps. Impacts and massive volcanism can both dramatically change the global climate on scales that far exceed the changes witnessed in the past few thousand years. These changes have dramatic influences on the evolution and extinction of species, and current ideas suggest that impacts and volcanism have been responsible for most of the great extinctions of geological time.

Impacts cause earthquakes of unimaginable magnitude, thousands of times stronger than any ever observed on Earth by humans. If a meteorite lands in the ocean, it can form giant tsunamis hundreds (if not thousands) of feet tall that sweep across ocean basins in minutes and run up hundreds of kilometers onto the continents. Impacts kick up tremendous amounts of dust and hot flaming gases that scorch the atmosphere and fill it with sun-blocking dust clouds for years. Global fires burn most organic matter, and these fires are followed by a period of dark, deep freeze, which is caused by the atmospheric dust blocking out warming sunlight. This may be followed rapidly by a warm period after the dust settles, which is caused by the extra CO_2 released in the atmosphere by the impact, and changes in atmospheric and oceanic chemistry kill off many of the remaining life forms in the oceans.

Other rare geologic events may be less dramatic than meteorite impacts, though still beyond the scale of anything we have seen on Earth in recent times. One example of this "extreme" geology is the formation of diatremes or kimberlite pipes, which are large volcanic pipes that come from deep in the earth and explode their way to the surface with such force that they may blow holes through the stratosphere. These kimberlite pipes carry diamonds from hundreds of kilometers in the earth and seem to form in places that were abnormally rich in fluids. The global effect of kimberlites is not known, but you certainly would not want to be in the vicinity when one punches through the crust.

METEORITE AND COMETARY IMPACT

Meteorites are rocky objects from space that strike Earth. When meteorites pass through Earth's atmosphere, they get heated and their surfaces become ionized, causing them to glow brightly and form a streak moving across the atmosphere known as a shooting star or fireball.

At certain times of the year, the earth passes through parts of our solar system that are rich in meteorites, and the night skies become filled with shooting stars and fireballs, sometimes as frequently as several per minute. These times of high-frequency meteorite encounters are known as meteor showers; they include the Perseid showers that appear around August 11 and the Leonid showers that appear about November 14.

There is ample evidence that many small meteorites have hit the earth frequently throughout time. Eyewitness accounts describe many events, and fragments of meteorites are regularly recovered from places like the Antarctic ice sheets, where rocky objects on the surface have no place to come from but space. Although meteorites may appear as flaming objects moving across the night skies, they are generally cold icy bodies when they land on Earth; after the deep freeze of space, only their outermost layers get heated during their short transit though the atmosphere.

There are several different main types of meteorites. Stony meteorites include chondrites, which are very primitive and ancient meteorites made of silicate minerals like those common in the earth's crust and mantle, but chondrites also contain small spherical objects known as chondrules. These chondrules contain frozen droplets of material that is thought to be remnants of the early solar nebula from which Earth and the other planets initially condensed. Achondrites are similar to chondrites in mineralogy except that they do not contain the chondritic spheres. Iron meteorites are made of an iron-nickel alloy with textures that suggest that they formed from slow crystallization inside a large asteroid or small planet, which has since been broken into billions of small pieces, probably due to an impact with another object. Stony-irons are meteorites that contain mixtures of stony and iron components, and they probably formed near the core-mantle boundary of the broken planet or asteroid. Almost all meteorites found on Earth are stony varieties.

Origin of Meteorites and Other Earth Orbit–Crossing Objects

Most meteorites originate in the asteroid belt situated between the orbits of Mars and Jupiter. There are at least a million asteroids in this belt with diameters greater than one kilometer, 1,000 with diameters greater than thirty kilometers, and 200 with diameters greater than 100 kilometers. Asteroids and meteorites are distinguished only by their size; asteroids are greater than 100 meters in diameter. Meteorites are referred to as meteors only after they enter Earth's atmosphere. These are thought to be either remnants of a small planet that was destroyed by a large impact event or perhaps fragments of rocky material that failed to coalesce into a planet, probably due to the gravitational effects of the nearby massive planet of Jupiter. Most scientists favor the second hypothesis but recognize that collisions between asteroids have fragmented a large body to expose a planet-like core and mantle now preserved in the asteroid belt.

Collisions between asteroids can alter their orbits and cause them to head into an Earth orbit–crossing path. At this point, the asteroid becomes hazardous to life on Earth and is known as an Apollo object. Presently, about 150 Apollo objects with diameters of greater than one kilometer are known, but there are bound to be many more. Only in 1996, an asteroid about one-quarter mile across nearly missed hitting Earth, speeding past at a distance about equal to the distance to the moon. The sobering reality of this near collision is that the asteroid was not even spotted until a few days before it sped past Earth. What if the object were bigger or slightly closer? Would it have been stoppable? If not, what would have been the consequences of its collision with Earth? A similar near-miss event was recorded again in 2001 when Asteroid 2001 YB5 passed Earth at a distance of twice that to the moon; it too was not recognized until two weeks before its near-miss. If YB5 hit Earth, it would have released energy equivalent to 350,000 times the energy released during the nuclear bomb blast in Hiroshima. The objects that are in Earth orbit–crossing paths could not have been in this path for very long because gravitational influences of Earth, Mars, and Venus would cause them to hit one of the planets or be ejected from the solar system within about 100 million years. The abundance of asteroids in an Earth orbit—crossing path demonstrates that ongoing collisions in the asteroid belt are replenishing the source of potential impacts on Earth. A few rare meteorites found on Earth have chemical signatures that suggest they originated on Mars and on the moon, probably being ejected toward Earth from giant impacts on those bodies.

Other objects from space may collide with Earth. Comets are masses of ice and carbonaceous material mixed with silicate minerals that are thought to originate in the outer parts of the solar system in a region called the Oort Cloud. Other comets have a closer origin in the Kuiper Belt just beyond the orbit of Neptune. There is considerable debate about whether small icy Pluto, long considered the small outermost planet, should actually be classified as a large Kuiper Belt object. Comets may be less common near Earth than meteorites, but they still may hit Earth with severe consequences. There are estimated to be more than a trillion comets in our solar system. Since they are lighter than asteroids and have water-rich and carbon-rich compositions, many scientists have speculated that cometary impact may have brought water, the atmosphere, and even life to Earth.

In 1908, a huge explosion rocked the Tunguska area of Siberia and devastated more than 3,000 square kilometers of forest. The force of the blast is estimated to have been equal to fifteen megatons and is thought to have been produced by the explosion, six miles above the surface of Earth, of an asteroid with a diameter of 200 feet. Shock waves were felt thousands of miles away, people located closer than sixty miles from the site of the explosion were knocked unconscious, and some were thrown into the air by the force of the explosion. Fiery clouds and deafening explosions were heard more than 300 miles from Tunguska.

Impact-Cratering Mechanics and Consequences

The collision of meteorites with Earth produces impact craters, which are generally circular bowl-shaped depressions (Figure 12.1). There are more than 200 known impact structures on Earth, although processes of weathering, erosion, volcanism, and tectonics have undoubtedly erased many thousands more. The moon and other planets show much greater densities of impact craters, and since Earth has a greater gravitational pull than the moon, it should have been hit by many more impacts than the moon.

Meteorite impact craters have a variety of forms but are of two basic types. Simple craters are circular bowl-shaped craters with overturned rocks around their edges, and they are generally less than five kilometers in diameter. They are thought to have been produced by impact with objects less than thirty meters in diameter. Examples of simple craters include the Barringer Meteor Crater in Arizona (Figure 12.1) and Roter Kamm in Namibia. Complex craters are larger and generally greater than three kilometers in diameter. They have an uplifted peak in the center of the crater and have a series of concentric rings around the excavated core of the crater. Examples of complex craters include Manicougan, Clearwater Lakes, and Sudbury in Canada; Chicxulub in Mexico; and Gosses Bluff in Australia.

The style of impact crater depends on the size of the impacting meteor, the speed at which it strikes the surface, and (to a lesser extent) the underlying geology and the angle at which the meteor strikes Earth. Most meteorites hit Earth with a velocity between 2.5 and twenty-five miles per second, releasing tremendous energy when they hit. Meteor Crater in Arizona was produced about 50,000 years ago by a meteorite 100 feet in diameter that released the equivalent of four megatons of TNT. The meteorite body and a large section of the ground at the site were suddenly melted by shock waves from the impact, which released about twice as much energy as the eruption of Mount St. Helens. Most impacts generate so much heat and shock pressure that the entire meteorite and a large amount of the rock it hits are melted and vaporized. Temperatures may exceed thousands of degrees in a fraction of a second as atmospheric pressures increase a million times during passage of the shock wave. These conditions cause the rock at the site of the impact to accelerate downward and outward, then the ground re-

Figure 12.1. Barringer Meteor Crater, Arizona. Note roads and trees on crater rim. Imagine the devastation that would occur if a meteorite were to hit a populated area of the earth. ©Charles O'Rear/CORBIS

bounds and tons of material are shot outward and upward into the atmosphere.

Impact cratering is a complex process. When the meteorite strikes, it explodes, vaporizes, and sends shock waves through the underlying rock; this compresses the rock, crushes it into breccia, and ejects material (conveniently known as *ejecta*) back up into the atmosphere from where it falls out as an ejecta blanket around the impact crater. Large impact events may melt the underlying rock forming an impact melt and may form distinctive minerals that only form at exceedingly high pressures.

After the initial stages of the impact crater–forming process, the rocks surrounding the excavated crater slide and fall into the deep hole, enlarging the diameter of the crater and typically making it much wider than it is deep. Many of the rocks that slide into the crater are brecciated or otherwise affected by the passage of the shock wave and may preserve these effects as brecciated rocks, high pressure mineral phases, shatter cones, or other deformation features.

Impact cratering was probably a much more important process in the early history of Earth than it is at present. The flux of meteorites from most parts of the solar system was much greater in early times, and it is likely that impacts totally disrupted the surface in the early Precambrian. At present, the meteorite flux is about 100 tons per day (somewhere between 10^7–10^9 kg/yr), but most of this material burns up as it enters the atmosphere. Meteorites that are about one-tenth of an inch to several feet in diameter make a flash of light (a shooting star) as they burn up in the atmosphere, and the remains fall to Earth as a tiny glassy sphere of rock. Smaller particles, known as cosmic dust, escape the effects of friction and slowly fall to Earth as a slow rain of extraterrestrial dust.

Meteorites must be greater than three feet in diameter to make it through the atmosphere without burning up from friction. Earth's surface is currently hit by about one small meteorite per year. The statistics of meteorite impact show that the larger events are the least frequent: Meteorites 300 feet in diameter hit once every 10,000 years, ones 3,000 feet in diameter hit Earth once every million years, and ones six miles in diameter hit every 100 million years. Meteorites of only hundreds of meters in diameter could create craters about 1–2 kilometers in diameter, or if they hit in the ocean, they would generate tsunami more than five meters tall over wide regions.

Hazards of Impacts

The chances of the earth being hit by a meteorite are small at any given time, but they are greater than the chances of winning a lottery. The chances of dying from an impact are about the same as dying in a plane crash for a person who takes one flight per year. These comparisons are statistical flukes, however, and reflect the fact that a meteorite impact is likely to kill so many people that it raises the statistical chances of dying by impact. The earth has been hit by a number of small impacts and by some very large impacts that have had profound effects on the life on Earth at those times. What would happen to humans if a huge impact were to hit the planet today?

A moderate-sized impact event, such as a collision with a meteorite with a 5–10 mile diameter that was moving at a moderate velocity of 12 km/sec, would release energy equivalent to 100 megatons or about 1,000 times the yield of all existing nuclear weapons on Earth. The meteorite would begin to glow brightly as it approached Earth and encountered the outer atmosphere. As this body entered the atmosphere, it would create a huge fireball that would crash with the Earth after about ten seconds (Figure 12.2). This collision would send out shock waves that would be felt globally as earthquakes of unimaginable size, destroying much of the surface of the planet and killing billions of people. It is estimated that the size of this earthquake would be at least a magnitude 11, more than a thousand times stronger than the Great Alaskan Quake of 1964. If the impact hit water, huge tsunamis would be formed that would be hundreds or perhaps thousands of feet tall (Figure 12.2). These would run up on coastlines, washing away the debris from the earthquakes of a few moments before. The force of the impact event would eject enormous quantities of superheated dust and gases into the atmosphere, some of which would fall back to Earth as flaming fireballs. Most of the dust would make it into the upper atmosphere, where it would encircle the entire planet. The energy from the impact would heat the atmosphere to such a degree that it would spontaneously ignite forests and much of the biomass, sending dark clouds of smoke into the atmosphere. This smoke and the dust from the impact would block out the sun, leading to a rapid plunge into a dark mini–ice age with most of the sun's energy blocked from reaching Earth, preventing photosynthesis and plant growth. This darkness and cold would last for several months as the dust slowly settled, forming a global layer of dust recording the chemical signature of the impact. This chemical signature of impacts includes a hallmark high concentration of the rare element iridium, which is produced by vaporization of the meteorite.

Gradually, rain would remove the sulfuric acid and dust from the atmosphere, but the chemical consequences include enhanced acid rain, which rapidly dissolves calcium carbonates from limestone and shells, releasing carbon dioxide to the atmosphere. Acid rain wreaks havoc on the ocean biosphere, changing the ocean chemistry to the point at which many life forms become extinct. The carbon dioxide released to the atmosphere heats the planet a few months after the impact, and the planet would enter an extended warm period caused by this greenhouse effect. Temperatures would be more than 10° hotter on average, shifting climate belts and leading to excessively long hot summers.

Impacts and Mass Extinction Events

The geologic record of life on Earth shows that there have been several sudden events that led to the extinction of large numbers of land and marine species in a very short interval of time. Many of the boundaries between geologic time periods have been selected based on these mass extinction events. Some of the major mass extinctions include that between the Cretaceous and Tertiary Periods, marking the boundary between the Mesozoic and Cenozoic Eras. At this boundary 66 million years ago, dinosaurs, ammonites, many marine reptile species, and a large number of marine invertebrates suddenly died off, and the planet lost about 26 percent of all biological families and numerous species. Two hundred and forty-five million years ago, at the boundary between the Permian and Triassic Periods and between the Paleozoic and Mesozoic Periods, 96 percent of all species became extinct. Lost were the rugose corals, trilobites, many types of brachiopods, and marine organisms including many foraminifer species. There are several other examples of mass extinctions, including one at the boundary between the Cambrian and Ordovician Periods 505 million years ago, during which more than half of all biological families disappeared forever.

These mass extinctions have several similar features that point to a common origin. Impacts have been implicated as the cause of many of the mass extinction events in Earth's history. The mass extinctions seem to have occurred on a geologically instantaneous time scale, with many species present in the rock record below a thin clay-rich layer and dramatically fewer species present immediately above the layer. In the case of the Cretaceous-Tertiary extinction, some organisms were dying off slowly before the dramatic die-off, but a clear sharp event occurred at the end of this time that resulted in environmental stress and gradual extinction. Iridium anomalies have been found along most of the clay layers, considered by many to be the "smoking gun" that indicates an impact origin for cause of the extinctions. One half-million tons of iridium are estimated to be in the Cretaceous-Tertiary boundary clay, equivalent to the amount that would be contained in a meteorite with a six-mile diameter. Other scientists argue that volcanic processes within the earth can produce iridium and impacts are not necessary. However, other rare elements and geochemical anomalies are present along the Cretaceous-Tertiary boundary, supporting the idea that a huge meteorite hit the earth at this time.

Other features have been found that support the impact origin for the mass extinctions. One of the most important is the presence of some high-pressure minerals formed at pressures not reachable in the outer layers of the earth. The presence of the high-pressure mineral equivalents of quartz, including coesite, stishovite, and an extremely high-pressure phase known as diaplectic glass, strongly implicates an impacting meteorite, which can produce tremendous pressures during the passage of shock waves related to the force of the impact. Many of the clay layers associated with the iridium anomalies also have layers of tiny glass spherules, which are thought to be remnants of melted rock produced during the impact that was thrown skyward, where it crystallized as tiny droplets that rained back on the planet's surface. The clay layers also have abundant microdiamonds similar to those produced during meteorite impact events. Layers of carbon-rich soot are also associated with some of the impact layers, and these are thought to represent remains of the global wildfires that were ignited by the impacts. Finally, some of the impact layers also record huge tsunami that swept across coastal regions.

Many of these features are found around and associated with an impact crater recently discovered on Mexico's Yucatán Peninsula. The Chicxulub Crater is about 66 million years old and lies half-buried beneath the waters of the Gulf of Mexico and half-buried on land. Tsunami deposits of the same age are found in inland Texas, much of the Gulf of Mexico, and the Caribbean, recording a huge tsunami perhaps several hundred feet high that was generated by the impact. The crater is at the center of a huge field of scattered spherules that extends across Central America and through the southern United States. It is a large structure and is the right age to be the crater that records the impact at the Cretaceous-Tertiary boundary, recording the extinction of the dinosaurs and other families. Other craters of similar age have been reported from elsewhere in the world, suggesting that Chicxulub may be one of several craters formed by the breakup of the meteorite before collision, forming several craters at virtually the same time.

An Example of Meteorite and Cometary Impact Events

On June 30, 1908, a huge explosion rocked the area of Tunguska in Siberia. At about 7:00 A.M. on June 30, a huge fireball was seen moving westward across Siberia. Next, an explosion was heard that was centered on the remote Tunguska region, with reports of the explosion and pressure waves being felt more than 600 miles away. People who were hundreds of miles from the site of the explosion were knocked down, and a huge twelve-mile-high column of fire was visible for more than 400 miles. Seismometers recorded the impact, and barometers around the world recorded the air pressure wave as it traveled two times around the globe. The impact caused a strange, bright, unexplained glow to light the night skies in Scotland and Sweden.

Several years after the impact, a scientific expedition to the remote region discovered that many trees were charred and knocked down in a 2,000-square-mile area, with near total destruction in the center 500 square miles. Many theories were advanced to explain the strange findings, and more than fifty years later in 1958, a new expedition to Tunguska found small melted globules of glass and metal, which were identified as pieces of an exploded meteorite or asteroid.

One of the biggest puzzles at Tunguska is the absence of an impact crater, despite all other evidence that point to an impact origin for this event. It is now thought that a piece of a comet, Comet Encke, broke off the main body as it was orbiting nearby Earth, and this fragment entered Earth's atmosphere and exploded about five miles above the Siberian plains at Tunguska. Comets are weaker than metallic or stony meteorites, and they more easily break up and explode in the atmosphere before hitting the earth's surface.

Mitigating the Dangers of Future Impacts

What would it take to wipe out the human race? A meteorite about a mile across would do enough damage to wipe out about one-quarter of the human race, and events of this magnitude may occur approximately every 3 million years. Larger events happen less frequently than smaller events: It is estimated that about fifty objects with diameters of 50–100 feet pass between Earth and the moon every day, though these rarely collide with Earth. Comets and stony meteorites of this size will typically break up upon entering the atmosphere, whereas iron meteorites tend to make it all the way to the planet's surface. Luckily, few of the near-Earth objects are iron meteorites.

What can humans do to try to prevent large meteorites from crashing into Earth and wiping out much of the population and biosphere? NASA has estimated that there are about 2,000 near-Earth objects greater than half a mile in diameter, and that about half of them may eventually hit the Earth. However, the time interval between individual impacts is greater than 100,000 years. If any of these objects hits the earth, the death toll will be tremendous, particularly if any of them hit populated areas or a major city.

It is now technically feasible to map and track many of the large objects that could be on an Earth-impacting trajectory. This would involve considerable expense to advanced societies, principally the taxpayers of the United States. NASA has mounted a preliminary program for mapping and tracking objects in near-Earth orbit and has already identified many significant objects. Lawmakers and the public must decide if greater expenses are worth the calculated risk of impacts hitting the earth. Is it more realistic to try to stop the spread of disease, crime, poverty, and famine and to prepare for other natural disasters, or should resources be spent looking for objects that might one day collide with the earth? What could be done if an object is identified that is likely to collide with the earth?

We have the technology to attempt to divert or blow up the meteorite using nuclear devices. Bombs could be exploded near the asteroid or meteorite in an attempt to move it out of Earth's orbit or to shatter it into small enough pieces that would break up upon entering the atmosphere. Alternatively, given enough time, rockets could be installed on the meteorite and fired to try to steer it out of its impact trajectory. However, if the object if very large, it is likely that even all of the nuclear weapons on the planet would not have a significant effect on altering the trajectory of the meteorite or asteroid.

MASSIVE GLOBAL VOLCANISM AND DEGASSING

At several times in Earth's history, vast outpourings of lava have accumulated and formed thick piles of basalt, which represent the largest known volcanic episodes on the planet in the past several hundred million years. These deposits include continental flood basalt provinces, anomalously thick and topographically high sea floors known as oceanic plateaus, and some volcanic rifted passive margins. These piles of volcanic rock represent times when the earth moved more material and energy from its interior than during intervals between the massive volcanic events. Such large amounts of volcanism also released large amounts of volcanic gases into the atmosphere with serious implications for global temperatures and climate, and they may have contributed to some global mass extinctions.

The largest continental flood basalt province in the United States is the Columbia River flood basalt in Washington, Oregon, and Idaho (Figure 12.2). The Columbia River flood basalt province is 6–17 million years old and contains an estimated 1,250 cubic miles of basalt. Individual lava flows erupted through fissures or cracks in the crust, then flowed laterally across the plain for up to 400 miles.

The 66-million-year-old Deccan flood basalts, also known as traps, cover a large part of western India and the Seychelles Islands. They are associated with the breakup of India from the Seychelles during the opening of the Indian Ocean. Slightly older flood basalts (83–90 million years old) are associated with the break away of Madagascar from India. The volume of the Deccan traps is estimated at 5,000,000 cubic miles, and the volcanics are thought to have been erupted during an approximately 1 million year long time period, starting slightly before the great Cretaceous–Tertiary extinction. Most workers now agree that the gases released during the flood basalt volcanism stressed the global biosphere to such an extent that

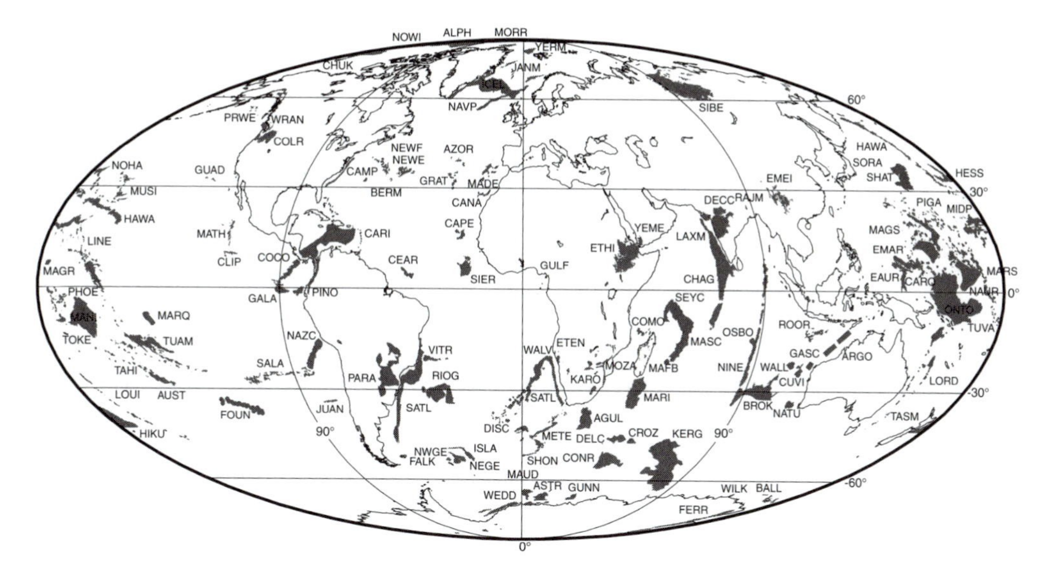

Figure 12.2. Map of world showing distribution of flood basalt provinces. Photo Credit: Institute of Geophysics, University of Texas, Austin

many marine organisms had gone extinct and many others were stressed. Then the planet was hit by the massive Chicxulub impact, causing the massive extinction that included the end of the dinosaurs.

The breakup of east Africa along the East African rift system and the Red Sea is associated with large amounts of Cenozoic (less than 30 million years old) continental flood basalts. Some of the older volcanic fields are located in east Africa in the Afar region of Ethiopia, south into Kenya and Uganda and north across the Red Sea and Gulf of Aden into Yemen and Saudi Arabia (Figure 12.2). These volcanic piles are overlain by younger (less than 15 million year old) flood basalts that extend both farther south into Tanzania, farther north through central Arabia where they are known as *harrats*, and into Syria, Israel, Lebanon, and Jordan.

An older volcanic province also associated with the breakup of a continent is known as the North Atlantic Igneous Province. It formed along with the breakup of the North Atlantic Ocean 55–62 million years ago and includes both onshore and offshore volcanic flows and intrusions in Greenland, Iceland, and the northern British Isles including most of the Rockall Plateau and Faeroe Islands. In the south Atlantic, a similar 129–134-million-year-old flood basalt was split by the opening of the ocean and now has two parts. In Brazil, the flood lavas are known as the Parana basalts, and in Namibia and Angola of West Africa as the Etendeka basalts.

The Caribbean Ocean floor represents one of the best examples of an oceanic plateau, with other major examples including the Ontong-Java Plateau, Manihiki Plateau, Hess Rise, Shatsky Rise, and Mid Pacific

Mountains (Figure 12.2). All of these oceanic plateaus contain between six- and twenty-five-mile-thick piles of volcanic and subvolcanic rocks representing huge outpourings of lava. The Caribbean sea floor preserves oceanic crust that is five to thirteen miles thick and that was formed before 85 million years ago in the eastern Pacific Ocean. This unusually thick ocean floor was transported eastward by plate tectonics, where pieces of the sea floor collided with South America as it passed into the Atlantic Ocean. Pieces of the Caribbean oceanic crust are now preserved in Colombia, Ecuador, Panama, Hispaniola, and Cuba, and some scientists estimate that the Caribbean oceanic plateau may have once been twice its present size. In either case, it represents a vast outpouring of lava that would have been associated with significant outgassing with possible consequences for global climate and evolution.

The western Pacific Ocean basin contains several large oceanic plateaus, including the twenty-mile-thick crust of the Alaskan-sized Ontong-Java Plateau (Figure 12.2), which is the largest outpouring of volcanic rocks on the planet. It apparently formed in two intervals 122 and 90 million years ago, entirely within the ocean, and represents magma that rose in a plume from deep in the mantle and erupted on the sea floor. It is estimated that the volume of magma erupted in the first event was equivalent to all of the magma being erupted at mid-ocean ridges at the present time. Sea levels rose by more than thirty feet in response to this volcanic outpouring. The gases released during these eruptions are estimated to have raised average global temperatures by 13° C (23° F).

Hazards of Flood Basalt Volcanism

The environmental impact of the eruption of large volumes of basalt in provinces, including those described above, can be severe. Huge volumes of sulfur dioxide, carbon dioxide, chlorine, and fluorine are released during large basaltic eruptions. Much of this gas may get injected into the upper troposphere and lower stratosphere during the eruption process, being released from eruption columns that reach 2–8 miles in height. Carbon dioxide is a greenhouse gas and can cause global warming, whereas sulfur dioxide and hydrogen sulfate have the opposite effect and can cause short-term cooling. Many of the episodes of volcanism preserved in these large igneous provinces were rapid, repeatedly releasing enormous quantities of gases over periods of less than 1 million years and releasing enough gas to significantly change the climate more rapidly than organisms could adapt. For instance, one eruption of the Columbia River basalts is estimated to have released 9,000 million tons of sulfur dioxide and thousands of millions of tons of other gases, whereas the eruption of Mount Pinatubo (see Chapter 3) in 1991 released 20 million tons of sulfur dioxide. The Columbia River basalts continued erupting for years at a time, for approximately a million years. During this time, the gases released would be equivalent to that of Mount Pinatubo, every week, over periods maintained for decades to thousands of years at a time. The atmospheric consequences are sobering. Sulfuric acid aerosols and acid from the fluorine and chlorine would form extensive poisonous acid rain, destroying habitats and making waters uninhabitable for some organisms. At the very least, the environmental consequences would be such that organisms were stressed to the point that they would not be able to handle any additional environmental stress, such as a global winter and subsequent warming, caused by a giant impact.

Faunal extinctions have been correlated with the eruption of the Deccan flood basalts at the Cretaceous-Tertiary (K/T) boundary, and with the Siberian flood basalts at the Permian-Triassic boundary. There is still considerable debate about the relative significance of flood basalt volcanism and impacts of meteorites for extinction events, particularly at the Cretaceous-Tertiary boundary. However, most scientists would now agree that the global environment was stressed shortly before the K/T boundary by volcanic-induced climate change, and then a huge meteorite hit the Yucatán Peninsula, forming the Chicxulub impact crater and causing the massive K/T boundary extinction and the death of the dinosaurs.

The Siberian flood basalts cover a large area of the Central Siberian Plateau northwest of Lake Baikal (Figure 12.2). They are more than half a mile thick over an area of 210,000 square miles, but have been significantly eroded from an estimated volume of 1,240,000 cubic miles. Two hundred and fifty million years ago, they were erupted over a remarkably short period of less than 1 million years at the Permian-Triassic boundary. They are remarkably coincident in time with the major Permian-Triassic extinction, implying a causal link. The Permian-Triassic boundary at 250 million years ago marks the greatest extinction in Earth's history, where 90 percent of marine species and 70 percent of terrestrial vertebrates became extinct. It has been postulated that the rapid volcanism and degassing released enough sulfur dioxide to cause a rapid global cooling, inducing a short ice age with an associated rapid fall of sea level. Soon after the ice age took hold, the effects of the carbon dioxide took over and the atmosphere heated, resulting in a global warming. The rapidly fluctuating climate postulated to have been caused by the volcanic gases is thought to have killed off many organisms, which were simply unable to cope with the wildly fluctuating climate extremes.

KIMBERLITES AND DIATREMES

Kimberlites and diatremes represent rare types of continental volcanic rock types that are produced by generally explosive volcanism with an origin deep within the mantle. They form pipe-like bodies extending vertically downward, and are the source of many of the world's diamonds. Kimberlites were first discovered in South Africa during diamond exploration and mining in 1869, when the source of many alluvial diamonds on the Vaal, Orange, and Riet Rivers was found to be circular mud "pans," later appreciated to be kimberlite pipes. In 1871, two very diamond-rich kimberlite pipes were discovered on the Vooruitzigt Farm, owned by Nicolas de Beer. These discoveries led to the establishment of several large mines and one of the most influential mining companies in history.

Kimberlites are very complicated volcanic rocks with mixtures of material derived from the upper mantle and with complex water-rich magma of several different varieties. A range of volcanic intrusive styles, including some extremely explosive events, characterizes kimberlites. True volcanic lavas are only rarely associated with kimberlites, so volcanic styles of typical volcanoes (see Chapter 3) are not typical of kimberlites. Most near-surface kimberlite rocks are pyroclastic deposits formed by explosive volcanism filling vertical pipes and surrounded by rings of volcanic tuff and related features. The pipes are typically a couple of hundred yards wide with the tuff ring extending another hundred yards or so beyond the pipes. The very upper part of many kimberlite pipes includes reworked pyroclastic rocks, which are deposited in lakes that filled the kimberlite pipes after the explosive volcanism blasted much of the kimberlite material out of the hole. Geologic studies of kimberlites have suggested that they intrude the crust suddenly and behave differently from typical volcanoes. Kimberlites intrude violently and catastrophically with the initial formation of a pipe filled with brecciated material from the mantle, reflecting the sudden and explosive character of the eruption. As the eruption wanes, a series of tuffs falls out of the eruption column and deposits the tuff ring around the pipes. Unlike most volcanoes, kimberlite eruptions are not followed by the intrusion of magma into the pipe. The pipes simply get eroded by near-surface processes, lakes form in the pipes, and nature tries to hide the very occurrence of the explosive event.

Below these upward-expanding craters are deep vertical pipes known as diatremes that extend down into the mantle source region of the kimberlites. Many diatremes have features that suggest that the brecciated mantle and crustal rocks were emplaced nonviolently at low temperature, presenting a great puzzle to geologists. How can a deep source of broken mantle rocks passively move up a vertical pipe to the surface, suddenly explode violently, and then disappear beneath a newly formed lake?

Early ideas for the intrusion and surface explosion of kimberlites suggested that they rose explosively and catastrophically from their origin in the mantle. Subsequent studies revealed that the early deep parts of their ascent did not seem to be explosive. It is likely that kimberlite magma rises from deep in the upper mantle along a series of cracks and fissures until it gets to shallow levels, where it mixes with water and becomes extremely explosive. Other diatremes may be more explosive from greater depths and may move as gas-filled bodies rising from the upper mantle. As the gases move into lower pressure areas they may expand, resulting in the kimberlite moving faster until it explodes at the surface. Still other ideas for the emplacement of kimberlites and diatremes invoke hydrovolcanism, or the interaction of the deep magma with near-surface water. Magma may rise slowly from depth until it encounters groundwater in fractures or other voids, then explode when the water mixes with the magma. The resulting explosion could produce the volcanic features and upward expanding pipe found in many kimberlites.

It is likely that some or all of the processes discussed here play a role in the intrusion of kimberlites and diatremes, the important consequences being a sudden, explosive volcanic eruption at the surface, far from typical locations of volcanism, and the relatively rapid removal of signs of this volcanism. The initial explosions are likely to be so explosive that they may blast material to the stratosphere, though other kimberlite eruptions may only form small eruptions and ash clouds. The hazards are thus similar to other volcanic hazards (see Chapter 3), but the location of kimberlites is more likely to be in the middle of a stable continental region.

MASS EXTINCTIONS AND CATASTROPHES

Most of the disasters discussed in this book have been about events such as volcanic eruptions, earthquakes, floods, and droughts that have resulted in tremendous loss of life and property, with death tolls in the thousands to hundreds of thousands of people in many cases. Although these disasters have been catastrophic and every effort should be taken to prevent such disasters from occurring again, the earth has experienced several major events that pale all of these catastrophes in comparison. This chapter has so far touched on the relationship between impacts with extraterrestrial objects and their relationship to mass extinctions, as well as the global effects of massive volcanism. In this section, mass extinctions are examined in more detail, since they represent the largest and most significant natural disasters in the history of the earth. Understanding mass extinctions may someday save the human race from the same fate.

Geologists and paleontologists study the history of life on Earth though detailed examination of that record as preserved in sedimentary rock layers laid down one upon the other. For hundreds of years, paleontologists have recognized that many organisms are found in a series of layers and then suddenly disappear at a certain horizon, never to reappear in the succeeding, progressively younger layers. These disappearances have been interpreted to mark extinctions of the organisms from the biosphere. After hundreds of years of work, many of these rock layers have been dated by using radioactive decay dating techniques on volcanic rocks in the sequences, and many of these sedimentary rock sequences have been correlated with each other on a global scale.

One of the more important results that have come from such detailed studies is that the rock record preserves a record of several extinction events that have occurred simultaneously on a global scale. Furthermore, these events do not just affect thousands or hundreds of thousands of members of a species, but they have also wiped out many species and families, each containing millions or billions of individuals.

THE ROLE OF NONCATASTROPHIC PROCESSES IN EVOLUTION AND EXTINCTION

Throughout this book, several examples of processes that strongly influence the progression of life, evolution, and extinction have been discussed. Variations in the style of plate tectonics or the positions of the continents as well as continental collisions can all affect life, evolution, and extinction. Plate tectonics also may cause glaciation and climate changes, which in turn influence evolution and extinction.

One of the primary ways that normal, noncatastrophic plate tectonic mechanisms drive evolution and extinction is through tectonic-induced changes to sea level. Fluctuating sea levels cause the global climate to fluctuate between warm periods, when shallow seas are easily heated, and cold periods, when glaciation draws cold water down to shorelines along the steep continental slopes. Many species can not tolerate such variations in temperature and drastic changes to their shallow shelf environments and thus become extinct. After organisms from a specific environment die off, their environmental niches are available for other species to inhabit.

Sea levels have risen and fallen dramatically in Earth's history, with water covering all but 5 percent of the land surface at times and with water falling so that continents occupy 40 percent more of the planet's surface at other times. The most important plate tectonic mechanism of changing sea level is to change the average depth of the sea floor by changing the volume of the mid-ocean ridge system. If the undersea ridges take up more space in the ocean basins, then the water will be displaced higher onto the land, much like dropping pebbles into a birdbath may cause it to overflow.

How can the volume of the mid-ocean ridge system be changed? There are several mechanisms, all of which may have the same effect. Young oceanic crust is hotter, more buoyant, and topographically higher than older crust. Thus, if the average age of the oceanic crust is decreased, then more of the crust will be at shallow depths, displacing more water onto the continents. If sea floor spreading rates are increased, then the average age of oceanic crust will be decreased, the volume of the ridges will be increased, the average age of the sea floor will be decreased, and sea levels will rise. This has happened at several times in Earth's history, including during the mid-Cretaceous between 85–110 million years ago when sea levels were 660 feet (200 meters) higher than they are today, covering much of the central United States and other low-lying continents with water. This also warmed global climates, because the sun easily warmed the abundant shallow seas. It has also been suggested that sea levels were consistently much higher in the Precambrian, when sea floor spreading rates were likely to have been generally faster.

Sea levels can also rise from additional magmatism on the sea floor. If the Earth goes through a period where sea floor volcanoes erupt more magma on the sea floor, then the space occupied by these volcanic deposits will be displaced onto the continents. The additional volcanic rocks may be erupted at hot spot volcanoes like Hawaii or along the mid-ocean ridge system; either way, the result is the same.

A third way for the mid-ocean ridge volume to increase sea level height is to simply have more ridges on the sea floor. At the present time, the mid-ocean ridge system is 40,000 miles long. If the Earth goes through a period where it needs to lose more heat, such as in the Precambrian, one of the ways it may do this is by increasing the length of the ridge system where magmas erupt and lose heat to the seawater. Ridge lengths were probably greater in the

Precambrian, which together with faster sea floor spreading and increased magmatism may have kept sea levels at high levels for millions of years.

Sea level may also be changed by glaciation, which may be induced by tectonic or astronomical causes (see Chapter 9). At present, glaciers cover much of Antarctica, Greenland, and mountain ranges in several regions. There is approximately 6 million cubic miles of ice locked up in glaciers. If this ice were to all melt, then sea levels would rise by 230 feet, covering many coastal regions, cities, and interior farmland with shallow seas. During the last glacial maximum in the Pleistocene ice ages (20,000 years ago), sea levels were 460 feet lower than today, with shorelines up to hundreds of miles seaward of their present locations along the continental slopes.

Continental collisions and especially the formation of supercontinents can cause glaciations. When continents collide, many of the carbonate rocks deposited on continental shelves are exposed to weathering. As the carbonates and other minerals weather, their weathering products react with atmospheric elements and tend to combine with atmospheric CO_2. Carbon dioxide is a greenhouse gas that keeps the climate warm, and steady reductions of CO_2 in the atmosphere by weathering or other processes lower global temperatures. Thus, times of continental collision and supercontinent formation tend to be times that draw CO_2 out of the atmosphere, plunging the Earth into a cold "icehouse" period. In the cases of supercontinent formation, this icehouse may remain in effect until the supercontinent breaks up and massive amounts of sea floor volcanism associated with new rifts and ridges add new CO_2 back into the atmosphere.

The formation and dispersal of supercontinent fragments, and of migrating land masses in general, also strongly influence evolution and extinction. When supercontinents break up, a large amount of shallow continental shelf area is created. Life forms tend to flourish in the diverse environments on the continental shelves, and many spurts in evolution have occurred in these shallow shelf areas. In contrast, when island areas are isolated, such as Australia and Madagascar today, life forms evolve independently on them. If plate tectonics brings these isolated islands into contact, the different species will compete for similar food and environments, and typically only the strongest will survive.

The position of continents relative to the spin axes (or poles) of the earth can also influence climate, evolution, and extinction. At times (like the present) when a continent is sitting on one or both of the poles, these continents tend to accumulate snow and ice and to become heavily glaciated. This causes ocean currents to become colder, lowers global sea levels, and reflects more of the sun's radiation back to space. Together, these effects can put a large amount of stress on species, inducing or aiding extinction.

The History of Life

Life on Earth has evolved from simple organisms known as archaea that appeared on Earth by 3.85 billion years ago (Figure 1.3). Life may have been here earlier, but the record is not preserved. The method by which life first appeared is also unknown and is the subject of much thought and research by scientists, philosophers, and religious scholars.

The archaea derive energy from breaking down chemical bonds of carbon dioxide, water, and nitrogen, and have survived to this day in environments where they are not poisoned by oxygen. They presently live around hot vents around mid-ocean spreading centers, deep in the ground in pore spaces between soil and mineral grains, and in hot springs. The archaea represent one of the three main branches of life, the other two branches being the bacteria and the eukarya. The plant and animal kingdoms are part of the eukarya.

Prokaryotic bacteria (single-celled organisms lacking a cell nucleus) were involved in photosynthesis by 3.5 billion years ago, gradually transforming atmospheric carbon dioxide to oxygen and setting the stage for the evolution of simple eukaryotes (containing a cell nucleus and membrane-bound organelles) in the Proterozoic. Two and half billion years later, by 1 billion years ago, cells began reproducing sexually. This long-awaited step allowed cells to exchange and share genetic material, speeding up evolutionary changes by orders of magnitude.

Oxygen continued to build in the atmosphere, and some of this oxygen was combined into O_3 to make ozone. Ozone forms a layer in the atmosphere that blocks ultraviolet rays of the sun, forming an effective shield against this harmful radiation. When the ozone shield became thick enough to block a large portion of the ultraviolet radiation, life began to migrate out of the deep parts of the ocean and out of deep inland soils, moving into shallow water and places exposed to the sun. Multicellular life evolved around 670 million years ago (Figure 1.3), around the same time that the supercontinent of Gondwana was forming near the equator. Most of the planet's land masses were joined together for a short while, and then began splitting up and drifting apart again by 550 million years ago. This breakup of the supercontinent of Gondwana is associated with the most remarkable diversification of life in the history of the planet. In a remarkably short period of no longer than 40 million years, life developed complex forms with hard shells, and an incredible number of species appeared for the first time. This period of change marked the transition from the Precambrian Era to the Cambrian Period, marking the beginning of the Paleozoic Era. The remarkable development of life in this period is known as the Cambrian explosion. In the past 540 million years since the Cambrian explosion, life has continued to diversify with many new species appearing.

The evolution of life forms is also punctuated with the disappearance or extinction of many species, some as isolated cases and others that die off at the same time as many other species in the rock record. There are a number of distinct horizons representing times when hundreds, thousands, and even more species suddenly died, being abundant in the record immediately before the formation of one rock layer and forever absent immediately above that layer. Mass extinctions are typically followed, after several million years, by the appearance of many new species and the expansion and evolution of old species that did not go extinct. These rapid changes are probably a response to availability of environmental niches vacated by the extinct organisms. The new species rapidly populate these available spaces.

Mass extinction events are thought to represent major environmental catastrophes on a global scale. In some cases, these mass extinction events can be tied to specific likely causes, such as meteorite impact or massive volcanism, but in others their cause is unknown. Understanding the triggers of mass extinctions has important and obvious implications for ensuring the survival of the human race.

Examples of Mass Extinctions

Most species are present on Earth for about 4 million years. Some species come and go during a relatively low rate of extinctions that characterize most of geological time, during which new species evolve from older ones, but the majority of changes occur during distinct mass dyings and repopulation of the environment. Earth's biosphere has experienced five major and numerous less significant mass extinctions in the past 500 million years (i.e., in the Phanerozoic Era, Figure 1.3). These events occurred at the end of the Ordovician, in the Late Devonian, at the Permian-Triassic boundary, the Triassic-Jurassic boundary, and at the Cretaceous-Tertiary (K/T) boundary (Figure 1.3).

The early Paleozoic saw many new life forms emerge in new environments for the first time. The Cambrian explosion led to the development of trilobites, brachiopods, conodonts, mollusks, echinoderms, and ostracods. Bryozoans, crinoids, and rugose corals joined the biosphere in the Ordovician, and reef-building stromatoporoids flourished in shallow seas. The end-Ordovician extinction is one of the greatest of all Phanerozoic time. About half of all species of brachiopods and bryozoans died off, and more than 100 other families of marine organisms disappeared forever.

The cause of the mass extinction at the end of the Ordovician appears to have been largely tectonic. The major landmass of Gondwana had been resting in equatorial regions for much of the Middle Ordovician, but migrated toward the South Pole at the end of the Ordovician. This caused global cooling and glaciation, lowering sea levels from the high stand they had been resting at for most of the Cambrian and Ordovician. The combination of cold climates with lower sea levels, leading to a loss of shallow shelf environments for habitation, probably were enough to cause the mass extinction at the end of the Ordovician.

The largest mass extinction in Earth's history occurred at the Permian-Triassic boundary over a period of about 5 million years (Figure 1.3). The Permian world included abundant corals, crinoids, bryozoans, and bivalves in the oceans; on land, amphibians wandered about amid lush plant life. Ninety percent of oceanic species became extinct and 70 percent of land vertebrates died off at the end of the Permian. This greatest catastrophe of Earth's history did not have a single cause but reflects the combination of various elements.

First, plate tectonics was again bringing many of the planet's land masses together in a supercontinent (this time, Pangea), causing greater competition for fewer environmental niches by Permian life forms. Drastically reduced were the rich continental shelf areas. As the continents collided, mountains were pushed up, reducing the effective volume of the continents available to displace the sea, so sea levels fell, putting additional stress on life by further limiting the availability of favorable environmental niches. The global climate became dry and dusty, and the supercontinent formation led to widespread glaciation. This lowered sea level even more, lowered global temperatures, and put many life forms on the planet in a very uncomfortable position. Many perished.

In the final million years of the Permian, the Northern Siberian plains let loose a final devastating blow. The Siberian flood basalts (Figure 12.2) began erupting 250 million years ago, becoming the largest known outpouring of continental flood basalts ever. Carbon dioxide was released in hitherto unknown abundance, warming the atmosphere and melting the glaciers. Other gases were also released, perhaps also including methane, as the basalts probably melted permafrost and vaporized thick accumulations of organic matter that accumulate in high latitudes like that at which Siberia was located 250 million years ago.

The global biosphere collapsed, and evidence suggests that the final collapse happened in less than 200,000 years and perhaps in less than 30,000 years. Entirely internal processes may have caused the end-Permian extinction, although some scientists now argue that an impact may have dealt the final death blow. After it was over, new life forms populated the seas and land, and these Mesozoic organisms tended to be more mobile and adept than their Paleozoic counterparts. The great Permian extinction created opportunities for new life forms to occupy now empty niches, and the most adaptable and efficient organisms took control. The toughest of the marine organisms survived, and a new class of land animals grew to new proportions and occupied the land and skies. The Mesozoic, time of the great dinosaurs, had begun.

The Triassic-Jurassic extinction is not as significant as the Permian-Triassic extinction. Mollusks were abundant in the Triassic shallow marine realm with fewer brachiopods, and ammonoids recovered from near total extinction at the Permian-Triassic boundary. Sea urchins became abundant, and new groups of hexacorals replaced the rugose corals. Many land plants survived the end-Permian extinction, including the ferns and seed ferns that became abundant in the Jurassic. Small mammals that survived the end-Permian extinction rediversified in the Triassic, many to only become extinct at the close of the Triassic. Dinosaurs evolved quickly in the late Triassic, starting off small and attaining sizes approaching twenty feet by the end of the Triassic. The giant pterosaurs were the first known flying vertebrate, appearing late in the Triassic. Crocodiles, frogs, and turtles lived along with the dinosaurs. The end of the Triassic is marked by a major extinction in the marine realm, including total extinction of the conodonts and a mass extinction of the placodont marine reptiles and the mammal-like reptiles known as therapsids. Although the causes of this major extinction event are poorly understood, the timing is coincident with the breakup of Pangea and the formation of major evaporite and salt deposits. It is likely that this was a tectonic-induced extinction, with the supercontinent's breakup initiating new oceanic circulation patterns and new temperature and salinity distributions.

After the Triassic-Jurassic extinction, dinosaurs became extremely diverse and many grew quite large. Birds first appeared at the end of the Jurassic. The Jurassic was the time of the giant dinosaurs, which experienced a partial extinction affecting the largest varieties of stegosauroids, sauropods, and the marine ichthyosaurs and plesiosaurs. This major extinction is also poorly explained but may be related to global cooling. The other abundant varieties of dinosaurs continued to thrive through the Cretaceous.

The Cretaceous-Tertiary (K/T) extinction is perhaps the most famous of mass extinctions because the dinosaurs perished during this event (Figure 1.3). The Cretaceous land surface of North America was occupied by bountiful species, including herds of dinosaurs both large and small, some herbivores and others carnivores. Other vertebrates included crocodiles, turtles, frogs, and several types of small mammals. The sky had flying dinosaurs including the vulture-like pterosaurs, and insects including giant dragonflies. The dinosaurs had dense vegetation to feed on, including the flowing angiosperm trees, tall grasses, and many other types of trees and flowers. Life in the ocean had evolved to include abundant bivalves including clams and oysters, ammonoids, and corals that built large reef complexes.

Near the end of the Cretaceous (Figure 1.3), though the dinosaurs and other life forms didn't know it, things were about to change. High sea levels produced by mid-Cretaceous rapid sea floor spreading were falling, decreasing environmental diversity, cooling global climates, and creating environmental stress. Massive volcanic outpourings in the Deccan traps and the Seychelles formed as the Indian Ocean rifted apart and magma rose from an underlying mantle plume. Massive amounts of greenhouse gases were released, raising temperatures and drastically changing the environment. Many marine species were going extinct, and others became severely stressed. Then, one bright day, a visitor from space about six miles wide slammed into the Yucatán Peninsula of Mexico, instantly forming a fireball 1,200 miles across that was followed by giant tsunamis perhaps thousands of feet tall. The dust from the fireball plunged the world into a dusty fiery darkness and months or years of freezing temperatures, followed by an intense global warming. Few species handled the environmental stress well, and more than a quarter of all the plant and animal kingdoms' families, including 65 percent of all species on the planet, became extinct forever. Gone were dinosaurs, mighty rulers of the Triassic, Jurassic, and Cretaceous. Oceanic reptiles and ammonoids died off, and 60 percent of marine planktonic organisms went extinct. The great K/T dyings affected not only the numbers of species, but also the living biomass—the death of so many marine plankton alone amounted to 40 percent of all living matter on Earth at the time. Similar punches to land-based organisms decreased the overall living biomass on the planet to a small fraction of what it was before the K/T 1-2-3 knockout blows.

Some evidence suggests that the planet is undergoing the first stages of a new mass extinction. In the past 100,000 years, the ice ages have led to glacial advances and retreats, sea level rises and falls, the appearance and rapid explosion of human (*Homo sapiens sapiens*) populations, and the mass extinction of many large mammals. In Australia, 86 percent of large (> 100 pounds) animals have become extinct in the past 100,000 years, and in South America, North America, and Africa the extinction is an alarming 79 percent, 73 percent, and 14 percent, respectively (see Martin and Klein, 1989). This ongoing mass extinction appears to be the result of cold climates and, more importantly, predation and environmental destruction by humans. The loss of large-bodied species in many cases has immediately followed the arrival of humans in the region, with the clearest examples being found in Australia, Madagascar, and New Zealand. Similar loss of races through disease and famine has accompanied many invasions and explorations of new lands by humans, as described beautifully in the book *Guns, Germs and Steel* by Jared Diamond.

THE FUTURE

Humans are experiencing a population explosion that can not be sustained by the planet's limited resources (Figure 12.3). Very soon, fresh water and food sources will not be able to sustain the global population, and hunger, famine, disease, and war will follow. Can this fate be controlled? People are migrating in huge numbers to hazard-prone areas including coastlines, river flood banks, and the flanks of active volcanoes. Natural geologic hazards in these areas will cause disasters, and thousands of people will die. Perhaps greater understanding of natural geologic hazards will cause people to move to safer areas, or to be better prepared for the hazards in the areas in which they live.

Why has the population of the human race exploded so dramatically as shown in Figure 12.3? About a million years ago, there are estimated to have been a few thousand migratory humans on Earth, and by about 10,000 ago, this number had increased to only 5–10 million. It wasn't until about 8,000 years ago, when humans began stable agricultural practices and domesticated some species of animals, that the population rate started to increase substantially. The increased standards of living and nutrition caused the population growth to soar to about 20 million by 2,000 years ago, and 100 million by 1,000 years ago. By the eighteenth century, humans began manipulating their environments more, began public health services, and began to recognize and seek treatments for diseases that were previously taking many lives. The average life span began to soar, and world population surpassed 1 billion in the year 1810. A mere 100 years later, world population doubled again to 2 billion, and it had reached 4 billion by 1974. World population is now close to 7 billion and climbing more rapidly than at any time in history, doubling every fifty years. As discussed in Chapter 1, this rate of growth is not sustainable—at this rate, in about 800 years, there will be one human for every three square feet on Earth!

What will humans do when the population exceeds the ability of the earth to

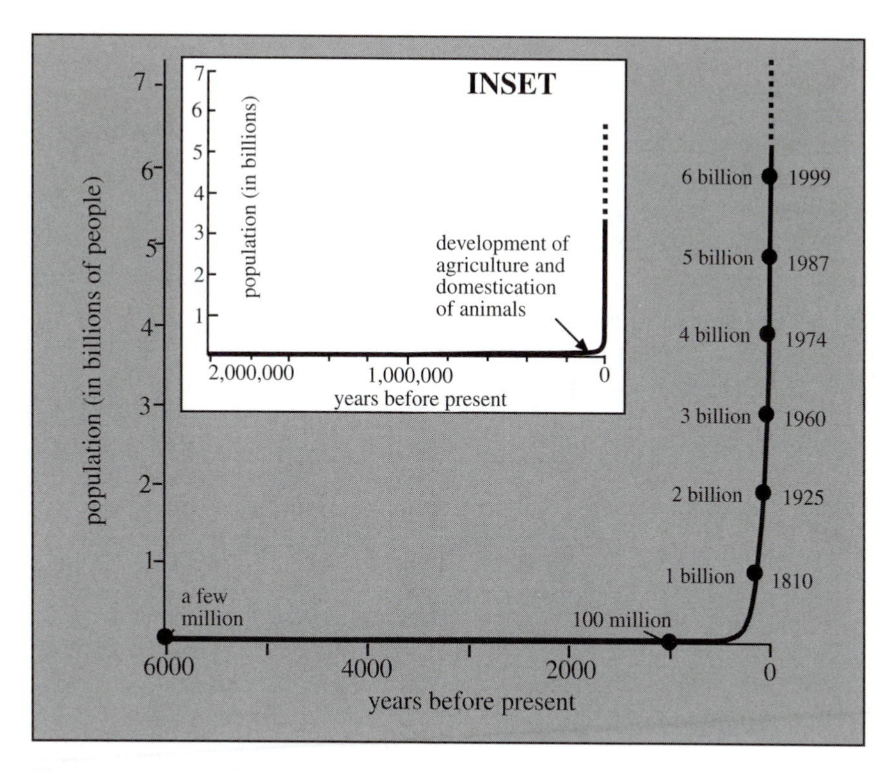

Figure 12.3. Population growth curves. Note the asymptotically growing population of the planet. Do we have enough resources and common sense to sustain this growth?

support us? Will we fight, like some past isolated societies that reached their capacity, to be supported by the land's natural resources? Will we succumb to disease and famine, perhaps induced by drought? Will a huge natural disaster such as an impact from outer space send us back to a small fraction of our current population? Will the population rate decrease in response to cultural changes? Some evidence suggests that the population growth rate is decreasing in developed countries and continues to rise in undeveloped and third world countries. This could be in response to increased medical care, which reduces the need to have many children to ensure that some survive to adulthood, or it could be a reflection of increased opportunities for women in developed societies.

We are causing mass extinction on a global scale that rivals the mass extinctions caused by massive flood basalt volcanism and meteorite impacts. Thousands of species are vanishing every year. Familiar to many is the rapid global decline in amphibians, particularly the frogs, in the past thirty years. Are frogs the proverbial canaries in the coal mine? We are destroying environments and, through uncontrolled population growth, we are competing for limited resources, causing a rapid decrease in global diversity. We are destroying many of the organisms that may be crucial to our own survival, perhaps without even knowing it because interrelationships between different organisms are complex. We may find ourselves at the point of someday realizing that that simple bacteria that went extinct, or the fungi that disappeared a few years back, was essential for some process we need to survive, such as removing some greenhouse gas from the atmosphere. Rarely does one change in the environment go without consequence. Death of one species may allow another to expand, and this new species may be a predator to a third or carry parasites that can wipe out large segments of the population. Exponential population growth is hazardous and could be the ultimate demise of the human race.

RESOURCES AND ORGANIZATIONS

Print Resources

Albritton, C. C., Jr. *Catastrophic Episodes in Earth History.* London: Chapman and Hale, 1989.

Alvarez, W. *T Rex and the Crater of Doom.* Princeton, N.J.: Princeton University Press, 1997, 236 pp.

Burke, K. "Tectonic Evolution of the Caribbean." *Annual Reviews of Earth and Planetary Sciences* 16 (1988), 201–30.

Chang, K. "Planet or No, It's on to Pluto." *The New York Times,* January 29, 2002, sec. D1, D4.

Chapman, C. R., and Morrison, D. "Impacts on the Earth by Asteroids and Comets: Assessing the Hazard." *Nature* 367 (1994), 33–39.

Cohen, J. E. *How Many People Can the World Support?* New York: W.W. Norton and Co., 1995.

Dawson, J. B. *Kimberlites and Their Xenoliths.* New York: Springer-Verlag, 1980.

Diamond, J. *Guns, Germs, and Steel: The Fates of Human Societies.* New York: W.W. Norton and Co., 1999, 480 pp.

Eldredge, N. *Fossils: The Evolution and Extinction of Species.* Princeton, N.J.: Princeton University Press, 1997, 240 pp.

Erwin, D. H. "The Permo-Traissic Extinction." *Nature* 367 (1994), 231–36.

MacDougall, J. D., ed. *Continental Flood Basalts.* Dordrecht, The Netherlands: Kluwer Academic Publishers, 1988.

Mahoney, J. J., and Coffin, M. F., eds. *Large Igneous Provinces, Continental, Oceanic, and Planetary Flood Volcanism.* Washington, D.C.: American Geophysical Union, Geophysical Monograph Series 100, 1997, 438 pp.

Mannard, G. W. "The Surface Expression of Kimberlite Pipes." Geological Association of Canada Proceedings 19, 1968.

Martin, P. S., and Klein, R. G., eds. *Quaternary Extinctions.* Tucson: University of Arizona Press, 1989.

McCormick, M. P., Thompson, L. W., and Trepte, C. R. "Atmospheric Effects of the Mt. Pinatubo Eruption." *Nature* 373 (1995), 399–404.

Melosh, H. J. *Impact Cratering: A Geologic Process.* New York: Oxford University Press, 1988.

Mitchell, R. H. *Kimberlites, Mineralogy, Geochemistry, and Petrology.* New York: Plenum Press, 1989, 442 pp.

Moores, E. "Pre-1 Ga (pre Rodinian) Ophiolites: Their Tectonic and Environmental Implications." *Geological Society of America Bulletin* 114 (2002), 80–95.

Poag, C. W. *Chesapeake Invader: Discovering America's Giant Meteorite Crater.* Princeton, N.J.: Princeton University Press, 1999, 168 pp.

Ponting, C. *A Green History of the World.* New York: St. Martin's Press, 1991.

Rampino, M. R., Self, S., and Stothers, R. B. "Volcanic Winters." *Annual Reviews of Earth and Planetary Science* 16 (1988), 73–99.

Renne, P. R., Zichao, Z., Richards, M. A., Black, M. T., and Basu, A. R. "Synchrony and Causal Relations between Permian Triassic Boundary Crises and Siberian Flood Volcanism." *Science* 269, 1413–16.

Self, S., Thordarson, T., and Keszthelyi, "Emplacement of Continental Flood Basalt Lava Flows," in J. J. Mahoney and M. F. Coffin, ed., *Large Igneous Provinces, Continental, Oceanic, and Planetary Flood Volcanism.* Washington, D.C.: American Geophysical Union, Geophysical Monograph Series 100, 1997, p. 381–410.

Sepkoski, J. J., Jr. "Mass Extinctions in the Phanerozoic Oceans: A Review," in *Patterns and Processes in the History of Life.* Amsterdam: Springer-Verlaag, 1982.

Sharpton, V. L., and Ward, P. D. *Global Catastrophes in Earth History.* Geological Society of America Special Paper 247, 1990.

Sheridan, M. F., and Wohletz, K. H. "Hydrovolcanism, Basic Considerations and Review." *Journal of Volcanology and Geothermal Research* 17 (1983), 1–29.

Stanley, S. M. *Extinction.* New York: Scientific American Library, 1987.

———. *Earth and Life through Time.* New York: W.H. Freeman and Co., 1986, 690 pp.

Wyllie, P. J. "The Origin of Kimberlite." *Journal of Geophysical Research* 85 (1980), 6902–10.

Non-Print Resources

Videos

Asteroids: Deadly Impact, National Geographic Society, 1997, 60 mins.

The Doomsday Asteroid, NOVA/BBC, 1995, 60 mins.

The Death of the Dinosaur, Public Broadcasting System (PBS), 1990, 60 mins.

Web Sites

Asteroid Introduction Web site:
http://www.solarviews.com/eng/asteroid.htm
This Web site by Calvin J. Hamilton offers views of the solar system, including the asteroid belt.

Dinosaur extinction Web site:
http://web.ukonline.co.uk/a.buckley/dino.htm
This site offers short summaries of some theories of dinosaur extinction, including meteorite impacts.

Ernstson impact structures Web site:
http://www.impact-structures.com/
This Web site describes many impact craters and structures.

Impact Web site:
http://personals.galaxyinternet.net/tunga/
This Web site describes the effects of a comet impact on Earth and what we might do to prevent or prepare for future disasters of this sort.

NASA's Ames Research Center:
http://impact.arc.nasa.gov//
This Web site describes various hazards associated with asteroid and comet impacts with Earth.

Organizations

National Aeronautics and Space Administration (NASA)
NASA Near-Earth Object Program
Jet Propulsion Laboratory
4800 Oak Grove Drive, Pasadena
California 91109
818–354–4321
http://neo.jpl.nasa.gov/
In 1998, NASA initiated a program called the "Near-Earth Object Program," whose aim is to catalog potentially hazardous asteroids that could present a hazard to Earth. This program uses five large telescopes to search the skies for asteroids that pose a threat to Earth and to calculate their mass and orbits. So far, the largest potential threat known is from asteroid 99AN10, which has a mass of 2.2 billion tons and may pass within the orbit of the moon at 7:10 A.M. on August 7, 2027. NASA has another related program called "Deep Impact," which is designed to collect data on the composition of a comet, Tempel 1, which will be passing beyond the orbit of Mars. The comet is roughly the size of midtown Manhattan, and the spacecraft will be shooting an object at the comet to determine its density by observing the characteristics of the impact.

Geological Survey of Canada
http://www.unb.ca/passc/ImpactDatabase/
The Geological Survey of Canada has compiled an Earth Impact database, available at the site above. The database is regularly maintained,

and contains maps and images of various impact craters. For more information, contact: Dr. John Spray (jgs@unb.ca), Director, PASSC Planetary and Space Science Centre, Department of Geology, University of New Brunswick, 2 Bailey Drive, Fredericton, NB. Canada. Telephone: (506) 453–3550 Email: impact@unb.ca.

University of Arizona, Lunar and Planetary Laboratory

http://seds.lpl.arizona.edu/nineplanets/nineplanets/meteorites.html

Lunar and Planetary Laboratory, University of Arizona. Web site has extensive list of information about meteors, meteorites, impacts, and links to other sites. Site is run by the Students for the Exploration and Development of Space (SEDS). UA SEDS, Box 174 Space Sciences Building, The University of Arizona, Tucson, Arizona 85721, E-mail: seds@seds.org, Phone: (520) 621–9790, Fax: (520) 621–4933.

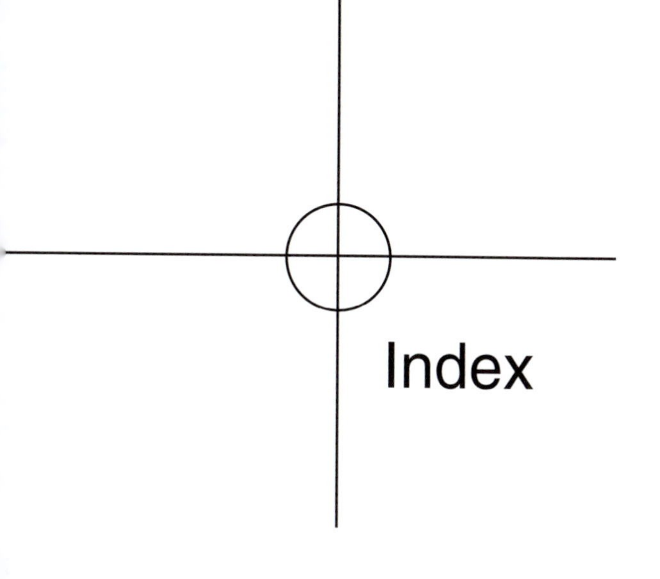

Index

Abrasion, 99, 175–77, 203, 260
Acid rain, 247, 273, 278
Aerosol, 65, 217, 278
Alluvial fan, 129, 175–76, 232
Aquiclude, 228, 232, 251
Aquifer, 153, 185, 193–97, 228, 232–33, 236, 252–56, 262
Archaea, 283–84
Arsenic, 221–23, 235–36, 242–43
Asbestos, 16, 222–26, 241–42
Ash, 1, 4–5, 15, 49–50, 56–57, 62, 65–71, 79, 112–14, 217, 281
Astenosphere. *See* Plate tectonics
Asteroid, 7, 82, 268–70, 275–76
Atacama Desert, 171, 173
Avalanche, 115. *See also* Mass wasting

Baldwin Hills dam and reservoir, 255–57
Barrier island, 145–48, 153–56, 159, 165–67
Basalt, 12, 17, 52–60, 64, 276–79, 286, 289–91
Base level, 33, 123–24, 129
Batholith, 51–52, 54
Beach, 15, 25, 80, 86–90, 95, 99, 122, 145–60, 165–67, 180–81, 214, 216, 260
Bicarbonate ion, 100, 128, 234
Black lung disease, 226–27

Calcium, 128, 207, 223, 234, 273
Caldera, 4, 52, 54, 58, 62, 156
Calving, 199–201, 211–12, 216
Carbon dioxide, 6, 64, 100–102, 217, 240, 273, 278–79, 283–86
Carbonic acid, 100, 247
Caves, 107, 233, 241, 246–53
Chicxulub Crater, 82, 267, 270, 274, 277–79
Cliff, 15, 31, 80, 88, 92–93, 99, 112, 116, 145–49, 152–58, 165–67, 174, 201, 250, 260
Climate change, 16, 102, 200–201, 279
Clinker, 239–40
Coal, 223–26, 235, 239–41
Coal mining, 226, 253
Coastal hazards: agencies, 166; examples, 156–62; human-induced, 146, 150–57, 164–65, 215; statistics, 165–66; what to do if you are in, 164–66
Coastline, 1, 3, 6, 29, 32–34, 39, 64, 77, 83, 88–89, 94–95, 140, 145–52, 163, 173, 217–18, 234, 245, 272, 288
Comet, 268–69, 274–75
Continental drift, 11–12, 209, 284
Continental shelf, 12, 95, 146, 210, 283
Convergent boundaries, 12–15, 50, 55–56, 59, 75, 84, 259
Coriolis force, 6, 150, 215
Crater, 82, 267, 270–75, 279–80
Creep, 1–2, 15–16, 31, 36, 56, 107–110, 116–201
Crevasse, 16, 201–2, 211

Dams, 107–8, 113, 119, 122–27, 133, 192–95, 255–57
Deccan Traps, 267, 276, 287

Deflation, 155, 175, 177
Deforestation, 103, 109, 128
Degassing, 64, 82, 276
Deltas, 97, 116, 127–29, 136–40, 150, 154–57, 161–63, 245, 260–63; submarine, 115
Deposition, 8, 114–16, 123–24, 128, 149, 203
Desert, 1, 7, 15–16, 169–97; and other hazards, 99, 103, 122, 128–31, 140, 207, 210, 216, 229, 232; atmospheric upwelling, 171, 187; dunes, 180–82; flash flooding, 15, 131–34, 170, 174–76; landforms, 174–83; types, 170–74
Desertification, 16, 170, 183–91, 207
Diamonds, 100, 205, 268, 274, 279
Diatreme, 268, 279–81
Dilation, 37
Dinosaurs, 3, 10, 82, 267, 273–74, 277, 279, 286–87
Drought, 3, 16, 19, 169–98; and other hazards, 112, 125, 207, 240, 245–49, 253, 256, 281, 289; causes, 185–96; Middle East, 190–95; United Nations Food program, 170, 189; U.S. southwest, 195–96
Dust Bowl, 182, 186, 189–90
Dust storm, 177, 179–80, 183, 189

Earth structure, 9–12
Earthquake, 1–5, 9, 14, 17, 21–48; agencies, 40–41; and other hazards, 49, 64, 69–70, 75–95, 97, 106–7, 115–16, 182, 228, 246, 258–59, 267, 272, 281; damage prevention, 38–39; examples, 42–46; hazards, 29–36; impacts, 267, 272; measuring, 25–29; origin, 23–25; predicting, 36–38; statistics, 41–46; subsidence, 246, 258–59; what to do if you are in, 39–41
Elastic rebound theory, 22
ENSO (El Nino Southern Oscillation), 186–88, 213, 217
EPA (Environmental Protection Agency), 225–28, 235, 241–42
Epicenter, 24–30, 38, 84, 87, 90, 94
Erosion, 3, 5, 8, 15, 175, 203, 259–60, 270; by ice, 214; by rain, 140; by streams and rivers, 116, 121–29, 175; by wind, 177, 184; coastal, 88, 118, 145–47, 153–59, 165, 187, 216; weathering, 98–99, 103, 112, 118–19
Estuary, 147, 153, 166, 260

Fault, faulting, 22–26, 29, 31, 36–37, 44–47, 78–81, 84, 87, 91, 106, 116, 194–95, 246, 257–59
Feldspar, 53, 100–101
FEMA (Federal Emergency Management Agency), 18–19, 40–42, 68–69, 155, 166–67
Fire, 17, 21, 25, 29–30, 35, 40–42, 45, 50–51, 55, 71, 82, 84, 90, 93–94, 112, 185, 189, 222–23, 235, 267–68, 272–77, 287; underground, 239–41
Fissure, 31, 57, 64, 240, 276, 280
Flood, 1–5, 15, 17, 121, 127–44; and other hazards, 32, 49–52, 60–62, 65, 79–80, 83, 86, 97, 107–8, 111, 121–47, 153, 159–65, 170, 173–76, 185–88, 191–94, 207, 215–16, 232, 241, 245, 255–56, 260–62, 276–81, 286–91; examples, 133–41; features, 128–30; flash flooding, 131–34; regional flooding, 134–37
Flood basalt, 52, 65, 276–79, 286
Floodplain, 122–24, 127–29, 136–38, 141, 232
Flows, 106–112
Fluid extraction, 253–56
Fracture, 37, 98–99, 101, 105–7, 113, 194–95, 232–33, 247–49, 251–53, 257, 280
Frost wedging, 174, 203

Geologic time scale, 9–10
Geyser, 54, 58, 64
Glacier, glaciation, 5, 6, 16, 199–220; ages, 208–211; and other hazards, 50, 102, 111–13, 116, 152, 201, 283, 286; causes, 206–11; cirque, 111, 199, 202–4; fiords, 82, 90, 199, 202–3; ice cap, 70, 199, 209, 213, 216–17; ice sheet, 199, 206–11, 216–19; landforms, 201–5; meltwater, 16, 201–3, 216; mountain glacier, 199, 212; movement, 201–5; piedmont glacier, 199, 210; study of, 205–6; tidewater glacier, 199, 202, 211–12, 216; zones of, 200–202
Global warming, 6, 102, 183, 200, 207, 210, 216–17, 223, 240, 278–79, 287
Granite, 53–54, 100, 113–15, 237–39
Gravity, 15, 26, 30, 97, 103–4, 107–10, 134, 150, 199–201, 231, 246, 249
Greenhouse effect, 207, 273
Greenhouse gases, 207, 211, 217, 287
Groundwater, 5, 16, 37, 58, 70, 80, 103, 128, 133, 142, 153, 170, 193–97, 218, 221–22, 228–38, 242, 245–49; discharge, 231–33; dissolu-

tion, 233–35, 246–48, 251; extraction, 253–56; recharge, 142, 231–33

Hadley cell, 170–72, 186
Hazard risk mapping, 38, 42, 66, 83–85, 118, 164–66, 235, 238–39, 263–64, 275–76
Herculaneum, Missouri, 227–28
Hot spot, 59, 80, 282
Hurricane, 2–3, 15, 18–19, 145–46, 150, 152, 159–67; and other hazards, 62, 131–32, 245; examples, 159–62
Hydrostatic pressure, 254

Ice age, 200, 208, 283
Ice drift, 214–15
Iceberg, 16, 199, 203, 213–15
Igneous rocks, 8–9, 30, 51–52, 100, 223, 254, 277
Impacts, 17, 267–76, 279, 281
Inertia, 26–27
Iodine, 223–24

Joints, 99, 101, 105–6, 115, 118

Karst, 247–52
Kilauea, Hawaii, 56, 59, 72
Kimberlite, 268, 279–81
Kobe, Japan, 35, 45
Krakatau, Indonesia, 60, 65, 68, 75, 79, 84, 88

Lagoon, 146–48, 150, 154–57
Lahar, 62, 67, 69–71, 112, 126, 159, 171
Landslide, 1, 5, 15, 97–98, 106–8, 116–20; and other hazards, 21, 29, 31, 33–34, 44, 75, 77–78, 80–82, 90–91, 173, 228
Lava, 1, 9, 12, 49–50, 54–62, 68, 71, 210, 246, 258, 267, 276–78, 280
Leaching, 100, 102–3, 223–24, 227
Lead, 221–24, 227–28, 236
Leeward slope, 169, 172, 180
Levee, 87, 121, 128, 134–39
Limestone, 99, 107, 207, 233–34, 245–51, 273
Liquefaction, 29, 32, 35, 45, 106, 246
Lithosphere. *See* Plate tectonics
Loess, 41, 107, 179, 182–83

Magma, 6, 8, 15, 21, 26, 49–59, 64–67, 70, 80, 246, 258, 278, 280–83, 287
Magnesium, 128, 223, 234
Mass extinction, 17, 273, 276, 281, 285–89
Mass wasting, 15, 97–120; and other hazards, 29, 31, 173; causes and controls, 103–6, 116–17; hazards to humans, 116–19; processes, 106–115; statistics, 117–19; what to do if you are in, 119
Mauna Loa, Hawaii, 50, 56, 58–59
Medical geology, 221, 224
Metamorphic rocks, 8, 30, 100, 254
Meteor Crater, Arizona, 270–71
Meteorite, 3, 17, 268–76, 279, 285
Methane, 116, 210, 217, 259, 286
Mexico City: earthquakes, 23, 30; subsidence, 255
Mines, 16, 205, 226, 240–41, 250, 253
Minoan civilization, 4, 76, 80
Mississippi River, 15, 97, 123, 129, 134–40, 162–63, 261
Mount Kilamanjaro, Tanzania, 50, 200
Mount Pelee, Martinique, 56, 60, 62, 68–69
Mount Pinatubo, Philippines, 59, 62, 65, 67, 71, 278, 290
Mount Saint Helens, Washington, 50, 57–59, 62, 66–70, 126, 270
Mount Vesuvius, Italy, 49, 62, 69

NASA, 275–76
New Madrid Fault, 22–23, 28, 47
New Orleans, Louisiana , 134–38, 162–63, 260–61
Nile River, 4, 80, 120, 123, 129, 137, 188, 190–94
Nitrogen, 217, 223, 284
Nuee ardent, 49, 56–57, 62, 66, 69

Oceanic plateaus, 276–79
OSHA (Occupational Safety Hazards Administration), 225, 242
Overland flow, 122, 140–43
Oxygen, 8, 100, 162, 223, 240–41, 259, 284
Ozone, 217, 284

Paleoclimate, 205, 219
Passive margin, 257, 260, 276
Permeability, 175, 195, 229–33, 238, 251
Petroleum: contamination, 179, 234; mining, 19, 90, 215–16, 245, 253, 255, 260–64

Plate tectonics, 2–3, 7–15, 21–22, 44, 46, 50, 55–56, 59, 70, 75, 78–79, 84, 89, 210, 217, 246, 257, 259, 276–85
Pompeii, Italy, 49, 62, 69
Population growth, 2–3, 16, 17, 117, 162, 166, 170, 184, 190–94, 216, 230, 241, 288–90
Porosity, 80, 98, 175, 195, 200, 229, 231–33, 238, 241, 254
Pull-apart basin, 257, 259
Pyroclast, 56–57, 62, 69–70, 280

Quartz, 100–101, 226, 260, 274

Radioactivity: alpha particles, 236–39; dating, 9, 281; decay, 13, 16, 221, 228, 236–37, 281
Radium, 236–37
Radon, 16, 37, 222–23, 236–39, 241–42
Red Cross, 40, 42, 185
Regolith, 8, 97–99, 104–9, 112–13, 174, 228, 230–33, 246, 254, 262–63
Rifts, rifting, 246, 257–59, 283
"Ring of Fire," 50, 55
River. *See* Stream
Rockfall, 97, 99, 101, 103, 106–7, 110–15, 119; and other hazards, 34, 80
Runoff, 118, 122, 156, 216, 230–32

Sahara Desert, 16, 169, 171–73, 177, 180–85, 188–89
Salalah Region, 249–50
Salt mining, 252–53
Saltation, 128, 169, 180
San Andreas Fault, 13–14, 21, 44–47
San Francisco, California, 30, 35, 42–46
Sand drift, 150, 153–59, 165, 193, 203
Sarawak Chamber, 247
Scientific method, 8–9
Sea ice, 213–15
Sea level change, 1, 16, 78–79, 145, 147, 152–53, 163–64, 216–18, 255–62, 282–88
Seawall, 146, 154–55, 160, 164–65, 215
Sedimentary rocks, 8–9, 129, 179, 201, 237, 254–55, 260, 281
Seiche waves, 4–5, 82–83, 90; and other hazards, 29, 32–35, 39, 41
Seismic waves, 24–30, 37–38, 42, 45, 67, 84
Seismograph, 25–28, 41, 47
Selenium, 221–24
Severe drought, 16, 186, 193, 153–54
Shrinking/swell clays, 229–30, 263
Silica, 52–55, 59, 100, 224–26, 229, 241, 268–69
Sinkholes, 3, 16, 233, 240, 246–53, 263–64
Slump, 32, 39, 44, 78–79, 82, 87, 97, 106, 109, 115–16, 119, 124, 146–47, 165
Snowball Earth, 199–200
Sodium, 128, 223–34
Soil: contamination, 222–27; formation, 102–3; mass wasting, 31, 33, 35, 97, 106, 109–112, 119, 140, 184, 189; mechanics, 228–30; water saturation, 131, 188–89, 228
Solar radiation, 183, 200, 206–211, 217
Storm surges, 34, 145, 152–55, 159–62, 245, 255, 260–61
Stream, 5–6, 16, 99, 107, 111, 116, 121–44, 174–76, 180, 182, 185–86, 203–5, 227, 230–32, 246–51, 259–60; braided, 125–26, 204–5; discharge, 123–28, 134, 136, 190, 218, 231–33; dynamics, 125–27; geometry, 123–25; meandering, 124–26
Stream load, 126–28
Streamflow, 122, 125
Striation, 102, 201, 205
Strike-slip faulting, 14, 45, 246, 259
Subduction zone, 11, 13–15, 44, 50, 79
Subsidence, 16, 33, 136, 204, 245–65; compaction-related, 35, 116, 245–46, 254–56, 260–63; deep, 246, 254; human-induced, 246, 250–56, 261–62; shallow, 246; tectonic, 246, 256, 260
Sulfate ion, 128, 234
Sulfur dioxide, 66, 68, 278–79
Supercontinent, 283; Gondwana, 208–9, 284–85; Pangea, 209–210, 285–86

Tephra, 56, 58, 68
Tides, 15, 32–34, 75, 83, 88, 146–53, 156–61, 166, 245, 255–56, 259–60
Total dissolved solids, 233–34
Trade wind, 7, 162, 170–73
Transform boundaries, 13, 14, 259
Tsunami, 1, 3–4, 17, 75–96; agencies, 83; and other hazards, 21, 32–34, 39, 42–44, 49, 57, 64–65, 116, 267, 272, 287; examples, 87–95; mechanics, 76–77; origin, 77–83; predicting, 84–86; statistics, 87–95; study of, 83; what if you are in, 86–87

Tunguska, Siberia, 270, 274–75
Turbidity current, 115–16

U.S. Army Corps of Engineers, 120, 136–38, 144, 155, 163
Undercutting, 116, 118, 146
Uranium, 221, 236–39
Urbanization, 112, 128, 133, 140, 142
USGS (United States Geological Survey), 19, 41, 42, 46, 48, 65, 67–68, 72–73, 83, 85, 96, 120, 143–44, 164–68, 196, 239, 241–43

Venice, Italy, 163–64, 245–46
Viscosity, 54, 59
Volcanism, 1–4, 12–15, 17, 49–73, 140; and other hazards, 23, 31–32, 75–80, 84, 106, 112–16, 126, 159, 209–10, 217, 223, 237, 246, 255, 258–59, 267–70, 276–90; examples, 68–71; hazards, 60–66; landforms, 55–59; origin, 50–55; plate tectonics, 59; predicting, 66–67; statistics, 68–71; what if you are in, 67–68

Water overuse, 170, 233, 256
Water table, 33, 133, 216, 231–33, 249–51, 254
Wave mechanics, 149–50
Weathering, 8, 97–102, 174, 224, 228, 247, 260, 270, 283; chemical, 98–102; factors, 101–2; mechanical, 99–102, 174
Wetlands, 146–47, 260–63
Wind, 6, 7, 15, 149–52, 155, 159–62, 169–96; and other hazards, 56, 65, 97–99, 131, 136, 149–50, 208, 211, 214–17, 256

Zinc, 221, 223

ABOUT THE AUTHOR

TIMOTHY M. KUSKY is a professor at St. Louis University in Missouri, where he teaches courses in geologic hazards, environmental geology, structural geology, remote sensing, tectonics, and Precambrian crustal evolution. He has an active, award-winning research program, including projects in Asia, Africa, the Middle East, the United States, Canada, and Australia. Kusky received his B.Sc. and M.Sc. in geological sciences from the State University of New York at Albany, and his M.S. and Ph.D. in earth and planetary sciences from Johns Hopkins University. He did post-doctoral studies in the Department of Mechanical Engineering at the University of California, Santa Barbara, and worked as a visiting professor in the Department of Geological Sciences and the Allied Geophysical Laboratories at the University of Houston. He also worked as an assistant professor at the Center for Remote Sensing and Department of Geology at Boston University.